AAOS
AMERICAN ACADEMY OF ORTHOPAEDIC SURGEONS

Ninth Edition

Emergency

Care and Transportation of the Sick and Injured

Student Workbook

JONES AND BARTLETT PUBLISHERS

Sudbury, Massachusetts

BOSTON   TORONTO   LONDON   SINGAPORE

Jones and Bartlett Publishers

World Headquarters
Jones and Bartlett Publishers
40 Tall Pine Drive, Sudbury, MA 01776
978-443-5000
info@jbpub.com
www.EMSzone.com

Jones and Bartlett Publishers Canada
6339 Ormindale Way
Mississauga, Ontario
L5V 1J2

Jones and Bartlett Publishers International
Barb House, Barb Mews
London W6 7PA
United Kingdom

AAOS
AMERICAN ASSOCIATION OF
ORTHOPAEDIC SURGEONS

Director, Department of Publications: Marilyn L. Fox, PhD
Managing Editor: Lynne Roby Shindoll
Managing Editor: Barbara A. Scotese

Jones and Bartlett's books and products are available through most bookstores and online booksellers. To contact Jones and Bartlett Publishers directly, call 800-832-0034, fax 978-443-8000, or visit our website www.jbpub.com.

Substantial discounts on bulk quantities of Jones and Bartlett's publications are available to corporations, professional associations, and other qualified organizations. For details and specific discount information, contact the special sales department at Jones and Bartlett via the above contact information or send an email to specialsales@jbpub.com.

Production Credits

Publisher, Public Safety: Kimberly Brophy
Managing Editor: Carol Brewer
Associate Editor: Janet Morris
Production Editor: Karen Ferreira
V.P., Manufacturing and Inventory Control: Therese Connell
Interior Design: Shepherd, Inc.
Cover Design: Kristin E. Ohlin
Composition: Shepherd, Inc.
Printing and Binding: Courier Westford
Chief Education Officer: Mark W. Wieting

Editorial Credits

Authors: Julie F. Chase, BS, BM, NREMT-P
Jay Keefauver, BS, EMT-P, CEMSI
Donald Royder, EMT-P

ISBN: 978-0-7637-7300-7
6048

This student workbook is intended solely as a guide to the appropriate procedures to be employed when rendering emergency care to the sick and injured. It is not intended as a statement of the standards of care required in any particular situation, because circumstances and the patient's physical condition can vary widely from one emergency to another. Nor is it intended that this Student Workbook shall in any way advise emergency personnel concerning legal authority to perform the activities or procedures discussed. Such local determinations should be made only with the aid of legal counsel.
Note: The patients described in *Ambulance Calls* are fictitious.

Printed in the United States of America
12 11 10 09 08   10 9 8 7

CONTENTS

Section 5: Trauma

Section 6: Special Populations

Section 7: Operations

Section 8: ALS Techniques

Answer Key

EMT-Basic Review Manual for National Certification

ISBN: 0-7637-1829-7

The *EMT-Basic Review Manual for National Certification* is designed to prepare you to sit for the national certification exam by including the same type of skill-based and multiple-choice questions that you are likely to see on the exam. The review manual will also evaluate your mastery of the material presented in your EMT-Basic training program. The *EMT-Basic Review Manual for National Certification* includes:

- Practice questions with answers and model exam
- Step-by-step walkthrough of skills, including helpful tips, commonly made errors, and sample scenarios
- Self-scoring guide and winning test-taking tips
- Coverage of the entire 1994 DOT EMT-Basic curriculum

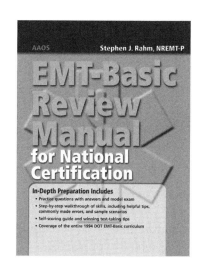

Student Review Manual

ISBN: 0-7637-2971-X (print)
ISBN: 0-7637-2970-1 (CD-ROM)
ISBN: 0-7637-2972-8 (online)

This *Review Manual* has been designed to prepare students for exams by including the same type of questions that they are likely to see on classroom and national examinations. The manual contains multiple-choice question exams with an answer key and page references. It is available in print, on CD-ROM, and online.

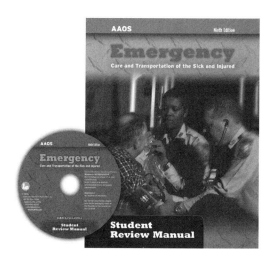

EMT-B Field Guide, Second Edition

ISBN: 0-7637-2214-6

This handy reference covers basic information from patient management tips to guidelines on helping a patient use medication. A few special features include a prescription medication reference and documentation tips. The *EMT-B Field Guide* is pocket-sized, spiral-bound, and water resistant for ready-reference in the field.

To order: Call 1-800-832-0034 or visit www.EMSzone.com

CHAPTER

1 Introduction to Emergency Medical Care

Workbook Activities

The following activities have been designed to help you. Your instructor may require you to complete some or all of these activities as a regular part of your EMT-B training program. You are encouraged to complete any activity that your instructor does not assign as a way to enhance your learning in the classroom.

Chapter Review

The following exercises provide an opportunity to refresh your knowledge of this chapter.

Matching

Match each of the items in the left column to the appropriate definition in the right column.

_____ **1.** ALS

_____ **2.** BLS

_____ **3.** EMT-B

_____ **4.** EMT-I

_____ **5.** EMT-P

_____ **6.** Medical control

_____ **7.** CQI

_____ **8.** EMS

_____ **9.** Continuing education

_____ **10.** Quality control

_____ **11.** Primary service area

_____ **12.** Medical Director

_____ **13.** Americans with Disabilities Act

A. EMS professional trained in ALS

B. a system of internal reviews and audits

C. a system to provide prehospital care to the sick and injured

D. the physician who authorizes the EMT to perform in the field

E. responsibility of the medical director to ensure appropriate care is delivered by an EMT

F. basic lifesaving interventions, such as CPR

G. EMS professional trained in some ALS interventions

H. advanced procedures such as drug administration

I. designated area in which the EMS service is responsible

J. protects disabled individuals from discrimination

K. physician instruction to EMS team

L. a required amount of training to maintain skills

M. EMS professional trained in BLS

Multiple Choice

Read each item carefully, and then select the best response.

_____ 1. Control of external bleeding, provision of oxygen and CPR are included in the "scope of practice" of the:
   A. EMT-P.
   B. EMT-B.
   C. EMT-I.
   D. EMT-D.

_____ 2. All of the following are true of medical control except:
   A. It is determined by the dispatcher.
   B. It may be written or "standing orders."
   C. It may require online radio or phone consultation.
   D. It describes the care authorized by the medical director.

_____ 3. All of the following are components of continuous quality control except:
   A. periodic run reviews.
   B. remedial training.
   C. internal reviews and audits.
   D. public seminars and meetings.

_____ 4. The major goal of quality improvement is to ensure that:
   A. quarterly audits of the EMS system are done.
   B. EMTs have received BLS/CPR training.
   C. the public receives the highest standard of care.
   D. the proper information is received in the billing department.

_____ 5. Your main concern while responding to a call should be the:
   A. safety of the crew and yourself.
   B. number of potential patients.
   C. request for mutual assistance.
   D. type of call.

Questions 6-10 are derived from the following scenario: After stocking the ambulance this morning, you and your partner go out for breakfast. While entering the restaurant, you see an older gentleman clutch his chest and collapse to the floor. When you get to him, he has no pulse and is not breathing.

_____ 6. The _____ _____ authorizes you as an EMT-B to provide emergency care to this patient.
   A. City Counsel
   B. Medical Control
   C. Medical Director
   D. Fire Chief

_____ 7. To treat this patient, you will follow:
   A. off-line medical control
   B. on-line medical control
   C. protocols
   D. all the above

_____ 8. The level of training that allows electrocardiogram and advanced life support to be given to this patient is:
   A. First responder
   B. EMT-B
   C. EMT-I
   D. EMT-P

_____ **9.** While you checked the patient's airway, breathing, and circulation, your partner followed local protocols and called for _____ back-up.
   A. ALS
   B. CQI
   C. PSA
   D. HIPAA

_____ **10.** While you performed CPR on this patient, your partner gathered the _____ that will deliver an appropriate electrical shock.
   A. EMD
   B. AED
   C. DEA
   D. GPS

Fill-in
Read each item carefully, and then complete the statement by filling in the missing word(s).

1. The training of the EMT-B should meet or exceed the guidelines of _____.

2. EMT-B training is divided into care of _____, _____, and important non-medical issues related to EMT abilities.

3. In some areas, EMT-Bs may provide selected ALS care such as automated external defibrillation, use of nonvisualized airways such as the Combitube or even endotracheal intubation, and assisting patients in taking prescription _____.

4. In most of the country, a communications center can be easily reached by dialing _____.

5. The appropriate care for injury or illness as described by the medical director either by radio or in written form is called _____ _____.

True/False
If you believe the statement to be more true than false, write the letter "T" in the space provided. If you believe the statement to be more false than true, write the letter "F."

_____ **1.** EMT-B personnel are the highest qualified members of the prehospital care team.
_____ **2.** The EMT-B scope of practice may include the use of an automated or semi-automated defibrillator.
_____ **3.** The purpose of continuous quality improvement (CQI) is to support discipline of personnel.
_____ **4.** A professional appearance and manner by the EMT-B will help a patient build confidence.
_____ **5.** Essential keys to being a good EMT-B include compassion, commitment, and desire.
_____ **6.** As a health care professional and an extension of physician care, you are bound by patient confidentiality—even in your own home.

Short Answer
Complete this section with short written answers using the space provided.

1. Describe the EMT-B's role in the EMS system.

_____

_____

2. What role has the U.S. Department of Transportation played in the development of EMS?

_____

_____

3. List five roles and/or responsibilities of being an EMT-B.

_____

_____

4. Describe the two basic types of medical direction that help the EMT-B provide care.

_____

_____

Word Fun

The following crossword puzzle is an activity provided to reinforce correct spelling and understanding of medical terminology associated with emergency care and the EMT-B. Use the clues in the column to complete the puzzle.

Across

2. System that provides prehospital emergency care
4. An EMT with training in BLS/CPR
7. Designed to protect individuals with disabilities
8. Section of the U.S. DOT that created programs to improve EMS
9. An individual trained to provide emergency care to the sick and injured
10. BLS intervention in cardiac arrest

Down

1. An EMT with extensive training in advanced life support
2. A type of 9-1-1 system that displays the caller's address
3. Simple lifesaving interventions, ie, CPR
5. Advanced lifesaving procedures, ie, defibrillation
6. Medical or Quality _____
10. A circular system of reviews and changes in an EMS system

Ambulance Calls

The following case scenarios provide an opportunity to explore the concerns associated with patient management. Read each scenario, and then answer each question in detail.

1. You are dispatched to a two car MVC. Upon arrival, you see minimal damage to both vehicles, as they were traveling less than 25 mph when the incident occurred. You and your partner examine all patients, and find no injuries. All patients are adults except for the driver of one vehicle. He is 16 years old, and has no complaints and no apparent injuries. He asks you if it's okay to leave.

   Do you let him go?

   _____

   _____

   _____

2. You are dispatched to a private residence. Several neighbors are gathered on the front lawn. There appears to be an argument taking place between two of them. A teenage boy is sitting on the doorstep, bleeding profusely from a cut above his left eye.

   How would you best manage this situation?

   _____

   _____

   _____

3. It's after midnight and you are dispatched to "laceration to the arm." You arrive to find a young man who tells you he was in a fight and his right arm was cut. You notice that his bleeding is controlled, and his vital signs are within acceptable ranges. As you apply dressings to his wound, another dispatch comes over the radio for "unconscious assault victim, police officers on the scene." You are one block away.

   What do you do?

   _____

   _____

   _____

CHAPTER

2 The Well-Being of the EMT-B

Workbook Activities

The following activities have been designed to help you. Your instructor may require you to complete some or all of these activities as a regular part of your EMT-B training program. You are encouraged to complete any activity that your instructor does not assign as a way to enhance your learning in the classroom.

Chapter Review

The following exercises provide an opportunity to refresh your knowledge of this chapter.

Matching

Match each of the terms in the left column to the appropriate definition in the right column.

_____ **1.** Cover
_____ **2.** Burnout
_____ **3.** Occupational Safety and Health Administration
_____ **4.** Posttraumatic stress disorder
_____ **5.** Body substance isolation
_____ **6.** Pathogen
_____ **7.** Transmission
_____ **8.** Tuberculosis
_____ **9.** Virulence
_____ **10.** Hepatitis
_____ **11.** Exposure
_____ **12.** Infection control
_____ **13.** Meningitis
_____ **14.** Contamination

A. chronic fatigue and frustration
B. assuming all body fluids are potentially infected
C. regulatory compliance agency
D. concealment for protection
E. capable of causing disease in a susceptible host
F. delayed stress reaction
G. chronic bacterial disease that usually affects the lungs
H. an inflammation of the meningeal coverings of the brain
I. an infection of the liver
J. contact with blood, body fluids, tissues, or airborne particles
K. the presence of infectious organisms in or on objects, or a patient's body
L. the strength or ability of a pathogen to produce disease
M. the way in which an infectious agent is spread
N. procedures to reduce transmission of infection among patients and health care personnel

Multiple Choice

Read each item carefully, then select the best response.

_____ **1.** From the age of 1 to the age of 34, _____ is the leading cause of death.

    A. cardiac arrest

    B. congenital disease

    C. trauma

    D. AIDS

_____ **2.** The stage of the grieving process where an attempt is made to secure a prize for good behavior or promise to change lifestyle is known as:

    A. denial.

    B. acceptance.

    C. bargaining.

    D. depression.

_____ **3.** The stage of the grieving process that involves refusal to accept diagnosis or care is known as:

    A. denial.

    B. acceptance.

    C. bargaining.

    D. depression.

_____ **4.** The stage of the grieving process that involves an open expression of grief, internalized anger, hopelessness, and/or the desire to die is:

    A. denial.

    B. acceptance.

    C. bargaining.

    D. depression.

_____ **5.** The stage of grieving where the person is ready to die is known as:

    A. denial.

    B. acceptance.

    C. bargaining.

    D. depression.

_____ **6.** When providing support for a grieving person, it is okay to say:

    A. "I'm sorry."

    B. "Give it time."

    C. "I know how you feel."

    D. "You have to keep on going."

_____ **7.** When grieving, family members may express:

    A. rage.

    B. anger.

    C. despair.

    D. all of the above

_____ **8.** _____ is a response to the anticipation of danger.

    A. Rage

    B. Anger

    C. Anxiety

    D. Despair

_____ **9.** Signs of anxiety include all of the following, except:

   A. diaphoresis.

   B. comfort.

   C. hyperventilation.

   D. tachycardia.

_____ **10.** Fear may be expressed as:

   A. anger.

   B. bad dreams.

   C. restlessness.

   D. all of the above

_____ **11.** If you find that you are the target of the patient's anger, make sure that you:

   A. are safe.

   B. do not take the anger or insults personally.

   C. are tolerant, and do not become defensive.

   D. all of the above

_____ **12.** When caring for critically ill or injured patients, _____ will be decreased if you can keep the patient informed at the scene.

   A. confusion

   B. anxiety

   C. feelings of helplessness

   D. all of the above

_____ **13.** When acknowledging the death of a child, reactions vary, but _____ is common.

   A. shock

   B. disbelief

   C. denial

   D. all of the above

_____ **14.** Factors influencing how a patient reacts to the stress of an EMS incident include all of the following, except:

   A. family history.

   B. age.

   C. fear of medical personnel.

   D. socioeconomic background.

_____ **15.** Negative forms of stress include all of the following, except:

   A. long hours.

   B. exercise.

   C. shift work.

   D. frustration of losing a patient.

_____ **16.** Stressors include _____ situations or conditions that may cause a variety of physiologic, physical, and psychologic responses.

   A. emotional

   B. physical

   C. environmental

   D. all of the above

_____ **17.** Prolonged or excessive stress has been proven to be a strong contributor to:

   A. heart disease.

   B. hypertension.

   C. cancer.

   D. all of the above

_____ **18.** _____ occur(s) when insignificant stressors accumulate to a larger stress-related problem.

    A. Negative stress

    B. Cumulative stress

    C. Psychological stress

    D. Severe stressors

_____ **19.** Events that can trigger critical incident stress include:

    A. mass-casualty incidents.

    B. serious injury or traumatic death of a child.

    C. death or serious injury of a coworker in the line of duty.

    D. all of the above

Questions 20-24 are derived from the following scenario: A 12-year-old boy told his grandmother he was going to collect the day's mail, located on the opposite side of the street for her. As he was returning with the mail, he was struck by a vehicle and was found lying lifeless in the middle of the street.

_____ **20.** What would be appropriate to say to the grandmother?

    A. "Don't worry, I'm sure he'll be fine."

    B. "What were you thinking? You will be reported!"

    C. "We're placing him on a backboard to protect his back, and we'll take him to the Columbus Community Hospital. Do you know who his doctor is?"

    D. It's best not to spend time talking to the grandmother.

_____ **21.** As an EMT-B, you know these types of calls are coming. How can you prepare to meet such stressful situations?

    A. Eat a balanced diet.

    B. Go for walks or other forms of exercise.

    C. Cut down on caffeine and sugars.

    D. All the above

_____ **22.** You or your partner may develop _____ after experiencing this call.

    A. Critical incident stress management

    B. Posttraumatic stress disorder

    C. Critical stress debriefing

    D. None of the above

_____ **23.** What signs of stress may you or your partner exhibit?

    A. Irritability towards coworkers, family, and friends.

    B. Loss of sexual activities

    C. Guilt

    D. All of the above

_____ **24.** When should you begin protecting yourself with BSI on this call?

    A. As soon as you are dispatched.

    B. As soon as you arrive.

    C. After you assess the victim, and know what you need.

    D. After speaking with the grandmother.

_____ **25.** The quickest source of energy is _____; however, this supply will last less than a day and is consumed in greater quantities during stress.

    A. glucose

    B. carbohydrate

    C. protein

    D. fat

_____ **26.** Stress management strategies include:
  A. changing work hours.
  B. changing your attitude.
  C. changing partners.
  D. all of the above

_____ **27.** _____ is a condition of chronic fatigue and frustration that results from mounting stress over time.
  A. Posttraumatic stress disorder
  B. Cumulative stress
  C. Critical incident stress
  D. Burnout

_____ **28.** The safest, most reliable sources for long-term energy production are:
  A. sugars.
  B. carbohydrates.
  C. fats.
  D. proteins.

_____ **29.** A _____ is any event that causes anxiety and mental stress to emergency workers.
  A. disaster
  B. mass-casualty incident
  C. critical incident
  D. stressor

_____ **30.** A CISD meeting is an opportunity to discuss your:
  A. feelings.
  B. fears.
  C. reactions to the event.
  D. all of the above

_____ **31.** Components of the CISM system include:
  A. preincident stress education.
  B. defusings.
  C. spouse and family support.
  D. all of the above

_____ **32.** Sexual harassment is defined as:
  A. any unwelcome sexual advance.
  B. unwelcome requests for sexual favors.
  C. unwelcome verbal or physical conduct of a sexual nature.
  D. all of the above

_____ **33.** Drug and alcohol use in the workplace can result in all of the following, except:
  A. absence from work more often.
  B. enhanced treatment decisions.
  C. an increase in accidents and tension among workers.
  D. lessened ability to render emergency medical care because of mental or physical impairment.

_____ **34.** You should begin protecting yourself:
  A. as soon as you arrive on the scene.
  B. before you leave the scene.
  C. as soon as you are dispatched.
  D. before any patient contact.

_____ 35. _____ is contact with blood, body fluids, tissues, or airborne droplets by direct or indirect contact.
   A. Transmission
   B. Exposure
   C. Handling
   D. all of the above

_____ 36. Modes of transmission for infectious diseases include:
   A. blood or fluid splash.
   B. surface contamination.
   C. needle stick exposure.
   D. all of the above

_____ 37. The spread of HIV and hepatitis in the health care setting can usually be traced to:
   A. careless handling of sharps.
   B. improper use of BSI precautions.
   C. not wearing PPE.
   D. sexual interaction with infected persons.

_____ 38. _____ is equipment that blocks entry of an organism into the body.
   A. Vaccination
   B. Body substance isolation
   C. Personal protective equipment
   D. Immunization

_____ 39. Recommended immunizations include:
   A. MMR vaccine.
   B. hepatitis B vaccine.
   C. influenza vaccine.
   D. all the above.

_____ 40. Why isn't tuberculosis more common?
   A. Absolute protection from infection with the tubercle bacillus does not exist.
   B. Everyone who breathes is at risk
   C. The vaccine for tuberculosis is only rarely used in the United States.
   D. Infected air is easily diluted with uninfected air.

_____ 41. You can use a bleach and water solution at a _____ dilution to clean the unit.
   A. 1:1
   B. 1:10
   C. 1:100
   D. 1:1000

_____ 42. Hazardous materials are classified according to _____, which dictate(s) the level of protection required.
   A. danger zones
   B. flammability
   C. toxicity levels
   D. all of the above

_____ 43. Factors to take into consideration for potential violence include:
   A. poor impulse control.
   B. substance abuse.
   C. depression.
   D. all of the above

Fill-in

Read each item carefully, then complete the statement by filling in the missing word.

1. The personal health, safety, and _____ of all EMT-Bs are vital to an EMS operation.

2. The struggle to remain calm in the face of horrible circumstances contributes to the _____

   _____ of the job.

3. Sixty percent of all deaths today are attributed to_____ _____.

4. Determination of the cause of death is the medical responsibility of a _____ .

5. In cases of hypothermia, the patient should not be considered dead until the patient is _____ and dead.

6. _____ is generally thought of in relation to the oncoming pain and the outcome of the damage.

7. Almost all dying patients feel some degree of _____ because of internalized anger and other factors.

8. You must also realize that the most _____ symptoms may be early signs of severe illness or injury.

9. EMS is a _____ job.

True/False

If you believe the statement to be more true than false, write the letter "T" in the space provided. If you believe the statement to be more false than true, write the letter "F."

_____ 1. Developing self-control is aided by proper training.
_____ 2. The primary mode of infection for herpes simplex is through close personal contact.
_____ 3. Denial is generally the first step in the grieving process.
_____ 4. Body fluids are generally not considered infectious substances.
_____ 5. Most EMT-Bs never suffer from stress.
_____ 6. Physical conditioning and nutrition are two factors the EMT-B can control in helping reduce stress.

Short Answer

Complete this section with short written answers using the space provided.

1. Describe the basic concept of body substance isolation (BSI).

   _____

   _____

2. List the five stages of the grieving process.

   _____

   _____

3. List five warning signs of stress.

   _____

   _____

4. List two strategies for managing stress.

   _____

   _____

5. Describe the process for proper hand washing.

_____

_____

6. Complete the following table on the toxicity of hazardous materials.

Level	Hazard	Protection Needed
0		
1		
2		
3		
4		

7. List the three layers of clothing recommended for cold weather.

_____

_____

8. List the four principal determinants of violence.

_____

_____

Word Fun emtb.com vocab explorer

The following crossword puzzle is an activity provided to reinforce correct spelling and understanding of medical terminology associated with emergency care and the EMT-B. Use the clues in the column to complete the puzzle.

Across

1. Impenetrable barrier for tactical use

2. Confronts responses and defuses them

5. Exposure by physical touching

7. Disease of the lungs

9. Delayed reaction to past incident

10. Infection of the liver

11. Capable of causing disease

Down

1. Disease spread from person to person

3. Chronic fatigue and frustration

4. Infection control process

6. Confidential discussion group

8. Federal workplace safety agency

9. Blocks entry of an organism into the body

12. May progress to AIDS

Ambulance Calls

The following case scenarios provide an opportunity to explore the concerns associated with patient management. Read each scenario, then answer each question in detail.

1. You are dispatched to a private residence for an "unknown medical problem." You arrive to find family members gathered around the patient's bed. Your patient is breathing and has a pulse, but is not conscious. The family tells you that she's a hospice patient who has cancer. They tell you that a relative from out of state called 9-1-1, and the patient has a DNR order. They tell you to leave.

   What do you do?

   _____

   _____

   _____

2. You are dispatched to a rollover vehicle crash with three known patients. All other ambulances are out on calls; you and your partner will likely not receive assistance for at least 15 minutes. One patient has obvious head trauma, but is still breathing. Your other two patients are conscious but have significant injuries. A crowd is gathering and is watching your every move.

   What do you do?

   _____

   _____

   _____

3. In the process of working a motor vehicle crash, your arm is gashed open and you are exposed to the blood of a patient who tells you that he is HIV positive. You have no water supply in which to wash. Your patient is stable and you are able to control his bleeding with direct pressure.

   How would you best manage this situation?

   _____

   _____

   _____

Skill Drills

Skill Drill 2-1: Proper Glove Removal Technique
Test your knowledge of skill drills by placing the photos below in the correct order. Number the first step with a "1," the second step with a "2," etc.

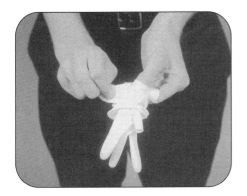

_____

Grasp both gloves with your free hand touching only the clean, interior surfaces.

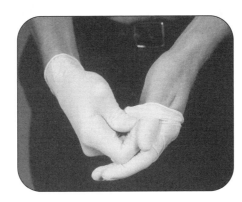

_____

Partially remove the first glove by pinching at the wrist. Be careful to touch only the outside of the glove.

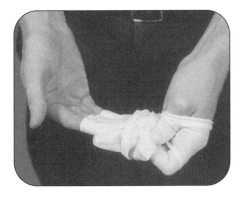

_____

Pull the second glove inside-out toward the fingertips.

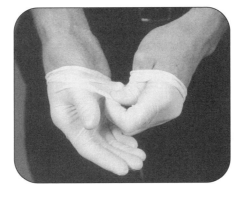

_____

Remove the second glove by pinching the exterior with the partially gloved hand.

CHAPTER

3 Medical, Legal, and Ethical Issues

Workbook Activities

The following activities have been designed to help you. Your instructor may require you to complete some or all of these activities as a regular part of your EMT-B training program. You are encouraged to complete any activity that your instructor does not assign as a way to enhance your learning in the classroom.

Chapter Review

The following exercises provide an opportunity to refresh your knowledge of this chapter.

Matching

Match each of the terms in the left column to the appropriate definition in the right column.

_____ 1. Assault

_____ 2. Abandonment

_____ 3. Advance directive

_____ 4. Battery

_____ 5. Certification

_____ 6. Competent

_____ 7. Consent

_____ 8. Duty to act

_____ 9. Expressed consent

_____ 10. Forcible restraint

_____ 11. Implied consent

_____ 12. Medicolegal

_____ 13. Negligence

_____ 14. Standard of care

A. able to make decisions

B. specific authorization to provide care expressed by the patient

C. confining a person from mental or physical action

D. granted permission

E. touching without consent

F. legal responsibility to provide care

G. written documentation that specifies treatment

H. unlawfully placing a patient in fear of bodily harm

I. unilateral termination of care

J. failure to provide standard of care

K. accepted level of care consistent with training

L. process that recognizes that a person has met set standards

M. legal assumption that treatment was desired

N. relating to law or forensic medicine

Multiple Choice

Read each item carefully, then select the best response.

_____ **1.** The care the EMT-B is able to provide, most commonly defined by state law, is:

A. duty to act.

B. competency.

C. scope of practice.

D. certification.

_____ **2.** How the EMT-B is required to act or behave is called the:

A. standard of care.

B. competency.

C. scope of practice.

D. certification.

_____ **3.** The process by which an individual, institution, or program is evaluated and recognized as meeting certain standards is called:

A. standard of care.

B. competency.

C. scope of practice.

D. certification.

_____ **4.** Negligence is based on the EMT-B's duty to act, cause, breach of duty, and:

A. expressed consent.

B. termination of care.

C. mode of transport.

D. real or perceived damages.

_____ **5.** While treating a patient with a suspected head injury, he becomes verbally abusive and tells you to "leave him alone." If you stop treating him you may be guilty of:

A. neglect.

B. battery.

C. abandonment.

D. slander.

_____ **6.** Good Samaritan laws generally are designed to offer protection to persons who render care in good faith. They do not offer protection from:

A. properly performed CPR.

B. acts of negligence.

C. improvising splinting materials.

D. providing supportive BLS to a DNR patient.

_____ **7.** Which of the following is generally NOT considered confidential?

A. assessment findings

B. patient's mental condition

C. patient's medical history

D. the location of the emergency

_____ **8.** An important safeguard against legal implication is:

A. responding to every call with lights and siren.

B. checking ambulance equipment once a month.

C. transporting every patient to an emergency department.

D. a complete and accurate incident report.

Questions 9-13 are derived from the following scenario: At 2 AM, a 17-year-old son accompanied by his 19-year-old girlfriend had driven to the bar to give his father (who had been drinking large amounts of alcohol) a ride home. On the way back, they were involved in a motor vehicle crash. The son has a large laceration with profuse bleeding on his forehead. His girlfriend is unconscious in the front passenger floor, and the father is standing outside the vehicle, appearing heavily intoxicated, and is refusing care.

_____ **9.** Should the father be allowed to refuse care?

    A. Yes – consent is required before care can be started.

    B. No – he is under the influence of drugs/alcohol and is therefore mentally incompetent.

    C. Yes – under implied consent.

    D. No – you would be guilty of abandonment.

_____ **10.** Why is it permissible for you to begin treatment on the girlfriend?

    A. consent is implied

    B. consent has been expressed

    C. consent was informed

    D. none of the above

_____ **11.** As you progress in your care for the patients, the father becomes unconscious. Can you begin/continue care for him now?

    A. Yes – consent in now implied.

    B. No – he made his wishes known before he fell unconscious.

    C. He just needs to sleep it off.

    D. none of the above

_____ **12.** With the son being a minor, what is the best way to gain consent to begin care, when his father has an altered mental status or is unconscious?

    A. Phone his mother for consent

    B. Call his grandparents for consent

    C. It's a true emergency, so consent is implied

    D. You're covered under the Good Samaritan Laws

_____ **13.** Your responsibility to provide patient care is called:

    A. Scope of practice

    B. Duty to act

    C. DNR

    D. Standard of care.

_____ **14.** Presumptive signs of death would not be adequate in cases of sudden death due to:

    A. hypothermia.

    B. acute poisoning.

    C. cardiac arrest.

    D. all of the above

_____ **15.** Definitive or conclusive signs of death that are obvious and clear to even nonmedical persons include all of the following, except:

    A. profound cyanosis.

    B. dependent lividity.

    C. rigor mortis.

    D. putrefaction.

_____ **16.** Medical examiner's cases include:

    A. violent death.

    B. suicide.

    C. suspicion of a criminal act.

    D. all of the above

_____ **17.** HIPAA is the acronym for the Health Insurance Portability and Accountability Act of 1996. It:

   A. makes ambulance services accountable for transporting patients in a safe manner.

   B. protects the privacy of health care information and to safeguard patient confidentiality.

   C. allows health insurers to transfer an insurance policy to another carrier if a patient does not pay his or her premium.

   D. enables emergency personnel to transfer a patient to a lower level of care when resources are scarce.

Fill-in

Read each item carefully, then complete the statement by filling in the missing word(s).

1. The _____ _____ _____ outlines the care you are able to provide.

2. The _____ _____ _____ is the manner in which the EMT must act when treating patients.

3. The legal responsibility to provide care is called the _____ _____ _____.

4. The determination of _____ is based on duty, breach of duty, damages, and cause.

5. Abandonment is _____ of care without transfer to someone of equal or higher training.

6. _____ consent is given directly by an informed patient, where _____ consent is assumed in the unconscious patient.

7. Unlawfully placing a person in fear of immediate harm is _____, while _____ is unlawfully touching a person without their consent.

8. A(n) _____ _____ is a written document that specifies authorized treatment in case a patient becomes unable to make decisions. A written document that authorizes the EMT not to attempt resuscitation efforts is a(n) _____ _____.

9. Mentally competent patients have the right to _____ _____.

10. Incidents involving child abuse, animal bites, childbirth, and assault have _____ _____ requirements in many states.

True/False

If you believe the statement to be more true than false, write the letter "T" in the space provided. If you believe the statement to be more false than true, write the letter "F."

_____ **1.** Failure to provide care to a patient once you have been called to the scene is considered negligence.

_____ **2.** For expressed consent to be valid, the patient must be a minor.

_____ **3.** If a patient is unconscious and a true emergency exists, the doctrine of implied consent applies.

_____ **4.** The EMT can legally restrain a patient against their will if the patient poses a threat to themselves or others.

_____ **5.** The best defense against legal action for the EMT is to always provide care in a manner consistent with ethical and moral standards.

Short Answer

Complete this section with short written answers using the space provided.

1. In many states, certain conditions allow a minor to be treated as an adult for the purpose of consenting to medical treatment. List three of these conditions.

   _____

   _____

2. When does your responsibility for patient care end?

   _____

   _____

3. There will be some instances when you will not be able to persuade the patient, guardian, conservator, or parent of a minor child or mentally incompetent patient to proceed with treatment. List five steps you should take to protect all parties involved.

   _____

   _____

4. List the two rules of thumb courts consider regarding reports and records.

   _____

   _____

5. List four steps to take when you are called to the scene involving a potential organ donor.

   _____

   _____

Word Fun

The following crossword puzzle is an activity provided to reinforce correct spelling and understanding of medical terminology associated with emergency care and the EMT-B. Use the clues in the column to complete the puzzle.

Across

1. Care that the EMT-B is authorized to provide
3. Relating to medical law
8. Failure to provide standard of care
10. Touching another person without their expressed consent
11. Direct permission to provide care
12. Responsibility to provide care

Down

2. Evaluation and recognition of meeting standards
4. Able to make rational decisions
5. Placing one in fear of bodily harm
6. Assumed permission to provide care
7. Unilateral termination of care
9. A serious situation, such as an injury or illness

Ambulance Calls

The following case scenarios provide an opportunity to explore the concerns associated with patient management. Read each scenario, then answer each question in detail.

1. You are dispatched to a woman complaining of abdominal pain. You arrive to find the 17-year-old female crying and holding her abdomen. She tells you that she fell down the stairs, and that she is pregnant. She is not sure how far along she is, but she is experiencing cramping and spotting. She asks you not to tell anyone, and says that if you tell her parents she will refuse transport to the hospital.

What do you do?

_____

_____

_____

2. You are off-duty when you see a child injured while riding his bike. You examine him and find abrasions on both knees, but no other injuries. He needs help getting to his house up the street. He tells you that his mother is not home, but his grandfather is (although he is bedridden). Looking through the window, you see the house is full of clothing, garbage, and papers.

What do you do?

_____

_____

_____

3. It is late in the night when police summon you to an auto crash. On arrival the officer directs you to the back of his patrol car. Sitting on the seat is your patient, snoring loudly with blood covering his face. The officer states that the patient was involved in a drunk driving accident in which he hit his head on the rear-view mirror. The patient initially refused care at the scene. You were called because his wound continues to bleed. Assessment reveals a sleeping 56-year-old man with a deep, gaping wound over the right eye with moderate venous bleeding. During assessment the patient wakes suddenly and pushes you away. He tells you to "leave him alone."

What actions are necessary in the management of this situation?

_____

_____

_____

CHAPTER

4 The Human Body

Workbook Activities

The following activities have been designed to help you. Your instructor may require you to complete some or all of these activities as a regular part of your EMT-B training program. You are encouraged to complete any activity that your instructor does not assign as a way to enhance your learning in the classroom.

Chapter Review

The following exercises provide an opportunity to refresh your knowledge of this chapter.

Matching

Match each of the terms in the left column to the appropriate definition in the right column.

_____ **1.** Anterior

_____ **2.** Capillary

_____ **3.** Anatomic position

_____ **4.** Superior

_____ **5.** Midline

_____ **6.** Carotid

_____ **7.** Medial

_____ **8.** Inferior

_____ **9.** Femoral

_____ **10.** Proximal

_____ **11.** Brachial

_____ **12.** Distal

_____ **13.** Midaxillary

_____ **14.** Radial

_____ **15.** Posterior

A. closer to the midline

B. farther from the midline

C. farther from the head; lower

D. standing, facing forward, palms facing forward

E. imaginary vertical line descending from the middle of the forehead to the floor

F. front surface of the body

G. closer to the head; higher

H. imaginary vertical line descending from the middle of the armpit to the ankle

I. back or dorsal surface of the body

J. closer to the midline

K. connects arterioles to venules

L. major artery that supplies blood to the head and brain

M. major artery that supplies blood to the lower extremities

N. major artery of the lower arm

O. major artery of the upper arm

For each of the bones listed in the left column, indicate whether it is an upper extremity bone (A) or a lower extremity bone (B).

_____ **16.** Acetabulum   **A.** upper extremity bone

_____ **17.** Patella   **B.** lower extremity bone

_____ **18.** Clavicle

_____ **19.** Fibula

_____ **20.** Calcaneus

_____ **21.** Ulna

_____ **22.** Acromion

For each of the muscle characteristics described in the left column, select the type of muscle from the right column.

_____ **23.** Attaches to the bone   **A.** skeletal

_____ **24.** Found in the walls of   **B.** smooth
          the gastrointestinal tract   **C.** cardiac

_____ **25.** Carries out much of the
          automatic work of the body

_____ **26.** Forms the major muscle mass of the body

_____ **27.** Under the direct control of the brain

_____ **28.** Found only in the heart

_____ **29.** Responds only to primitive stimulus, such as heat

_____ **30.** Can tolerate blood supply interruption for only a very short period

_____ **31.** Responsible for all bodily movement

_____ **32.** Has its own blood supply and electrical system

For each of the parts of the nervous system in the left column, select the phrase in the right column with which it is associated.

_____ **33.** Spinal cord   **A.** exits the brain through an opening at the base of the skull

_____ **34.** Central nervous system   **B.** transmits electrical impulses to the muscles, causing them to contract

_____ **35.** Sensory nerves   **C.** brain and spinal cord

_____ **36.** Motor nerves   **D.** links the central nervous system to various organs in the body

_____ **37.** Brain   **E.** carries sensations of taste and touch to the brain

_____ **38.** Peripheral nervous   **F.** controlling organ of the body system

Multiple Choice

Read each item carefully, then select the best response.

_____ **1.** The topographic term used to describe the location of an injury that is toward the midline center of the body is:

    A. lateral.

    B. medial.

    C. midaxillary.

    D. midclavicular.

_____ **2.** Topographically, the term distal means:

    A. near the trunk.

    B. near a point of reference.

    C. below a point of reference.

    D. toward the center of the body.

_____ 3. The leaf-shaped flap of tissue that prevents food and liquid from entering the trachea is called the:

     A. uvula.

     B. epiglottis.

     C. laryngopharynx.

     D. cricothyroid membrane.

_____ 4. Which of the following systems is responsible for releasing chemicals that regulate body activities?

     A. nervous

     B. endocrine

     C. cardiovascular

     D. skeletal

_____ 5. Which of the following vessels does NOT carry blood to the heart?

     A. inferior venae cavae

     B. superior venae cavae

     C. pulmonary vein

     D. pulmonary artery

Questions 6-10 are derived from the following scenario: Kory, a 16-year-old skateboarder, attempted to "gap" (jumping down a flight of stairs) with his skateboard, landing on his right side. He hit the pavement with his unprotected head, and had two bones sticking out by his right ankle.

_____ 6. His open wound would be known as a(n):

     A. ulna/radial fracture

     B. acromion/humerus fracture

     C. tib/fib fracture

     D. patella/fibula fracture

_____ 7. You would want to keep what part of his spinal column immobilized, so as not to move all 7 vertebrae?

     A. Cervical

     B. Thoracic

     C. Sacrum

     D. Coccyx

_____ 8. If he were to develop pain in his upper-right quadrant, what organ may be causing the pain?

     A. liver

     B. stomach

     C. spleen

     D. appendix

_____ 9. Kory was found in what position?

     A. prone

     B. supine

     C. shock position

     D. recovery position

_____ 10. In order to keep his spinal column straight, in what position would you place him on the cot?

     A. prone

     B. supine

     C. Fowler's position

     D. Trendelenburg position

Labeling

Label the following diagrams with the correct terms.

1. Directional Terms

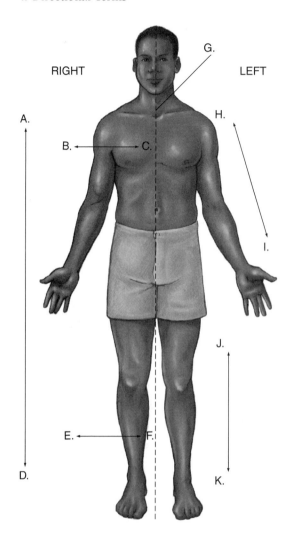

RIGHT   LEFT

A. _____

B. _____

C. _____

D. _____

E. _____

F. _____

G. _____

H. _____

I. _____

J. _____

K. _____

2. Anatomic Positions

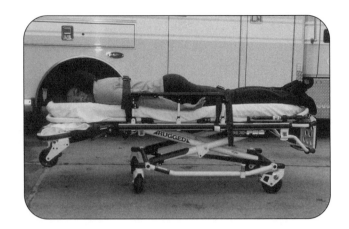
_____

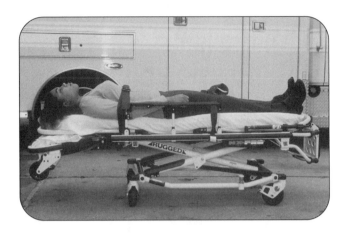
_____

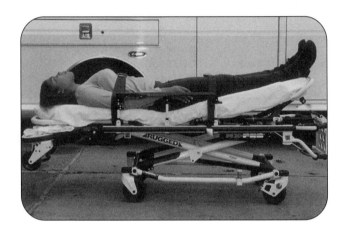
_____

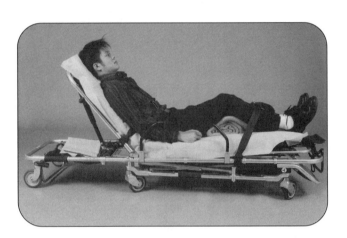
_____

3. The Skeletal System

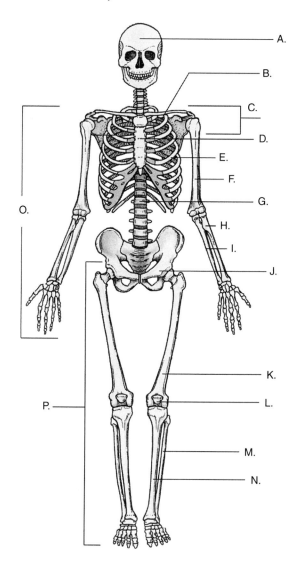

A. _____

B. _____

C. _____

D. _____

E. _____

F. _____

G. _____

H. _____

I. _____

J. _____

K. _____

L. _____

M. _____

N. _____

O. _____

P. _____

4. The Skull

A. _____

B. _____

C. _____

D. _____

E. _____

F. _____

G. _____

H. _____

I. _____

J. _____

K. _____

L. _____

M. _____

N. _____

O. _____

P. _____

Q. _____

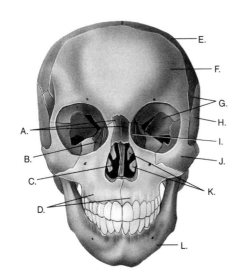

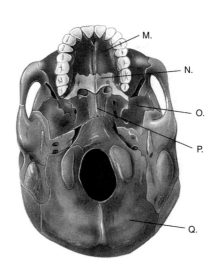

5. The Spinal Column

A. _____

B. _____

C. _____

D. _____

E. _____

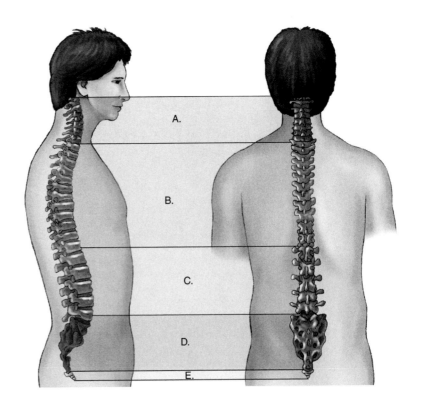

6. The Thorax

A. _____

B. _____

C. _____

D. _____

E. _____

F. _____

G. _____

H. _____

I. _____

J. _____

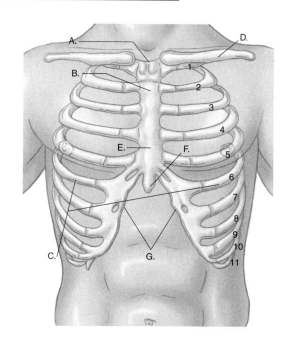

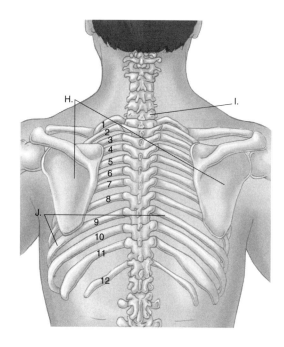

7. The Pelvis

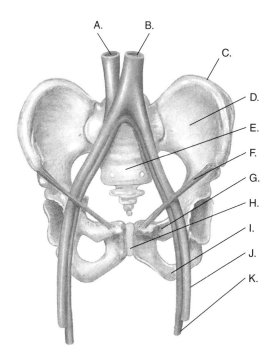

A. _____

B. _____

C. _____

D. _____

E. _____

F. _____

G. _____

H. _____

I. _____

J. _____

K. _____

8. The Lower Extremity

A. _____

B. _____

C. _____

D. _____

E. _____

F. _____

G. _____

H. _____

I. _____

J. _____

K. _____

L. _____

M. _____

N. _____

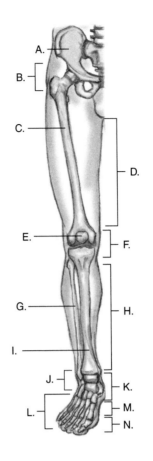

9. The Shoulder Girdle

A. _____

B. _____

C. _____

D. _____

E. _____

F. _____

G. _____

H. _____

I. _____

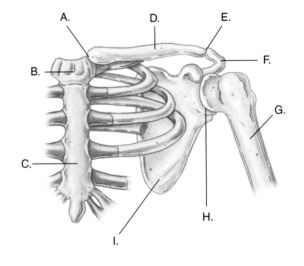

10. The Upper Extremity

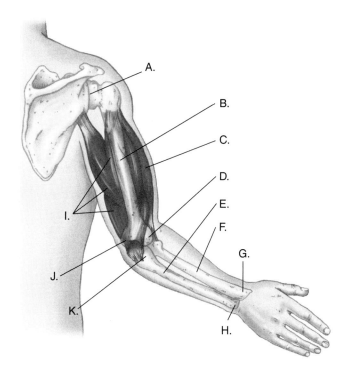

A. _____

B. _____

C. _____

D. _____

E. _____

F. _____

G. _____

H. _____

I. _____

J. _____

K. _____

11. Wrist and Hand

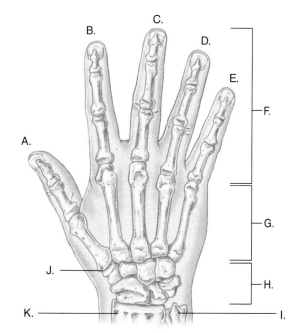

A. _____

B. _____

C. _____

D. _____

E. _____

F. _____

G. _____

H. _____

I. _____

J. _____

K. _____

12. The Respiratory System

A. _____

B. _____

C. _____

D. _____

E. _____

F. _____

G. _____

H. _____

I. _____

J. _____

K. _____

L. _____

M. _____

N. _____

O. _____

P. _____

Q. _____

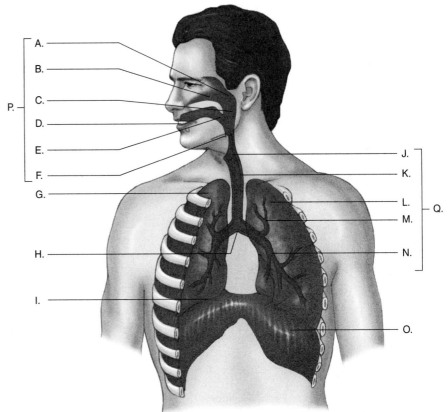

13. The Circulatory System

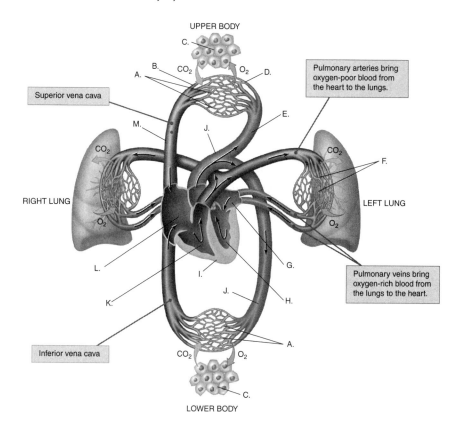

UPPER BODY

Superior vena cava

Pulmonary arteries bring oxygen-poor blood from the heart to the lungs.

RIGHT LUNG

LEFT LUNG

Pulmonary veins bring oxygen-rich blood from the lungs to the heart.

Inferior vena cava

LOWER BODY

A. _____
B. _____
C. _____
D. _____
E. _____
F. _____
G. _____
H. _____
I. _____
J. _____
K. _____
L. _____
M. _____

14. Electrical Conduction

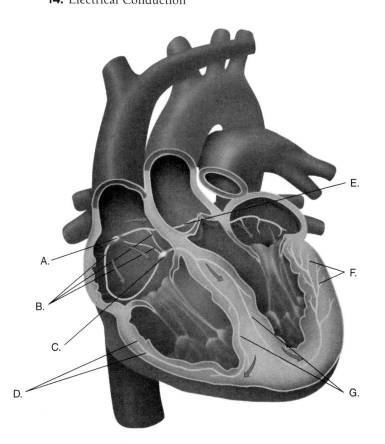

A. _____
B. _____
C. _____
D. _____
E. _____
F. _____
G. _____

15. Central and Peripheral Pulses

A. _____

B. _____

C. _____

D. _____

E. _____

F. _____

G. _____

H. _____

I. _____

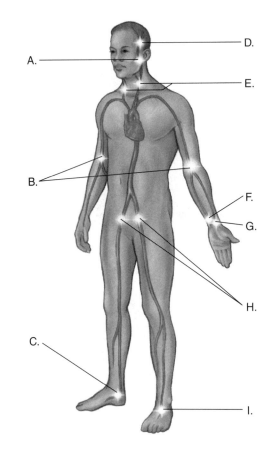

16. The Brain

A. _____

B. _____

C. _____

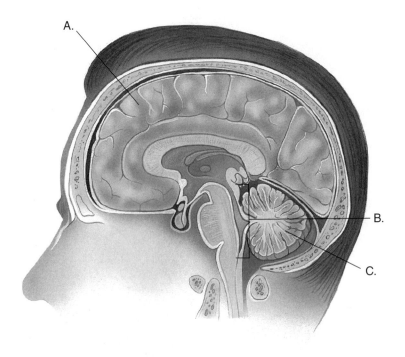

17. Anatomy of the Skin

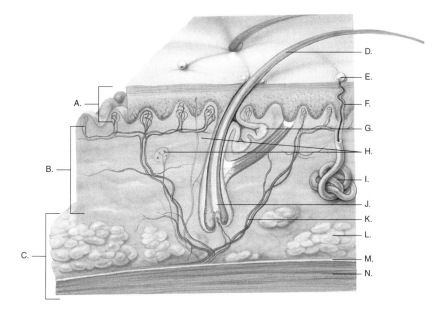

A. _____

B. _____

C. _____

D. _____

E. _____

F. _____

G. _____

H. _____

I. _____

J. _____

K. _____

L. _____

M. _____

N. _____

18. The Male Reproductive System

A. _____   G. _____   M. _____

B. _____   H. _____   N. _____

C. _____   I. _____   O. _____

D. _____   J. _____   P. _____

E. _____   K. _____   Q. _____

F. _____   L. _____

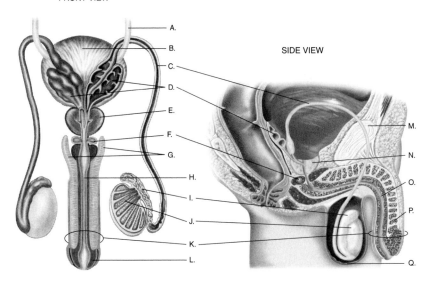

FRONT VIEW

SIDE VIEW

19. The Female Reproductive System

A. _____

B. _____

C. _____

D. _____

E. _____

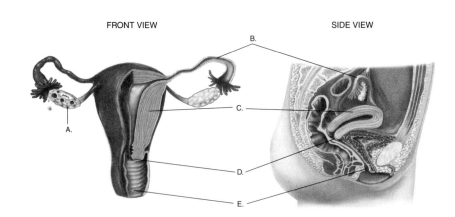

FRONT VIEW   SIDE VIEW

Fill-in
Read each item carefully, then complete the statement by filling in the missing word.

1. There are _____ cervical vertebrae.

2. The movable bone in the skull is the _____ .

3. There is a total of _____ lobes in the right and left lungs.

4. There are _____ pairs of ribs that attach posteriorly to the thoracic vertebrae.

5. The spinal column has _____ vertebrae.

6. The ankle bone is known as _____ .

7. The 11th and 12th rib are called _____ _____.

True/False
If you believe the statement to be more true than false, write the letter "T" in the space provided. If you believe the statement to be more false than true, write the letter "F."

_____ **1.** The aorta is the major artery that supplies the groin and lower extremities with blood.
_____ **2.** The largest joint in the body is the knee.
_____ **3.** The phalanges are the bones of the finger and toes.
_____ **4.** The right atrium receives blood from the pulmonary veins.
_____ **5.** There are 12 ribs that attach to the sternum.

Short Answer
Complete this section with short written answers using the space provided.

1. List the four components of blood and each of their functions.

_____

_____

2. List the five sections of the spinal column and indicate the number of vertebrae in each.

_____

_____

3. What organs are in each of the quadrants of the abdomen?

RUQ _____

_____

LUQ _____

_____

RLQ_____

_____

LLQ_____

_____

4. List in the proper order the parts of the heart that blood flows through.

1. _____ 6. _____

2. _____ 7. _____

3. _____ 8. _____

4. _____ 9. _____

5. _____

Word Fun

The following crossword puzzle is an activity provided to reinforce correct spelling and understanding of medical terminology associated with emergency care and the EMT-B. Use the clues in the column to complete the puzzle.

Across

1. Inner layer of skin

8. Nearer to the feet

10. Lower chamber of heart

13. Front surface of the body

14. Away from the midline, sides

15. Slow, dying respirations or pulse

16. Bone on thumb side of forearm

Down

1. Nearer the end

2. Behind the abdomen

3. Lower jawbone

4. Adequate circulation of blood

5. Upper chamber of the heart

6. Back surface of the body

7. Appears on both sides

9. Windpipe

11. Large solid organ in RUQ

12. Sitting up with knees bent

Ambulance Calls

The following case scenarios provide an opportunity to explore the concerns associated with patient management. Read each scenario, then answer each question in detail.

1. You are dispatched to the scene of a bar fight. A 34-year-old man has been stabbed in the right upper quadrant of the abdomen with a knife.

Using medical terminology, indicate which organ(s) might be affected. How would you describe this patient's injuries?

_____

_____

_____

2. You are dispatched to a one car MVC, car versus telephone pole. You arrive to find an unrestrained driver who is complaining of chest pain. You notice the steering wheel is deformed.

Based on the MOI and his chief complaint, what are his potential injuries?

_____

_____

_____

3. You are dispatched to a local BMX bike track just down the road from the fire station. You arrive to find 14-year-old male patient walking toward you, holding his left arm in place. He tells you that as he was turning a corner on the track, he fell off his bike and landed on his shoulder.

What possible injuries does this patient have?

_____

_____

_____

CHAPTER

5 Baseline Vital Signs and SAMPLE History

Workbook Activities

The following activities have been designed to help you. Your instructor may require you to complete some or all of these activities as a regular part of your EMT-B training program. You are encouraged to complete any activity that your instructor does not assign as a way to enhance your learning in the classroom.

Chapter Review

The following exercises provide an opportunity to refresh your knowledge of this chapter.

Matching

Match each of the terms in the left column to the appropriate definition in the right column.

_____ 1. Pulse

_____ 2. Auscultation

_____ 3. Palpation

_____ 4. Perfusion

_____ 5. Blood pressure

_____ 6. SAMPLE history

_____ 7. Sphygmomanometer

_____ 8. Capillary refill

_____ 9. Bradycardia

_____ 10. Cyanosis

_____ 11. Diaphoretic

_____ 12. Hypertension

_____ 13. Hypotension

_____ 14. Sclera

_____ 15. Tachycardia

A. the pressure of the circulating blood against the walls of the arteries

B. characterized by profuse sweating

C. white portion of the eye

D. examination by touch

E. listening to sounds within the organs, usually with a stethoscope

F. the ability of the circulatory system to restore blood to the capillary blood vessels after it is squeezed out

G. blood pressure that is higher than normal range

H. a patient's history consisting of signs/symptoms, allergies, medications, pertinent past history, last oral intake, and events leading to the illness/injury

I. blood pressure that is lower than normal range

J. rapid heart rate

K. a blood pressure cuff

L. the process whereby blood enters an organ or tissue through its arteries and leaves through its veins, providing nutrients and oxygen and removing waste

M. slow heart rate

N. the pressure wave that is felt with the expansion and contraction of an artery

O. bluish-gray skin color caused by reduced blood-oxygen levels

Multiple Choice
Read each item carefully, then select the best response.

_____ **1.** The reason that a patient or other individual calls 9-1-1 is called the:
   A. signs.
   B. symptoms.
   C. chief complaint.
   D. primary problem.

_____ **2.** Signs include all of the following, except:
   A. dizziness.
   B. marked deformities.
   C. external bleeding.
   D. wounds.

_____ **3.** The first set of vital signs that you obtain is called the:
   A. original vital signs.
   B. baseline vital signs.
   C. actual vital signs.
   D. real vital signs.

_____ **4.** Besides pulse, respirations, and blood pressure, you should also include evaluation of:
   A. level of consciousness.
   B. pupillary reaction.
   C. capillary refill in children.
   D. all of the above

_____ **5.** During inspiration:
   A. the chest rises up and out.
   B. the phase is passive.
   C. carbon dioxide is released.
   D. all of the above

_____ **6** Assess breathing by:
   A. watching for chest rise and fall.
   B. feeling for air through the mouth and nose during exhalation.
   C. listening for breath sounds with a stethoscope.
   D. all of the above

_____ **7.** The normal range for adult respirations is _____ breaths/min.
   A. 8 to 20
   B. 15 to 30
   C. 25 to 50
   D. none of the above

_____ **8.** In the _____ position, the patient sits leaning forward on outstretched arms with the head and chin thrust slightly forward.

    A. Fowler's

    B. tripod

    C. sniffing

    D. lithotomy

_____ **9.** In this position, the patient sits upright with the head and chin thrust slightly forward.

    A. Fowler's

    B. tripod

    C. sniffing

    D. lithotomy

_____ **10.** Signs of labored breathing include all of the following, except:

    A. accessory muscle use.

    B. dyspnea.

    C. retractions.

    D. gasping.

_____ **11.** The _____ is the pressure wave that occurs as each heartbeat causes a surge in the blood circulating through the arteries.

    A. systolic pressure

    B. diastolic pressure

    C. pulse

    D. ventricular pressure

_____ **12.** In responsive patients who are older than 1 year, you should palpate a pulse at the _____ artery.

    A. carotid

    B. femoral

    C. radial

    D. brachial

_____ **13.** In unresponsive patients who are older than 1 year, you should palpate a pulse at the _____ artery.

    A. carotid

    B. femoral

    C. radial

    D. brachial

_____ **14.** A pulse that is weak and _____ should be palpated and counted for a full minute.

    A. difficult to palpate

    B. irregular

    C. extremely slow

    D. all of the above

_____ **15.** When assessing the skin, you should evaluate:

    A. color.

    B. temperature.

    C. moisture.

    D. all of the above

_____ **16.** Perfusion may be assessed in the:

    A. fingernail beds.

    B. lips.

    C. conjunctiva.

    D. all of the above

_____ **17.** Poor peripheral circulation will cause the skin to appear:

  A. pale.

  B. ashen.

  C. gray.

  D. all of the above

_____ **18.** Liver disease or dysfunction may cause _____, resulting in the patient's skin and sclera turning yellow.

  A. cyanosis

  B. jaundice

  C. diaphoresis

  D. lack of perfusion

_____ **19.** The skin will feel cool when the patient:

  A. is in early shock.

  B. has mild hypothermia.

  C. has inadequate perfusion.

  D. all of the above

_____ **20.** Capillary refill reflects the patient's perfusion and is often affected by the patient's:

  A. body temperature.

  B. position.

  C. medications.

  D. all of the above

_____ **21.** With adequate perfusion, the color in the nail bed should be restored to its normal pink within _____ seconds when checking capillary refill.

  A. 1½

  B. 2

  C. 2½

  D. 3

_____ **22.** A decrease in blood pressure may indicate:

  A. loss of blood.

  B. loss of vascular tone.

  C. a cardiac pumping problem.

  D. all of the above

_____ **23.** When blood pressure drops, the body compensates to maintain perfusion to the vital organs by:

  A. decreasing pulse rate.

  B. decreasing the blood flow to the skin and extremities.

  C. decreasing respiratory rate.

  D. dilating the arteries.

_____ **24.** Blood pressure is usually measured through:

  A. auscultation.

  B. palpation.

  C. visualization.

  D. rationalization.

_____ **25.** When obtaining a blood pressure by palpation, you should place your fingertips on the _____ artery.

  A. carotid

  B. brachial

  C. radial

  D. posterior tibial

_____ **26.** You must assume that a patient who has a critically _____ can no longer compensate sufficiently to maintain adequate perfusion.

    A. low blood pressure

    B. high blood pressure

    C. low pulse rate

    D. high pulse rate

_____ **27.** The patient's level of consciousness reflects the status of the:

    A. peripheral nervous system.

    B. central nervous system.

    C. peripheral perfusion.

    D. distal perfusion.

_____ **28.** The diameter and reactivity to light of the patient's pupils reflect the status of the brain's:

    A. perfusion.

    B. oxygenation.

    C. condition.

    D. all of the above

_____ **29.** You should reassess vital signs:

    A. every 15 minutes in a stable patient.

    B. every 5 minutes in an unstable patient.

    C. after every medical intervention.

    D. all of the above

_____ **30.** In the mnemonic "SAMPLE," the "P" stands for:

    A. pupillary response.

    B. pulse rate.

    C. pertinent past history.

    D. pain level.

Questions 31-35 are derived from the following scenario: You've responded to a state wrestling tournament where several wrestling matches are taking place at the same time. The room is extremely noisy with fans cheering, and you are led to a 17-year-old male who appears unconscious and is surrounded by his coaches.

_____ **31.** His unconsciousness is a:

    A. sign

    B. symptom

    C. both A and B

    D. none of the above

_____ **32.** What method should be used to take his blood pressure?

    A. auscultation

    B. palpation

    C. by using the carotid artery

    D. by using the femoral artery

_____ **33.** By using a painful stimulus, the patient moans. Which level of unconsciousness is the patient exhibiting?

    A. A

    B. V

    C. P

    D. U

_____ **34.** Your patient is also exhibiting unequal pupils. This is likely caused by:

    A. internal bleeding

    B. cataracts

    C. brain injury

    D. none of the above

_____ **35.** His skin color, temperature, and moisture are signs of:

A. level of consciousness

B. perfusion

C. brain injury

D. none of the above

Fill-in

Read each item carefully, then complete the statement by filling in the missing word.

1. _____ _____ is the amount of air that is exchanged with each breath.

2. The _____ is the delicate membrane lining the eyelids and covering the exposed surface of the eye.

3. By using your _____ powers, you will be able to interpret the meaning and implications of your findings and the information that you have gathered while assessing the patient.

4. The severity of a _____ is subjective because it is based on the patient's interpretation and tolerance.

5. A patient who is breathing without assistance is said to have _____ _____.

6. _____ _____ are the key signs that are used to evaluate the patient's initial general condition.

7. When assessing respirations, you must determine the rate, _____, and depth of the patient's breathing.

8. When you can actually see the effort of the patient's breathing, it is described as _____ _____.

9. If you can hear bubbling or gurgling, the patient probably has _____ in the airway.

10. The condition of the patient's skin can tell you a lot about the patient's peripheral circulation and _____, blood oxygen levels, and body temperature.

True/False

If you believe the statement to be more true than false, write the letter "T" in the space provided. If you believe the statement to be more false than true, write the letter "F."

_____ **1.** A pulse is an indicator of the condition of the heart.

_____ **2.** A blood pressure determined by palpation is less accurate than if determined by auscultation.

_____ **3.** Only the diastolic pressure can be measured by the palpation method.

_____ **4.** Labored breathing can be described as increased breathing effort, grunting, and use of accessory muscles.

_____ **5.** A conscious patient is likely to alter his or her breathing if he or she is aware that you are evaluating it.

_____ **6.** The pulse is most commonly palpated at the femoral artery.

_____ **7.** The normal pulse range for a newborn is 140 to 160 beats/min.

_____ **8.** To assess skin color in an infant, you should look at the palms of the hands and soles of the feet.

_____ **9.** Cyanosis indicates a need for oxygen.

_____ **10.** BSI precautions should be followed when a patient is jaundiced.

_____ **11.** The skin will feel hot when the patient is in profound shock or has hypothermia.

_____ **12.** Normal reaction to a bright light shone in one eye is pupil constriction in only that eye.

_____ **13.** Normal respirations in an adult are 15 to 30 breaths/min.

Short Answer

Complete this section with short written answers using the space provided.

1. Name the seven basic vital signs.

_____

_____

2. List four abnormal skin colors.

_____

_____

3. Name four factors to be considered when assessing adequate or inadequate breathing.

_____

_____

4. Define systolic blood pressure.

_____

_____

5. Define diastolic blood pressure.

_____

_____

6. Name the three factors to consider when assessing a patient's pulse.

_____

_____

7. Name the three factors to consider when assessing a patient's skin.

_____

_____

8. Describe the process for assessing capillary refill.

_____

_____

9. Define the acronym PEARRL.

_____

_____

10. Explain the difference between a sign and a symptom.

_____

_____

Word Fun

The following crossword puzzle is an activity to reinforce correct spelling and understanding of medical terminology associated with emergency care and the EMT-B. Use the clues in the column to complete the puzzle.

Across

3. Acronym used to assess LOC
4. Circulation of blood within an organ or tissue
6. Pressure against the walls of the arteries
8. BP lower than normal
9. Bluish-gray skin color

Down

1. BP that is higher than normal
2. Harsh, high-pitched inspiratory sound
3. Listening
5. Subjective finding
7. Cardiac pressure wave

Ambulance Calls

The following case scenarios provide an opportunity to explore the concerns associated with patient management. Read each scenario, then answer each question in detail.

1. It's Thanksgiving and you are dispatched to a private residence for "female choking." You arrive to find an older woman who is alert and oriented but is coughing and drooling. Family members tell you that the patient was eating a piece of strip steak when she suddenly clutched her throat and started coughing. She keeps repeating, "It's stuck in my throat!" Every time she tries to swallow, she feels that her airway becomes blocked.

   How would you best manage this patient?

   _____

   _____

   _____

2. It's the first day of a major snowfall in your area. You are dispatched to a man complaining of chest pain. You arrive to find a 50-year-old male patient sitting on the steps of his home. He tells you that he was shoveling snow from his sidewalk when suddenly he felt pain and pressure in his chest. He says that it is slightly better now, but there is still a feeling of "heaviness," and he also feels sick to his stomach.

   How would you best manage this patient?

   _____

   _____

   _____

3. You are called to a residence for a possible cardiac arrest. Family members tell you that the patient, a 57-year-old man, stopped breathing and had no pulse. They also tell you he has a cardiac history. They stopped CPR when he resumed breathing and regained a pulse, approximately 2 minutes before your arrival.

   How would you best manage this patient?

   _____

   _____

   _____

Skill Drills

Skill Drill 5-1: Obtaining a Blood Pressure by Auscultation or Palpation

Test your knowledge of this skill drill by placing the photos below in the correct order. Number the first step with a "1," the second step with a "2," etc.

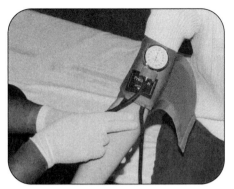

_____

Palpate the brachial artery.

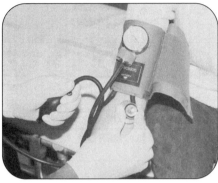

_____

Close the valve and pump to 20 mm Hg above the point at which you stop hearing pulse sounds. Note the systolic and diastolic pressures as you let air escape slowly.

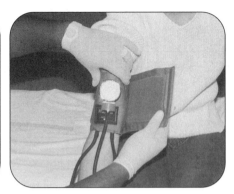

_____

Apply the cuff snugly.

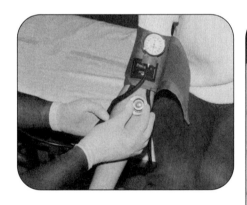

_____

Place the stethoscope over the brachial artery and grasp the ball-pump and turn-valve.

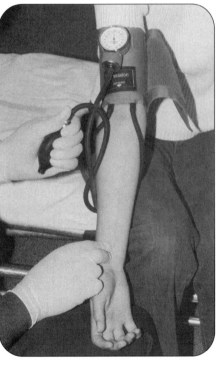

_____

When using the palpatation method, you should place your fingertips on the radial artery so that you feel the radial pulse.

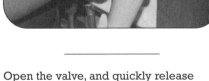

_____

Open the valve, and quickly release remaining air.

CHAPTER

6 Lifting and Moving Patients

Workbook Activities

The following activities have been designed to help you. Your instructor may require you to complete some or all of these activities as a regular part of your EMT-B training program. You are encouraged to complete any activity that your instructor does not assign as a way to enhance your learning in the classroom.

Chapter Review

The following exercises provide an opportunity to refresh your knowledge of this chapter.

Matching
Match each of the terms in the left column to the appropriate definition in the right column.

_____ 1. Extremity lift

_____ 2. Flexible stretcher

_____ 3. Stair chair

_____ 4. Basket stretcher

_____ 5. Scoop stretcher

_____ 6. Backboard

_____ 7. Direct ground lift

_____ 8. Portable stretcher

_____ 9. Wheeled ambulance stretcher

A. separates into two or four pieces

B. tubular framed stretcher with rigid fabric stretched across it

C. used for patients without a spinal injury who are supine

D. specifically designed stretcher that can be rolled along the ground

E. used to carry patients across uneven terrain

F. used for patients who are found lying supine with no suspected spinal injury

G. can be folded or rolled up

H. used to carry patients up and down stairs

I. spine board or longboard

Multiple Choice

Read each item carefully, then select the best response.

_____ **1.** _____ safety depends on the use of proper lifting techniques and maintaining a proper hold when lifting or carrying a patient.

   A. Your

   B. Your team's

   C. The patient's

   D. all of the above

_____ **2.** You should perform an urgent move:

   A. if a patient has an altered level of consciousness.

   B. if a patient has inadequate ventilation or shock.

   C. in extreme weather conditions.

   D. all of the above

_____ **3.** You may injure your back if you lift:

   A. with your back curved.

   B. with your back straight, but bent significantly forward at the hips.

   C. with the shoulder girdle anterior to the pelvis.

   D. all of the above

_____ **4.** When lifting you should:

   A. spread your legs shoulder width apart.

   B. never lift a patient while reaching any significant distance in front of your torso.

   C. keep the weight that you are lifting as close to your body as possible.

   D. all of the above

_____ **5.** When carrying a patient on a cot, be sure:

   A. to flex at the hips.

   B. to bend at the knees.

   C. that you do not hyperextend your back.

   D. all of the above

_____ **6.** In lifting with the palm down, the weight is supported by the _____ rather than the palm.

   A. fingers

   B. forearm

   C. lower back

   D. wrist

_____ **7.** When you must carry a patient up or down a flight of stairs or other significant incline, use a _____ if possible.

   A. backboard

   B. stair chair

   C. stretcher

   D. short spine board

_____ **8.** If you need to lean to either side to compensate for a weight imbalance, you have probably _____ your weight limitation.

   A. met

   B. exceeded

   C. increased

   D. countered

_____ **9.** A backboard is a device that provides support to patients who you suspect have:

    A. hip injuries.

    B. pelvic injuries.

    C. spinal injuries.

    D. all of the above

_____ **10.** The team leader should do all of the following, except _____, before any lifting is initiated.

    A. give a command of execution

    B. indicate where each team member is to be located

    C. rapidly describe the sequence of steps that will be performed

    D. give a brief overview of the stages

_____ **11.** Special _____ are usually required to move any patient who weighs more than 300 lb to an ambulance.

    A. techniques

    B. equipment

    C. resources

    D. all of the above

_____ **12.** When carrying a patient in a stair chair, always remember to:

    A. keep your back in a locked-in position.

    B. flex at the hips, not at the waist.

    C. keep the patient's weight and your arms as close to your body as possible.

    D. all of the above

_____ **13.** When you use a body drag to move a patient:

    A. your back should always be locked and straight.

    B. you should avoid any twisting so that the vertebrae remain in normal alignment.

    C. avoid hyperextending.

    D. all of above

_____ **14.** When pulling a patient, you should do all of the following, except:

    A. extend your arms no more than about 15 to 20 inches.

    B. reposition your feet so that the force of pull will be balanced equally.

    C. when you can pull no farther, lean forward another 15 to 20 inches.

    D. pull the patient by slowly flexing your arms.

_____ **15.** When log rolling a patient:

    A. kneel as close to the patient's side as possible.

    B. lean solely from the hips.

    C. use your shoulder muscles to help with the roll.

    D. all of the above

_____ **16.** If the weight you are pushing is lower than your waist, you should push from:

    A. the waist.

    B. a kneeling position.

    C. the shoulder.

    D. a squatting position.

_____ **17.** If you are alone and must remove an unconscious patient from a car, you should first move the patient's:

    A. legs.

    B. head.

    C. torso.

    D. pelvis.

_____ **18.** Situations in which you should use an emergency move include those where:

    A. there is the presence of fire, explosives, or hazardous materials.

    B. you are unable to protect the patient from other hazards.

    C. you are unable to gain access to others in a vehicle who need lifesaving care.

    D. all of the above

_____ **19.** You can move a patient on his or her back along the floor or ground by using all of the following methods, except:

    A. pulling on the patient's clothing in the neck and shoulder area.

    B. placing the patient on a blanket, coat, or other item that can be pulled.

    C. pulling the patient by the legs if they are the most accessible part.

    D. placing your arms under the patient's shoulders and through the armpits, while grasping the patient's arms, drag the patient backward.

_____ **20.** An urgent move may be necessary for moving a patient with:

    A. an altered level of consciousness.

    B. inadequate ventilation.

    C. shock.

    D. all of the above

_____ **21.** Use the rapid extrication technique in the following situation(s):

    A. The vehicle on the scene is unsafe.

    B. The patient's condition cannot be properly assessed before being removed from the car.

    C. The patient blocks access to another seriously injured patient.

    D. all of the above

_____ **22.** To avoid the strain of unnecessary lifting and carrying, you should use _____ or assist an able patient to the cot whenever possible.

    A. the direct ground lift

    B. the extremity lift

    C. the draw sheet method

    D. a scoop stretcher

_____ **23.** You should use a rigid _____, often called a Stokes litter, to carry a patient across uneven terrain from a remote location that is inaccessible by ambulance or other vehicle.

    A. basket stretcher

    B. scoop stretcher

    C. molded backboard

    D. flotation device

_____ **24.** Basket stretchers can be used:

    A. for technical rope rescues and some water rescues.

    B. to carry a patient across fields on an all-terrain vehicle.

    C. to carry a patient on a toboggan.

    D. all of the above

Questions 25-29 are derived from the following scenario: You've been called to the scene of a high-speed motor-vehicle accident involving two compact cars. The first vehicle was a roll-over, ejecting the driver. The second vehicle contained both a driver and a front seat passenger who can't be reached as their door is up against a building.

_____ **25.** What device will you use to put the roll-over victim onto the wheeled ambulance stretcher?

    A. extremity lift

    B. scoop stretcher

    C. short backboard

    D. backboard

_____ **26.** For the passenger in vehicle number 2, you may need to perform a _____ on the driver in order to reach the patient.

    A. extremity lift

    B. emergency move

    C. short backboard

    D. You should do nothing different, treating each patient the same.

_____ **27.** An advantage of using the diamond carry is that:

    A. It uses an even number of people (less likely to drop)

    B. It can be done with one person, freeing up others for patient care

    C. They can be slid along the ground

    D. It provides the best means of spinal immobilization

_____ **28.** You'll likely use the _____ to transfer the patient from your cot to the hospital bed.

    A. diamond carry

    B. scoop stretcher

    C. portable cot

    D. draw sheet method

_____ **29.** After the doctor examines the patient, you are asked to transfer the patient to a specialized facility. What method will be likely used to get the patient back onto your cot?

    A. extremity lift

    B. draw sheet method

    C. direct ground lift

    D. scoop stretcher

_____ **30.** Bariatrics is:

    A. The branch of medicine concerned with the elderly.

    B. The branch of medicine concerned with the obese.

    C. The branch of medicine concerned with babies.

    D. The method used to take blood pressures.

Fill-in

Read each item carefully, then complete the statement by filling in the missing word.

1. To avoid injury to you, the patient, or your partners, you will have to learn how to lift and carry the patient properly, using proper _____ _____ and a power grip.

2. The key rule of lifting is to always keep the back in a straight, _____ position and to lift without twisting.

3. The safest and most powerful way to lift, lifting by extending the properly placed flexed legs, is called a _____ _____.

4. The arm and hand have their greatest lifting strength when facing _____ up.

5. Be sure to pick up and carry the backboard with your back in the _____ position.

6. You should not attempt to lift a patient who weighs more than _____ pounds with fewer than four rescuers, regardless of individual strength.

7. During a body drag where you and your partner are on each side of the patient, you will have to alter the usual pulling technique to prevent pulling _____ and producing adverse lateral leverage against your lower back.

8. When you are rolling the wheeled ambulance stretcher, your back should be _____, straight, and untwisted.

9. Be careful that you do not push or pull from a(n) _____ position.

10. Remember to always consider whether there is an option that will cause _____ _____ to you and the other EMT-Bs.

11. The manual support and immobilization that you provide when using the rapid extrication technique produce a greater risk of _____ _____.

12. The _____ _____ _____ is used for patients with no suspected spinal injury who are found lying supine on the ground.

13. The _____ _____ may be especially helpful when the patient is in a very narrow space or there is not enough room for the patient and a team of EMTs to stand side by side.

14. The mattress on a stretcher must be _____ _____ so that it does not absorb any type of potentially infectious material, including water, blood, or other body fluid.

15. A _____ _____ may be used for patients who have been struck by a motor vehicle.

True/False

If you believe the statement to be more true than false, write the letter "T" in the space provided. If you believe the statement to be more false than true, write the letter "F."

_____ **1.** A portable stretcher is typically a lightweight folding device that does not have the undercarriage and wheels of a true ambulance stretcher.

_____ **2.** The term "power lift" refers to a posture that is safe and helpful for EMT-Bs when they are lifting.

_____ **3.** If you find that lifting a patient is a strain, try to move to the ambulance as quickly as possible to minimize the possibility of back injury.

_____ **4.** The use of adjunct devices and equipment, such as sheets and blankets, may make the job of lifting and moving a patient more difficult.

_____ **5.** One-person techniques for moving patients should only be used when immediate patient movement is necessary due to a life-threatening hazard and only one EMT-B is available.

_____ **6.** A scoop stretcher may be used alone for a standard immobilization of a patient with a spinal injury.

_____ **7.** When carrying a patient down stairs or on an incline, make sure the stretcher is carried with the head end first.

_____ **8.** The rapid extrication technique is the preferred technique to use on all sitting patients with possible spinal injuries.

_____ **9.** It is unprofessional for you to discuss and plan a lift at the scene in front of the patient.

_____ **10.** Americans are becoming so large that a new field of medicine has been named for the care of the obese.

Short Answer

Complete this section with short written answers using the space provided.

1. List the one-rescuer drags, carries, and lifts.

_____

_____

2. List the situations where the rapid extrication technique is used.

_____

_____

3. List the three guidelines for loading the cot into the ambulance.

_____

_____

4. List the five guidelines for carrying a patient on a cot.

_____

_____

5. Describe the key rule of lifting.

_____

_____

Word Fun emtb.com vocab explorer

The following crossword puzzle is an activity provided to reinforce correct spelling and understanding of medical terminology associated with emergency care and the EMT-B. Use the clues in the column to complete the puzzle.

Across

3. Stretcher used with technical rescues, particularly when water is involved

5. Safest way to lift

7. Stretcher that becomes rigid when secured around patient

10. Tubular framed stretcher with fabric across

11. Four-rescuer carry with one at the head, one at the foot, and one on each side

Down

1. A ground lift used with no suspected spinal injury

2. Used to support patient with hip, pelvic, or spine injury

4. Folding device for moving seated patients up or down floors

6. Stretcher designed to roll along ground

8. Using the patient's limbs to lift

9. Stretcher that can be split into two or four sections

Ambulance Calls

The following case scenarios provide an opportunity to explore the concerns associated with patient management. Read each scenario, then answer each question in detail.

1. You are dispatched to a construction site for a 26-year-old man who fell into a ravine. He is approximately 35 feet down a rocky ledge. He is alert with an unstable pelvis and weak radial pulses. You have all the help you need from the construction crew and the volunteer fire department.

How would you best manage this patient?

_____

_____

_____

2. You are dispatched to "unknown medical problem" at a local residence. You are met at the door by the wife of the patient who tells you that her husband is in the bathroom and is not acting right. You find the 350 pound patient lying in the bathroom, stuck between the bathroom toilet and the wall. He is not breathing and has no pulse.

What do you do?

_____

_____

_____

3. You are dispatched to "difficulty breathing" at a nearby apartment complex. The patient's apartment is located on the top floor of a three-story building, is accessed through an exterior entryway and no elevators are available. Your patient is morbidly obese and cannot walk.

What do you do?

_____

_____

_____

Skill Drills

Skill Drill 6-1: Performing the Power Lift
Test your knowledge of this skill drill by filling in the correct words in the photo captions.

1. Lock your back into a(n) _____, inward curve. _____ and bend your legs. Grasp the backboard, palms up and just in front of you. _____ and _____ the weight between your arms.

2. Position your feet, _____ the object, and _____ weight.

3. _____ your legs and lift, keeping your back locked in.

Skill Drill 6-2: Performing the Diamond Carry

Test your knowledge of this skill drill by placing the photos below in the correct order. Number the first step with a "1," the second step with a "2," etc.

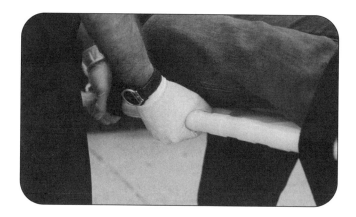

_____

EMT-Bs at the side each turn the head-end hand palm down and release the other hand.

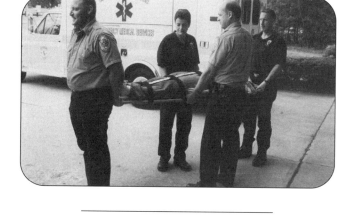

_____

After the patient has been lifted, the EMT-B at the foot turns to face forward.

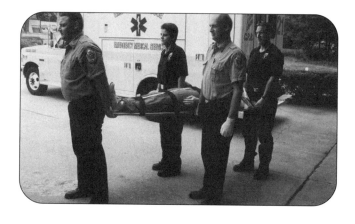

_____

EMT-Bs at the side turn toward the foot end.

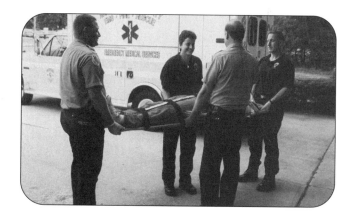

_____

Position yourselves facing the patient.

Skill Drill 6-3: Performing the One-Handed Carrying Technique
Test your knowledge of this skill drill by filling in the correct words in the photo captions.

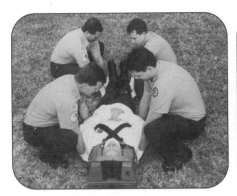 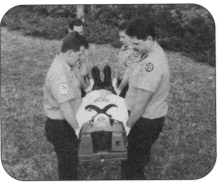 

1. _____ each other and use both _____.

2. Lift the backboard to _____ _____.

3. _____ in the direction you will walk and _____ to using one hand.

Skill Drill 6-4: Carrying a Patient on Stairs
Test your knowledge of this skill drill by filling in the correct words in the photo captions.

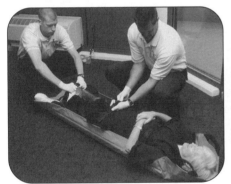  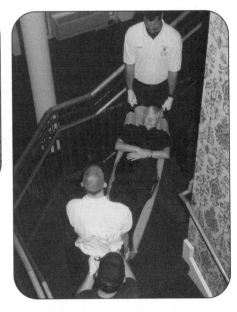

1. _____ the patient securely. Make sure one strap is tight across the _____ _____, under the arms, and secured to the handles to prevent the patient from _____.

2. Carry a patient down stairs with the _____ end first, _____ elevated.

3. Carry the _____ end first going up stairs, always keeping the head elevated.

Skill Drill 6-5: Using a Stair Chair
Test your knowledge of this skill drill by filling in the correct words in the photo captions.

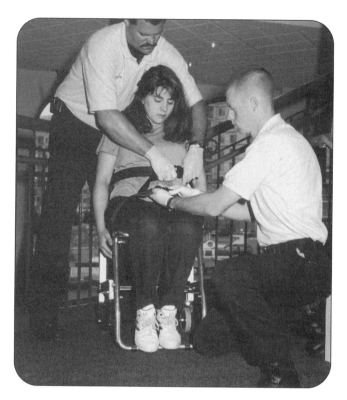

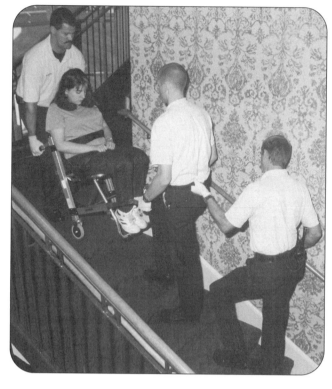

1. Position and secure the patient on the chair with _____ .

2. Take your places at the _____ and _____ of the chair.

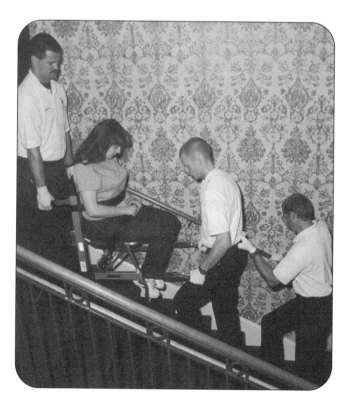

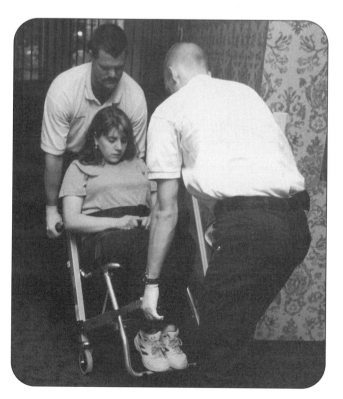

3. A third _____ "backs up" the rescuer carrying the _____ .

4. _____ the chair to roll on landings, or for transfer to the cot.

Skill Drill 6-6: Performing the Rapid Extrication Technique

Test your knowledge of this skill drill by placing the photos below in the correct order. Number the first step with a "1," the second step with a "2," etc.

Second EMT-B supports the torso.

Third EMT-B frees the patient's legs from the pedals and moves the legs together, without moving the pelvis or spine.

First EMT-B provides in-line manual support of the head and cervical spine.

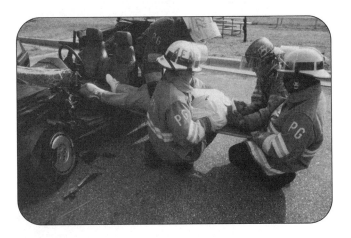

Third EMT-B exits the vehicle, moves to the backboard opposite Second EMT-B, and they continue to slide the patient until patient is fully on the board.

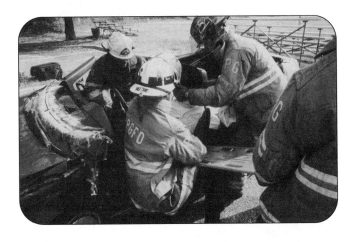

First (or Fourth) EMT-B places the backboard on the seat against patient's buttocks.

(continued)

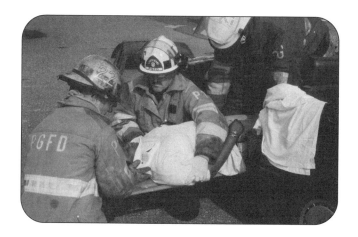

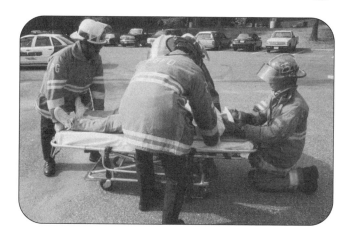

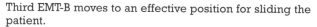

Third EMT-B moves to an effective position for sliding the patient.

Second and Third EMT-Bs slide the patient along the backboard in coordinated, 8" to 12" moves until the hips rest on the backboard.

First (or Fourth) EMT-B continues to stabilize the head and neck while Second and Third EMT-Bs carry the patient away from the vehicle.

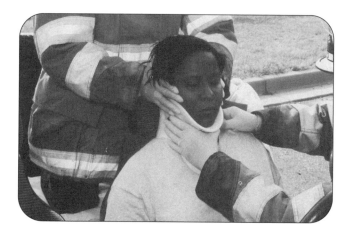

Second and Third EMT-Bs rotate the patient as a unit in several short, coordinated moves.

First EMT-B (relieved by Fourth EMT-B or bystander as needed) supports the head and neck during rotation (and later steps).

Second EMT-B gives commands, applies a cervical collar, and performs the initial assessment.

Skill Drill 6-7: Extremity Lift

Test your knowledge of this skill drill by filling in the correct words in the photo captions.

1. Patient's hands are

   _____ over the chest.

   First EMT-B grasps patient's wrists

   or _____ and pulls

   patient to a _____

   position.

2. When the patient is sitting, First

   EMT-B passes his or her arms through

   patient's _____ and grasps the

   patient's opposite (or his or her own)

   _____ or _____.

   Second EMT-B kneels between the

   _____, facing in the same

   direction as the patient, and places

   his or her hands under the

   _____.

3. Both EMT-Bs rise to _____.

   On _____, both lift and begin

   to move.

Skill Drill 6-8: Using a Scoop Stretcher

Test your knowledge of this skill drill by filling in the correct words in the photo captions.

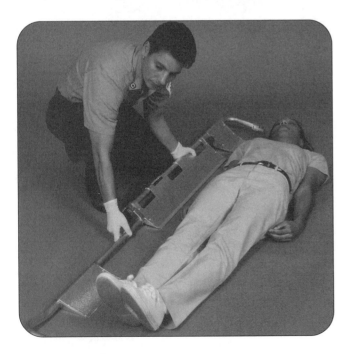

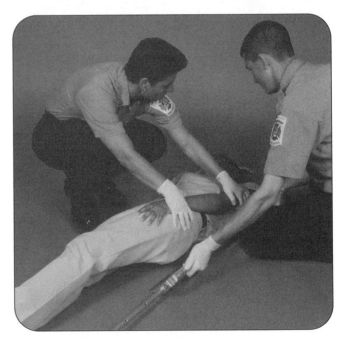

1. Adjust stretcher _____.

2. _____ the patient slightly and _____

   stretcher into place, one side at a time.

(continued)

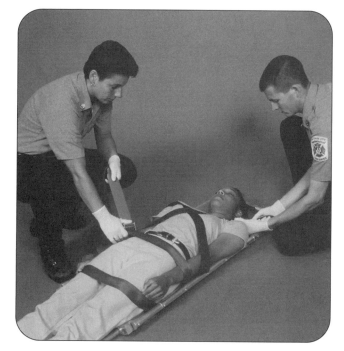

3. _____ the stretcher ends together, avoiding _____ .

4. _____ the patient to the scoop stretcher and _____ it to the cot.

Skill Drill 6-9: Loading a Cot into an Ambulance

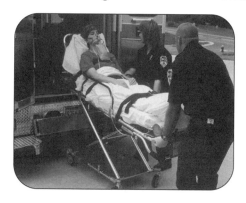

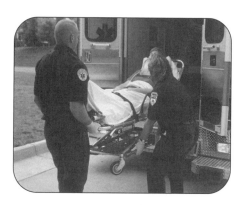

1. Tilt the head of the cot _____, and place it into the patient compartment with the wheels on the floor.

2. Second rescuer on the side of the cot releases the _____ lock and lifts the undercarriage.

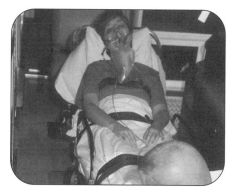

3. _____ the cot into the back of the ambulance.

4. Secure the cot to the _____ mounted in the ambulance.

CHAPTER

7 Airway

Workbook Activities

The following activities have been designed to help you. Your instructor may require you to complete some or all of these activities as a regular part of your EMT-B training program. You are encouraged to complete any activity that your instructor does not assign as a way to enhance your learning in the classroom.

Chapter Review

The following exercises provide an opportunity to refresh your knowledge of this chapter.

Matching

Match each of the terms in the left column to the appropriate definition in the right column.

_____ **1.** Inhalation	**A.** moves down slightly when it contracts
_____ **2.** Exhalation	**B.** irregular breathing pattern with increased rate and depth, followed by apnea
_____ **3.** Alveoli	**C.** active part of breathing
_____ **4.** Automatic function	**D.** voice box
_____ **5.** Hypoxic drive	**E.** amount of air moved during one breath
_____ **6.** Tidal volume	**F.** raises ribs when it contracts
_____ **7.** Diaphragm	**G.** controls breathing when we sleep
_____ **8.** Intercostal muscle	**H.** site of oxygen diffusion
_____ **9.** Ventilation	**I.** thorax size decreases
_____ **10.** Larynx	**J.** insufficient oxygen for cells and tissues
_____ **11.** Hypoxia	**K.** backup system to control respiration
_____ **12.** Cheyne-Stokes	**L.** exchange of air between lungs and respirations environment

Multiple Choice

Read each item carefully, then select the best response.

_____ **1.** What percentage of the air we breathe is made up of oxygen?
   A. 78%
   B. 12%
   C. 16%
   D. 21%

_____ **2.** Regarding the maintenance of the airway in an unconscious adult, which of the following is false?
   A. Insertion of an oropharyngeal airway helps keep the airway open.
   B. The head tilt-chin lift maneuver should always be used to open the airway.
   C. Secretions should be suctioned from the mouth as necessary.
   D. Inserting a rigid suction catheter beyond the tongue may cause gagging.

_____ **3.** The normal respiratory rate for an adult is:
   A. about equal to the person's heart rate.
   B. 12 to 20 breaths/min.
   C. faster when the person is sleeping.
   D. the same as in infants and children.

_____ **4.** All of the following conditions are associated with hypoxia, except:
   A. heart attack.
   B. altered mental status.
   C. chest injury.
   D. hyperventilation syndrome.

_____ **5.** The brain stem normally triggers breathing by increasing respirations when:
   A. carbon dioxide levels increase.
   B. oxygen levels increase.
   C. carbon dioxide levels decrease.
   D. nitrogen levels decrease.

_____ **6.** Which of the following is not a sign of abnormal breathing?
   A. Warm, dry skin
   B. Speaking in two- or three-word sentences
   C. Unequal breath sounds
   D. Skin pulling in around the ribs during inspiration

_____ **7.** The proper technique for sizing an oropharyngeal airway before insertion is to:
   A. measure the device from the tip of the nose to the earlobe.
   B. measure the device from the bridge of the nose to the tip of the chin.
   C. measure the device from the corner of the mouth to the earlobe.
   D. measure the device from the center of the jaw to the earlobe.

_____ **8.** What is the most common problem you may encounter when using a BVM device?
   A. volume of the BVM device
   B. positioning of the patient's head
   C. environmental conditions
   D. maintaining an airtight seal

_____ **9.** When ventilating a patient with a BVM device, you should:
   A. look for inflation of the cheeks.
   B. look for signs of the patient breathing on his or her own.
   C. look for rise and fall of the chest.
   D. listen for gurgling.

_____ **10.** Suctioning the oral cavity of an adult should be accomplished within:

   A. 5 seconds.

   B. 10 seconds.

   C. 15 seconds.

   D. 20 seconds.

Questions 11-15 are derived from the following scenario: You respond to a construction site and find a worker lying supine in the dirt. He has been hit by a heavy construction vehicle and flew over 15 feet of distance before landing in this position. There is discoloration and distention of the abdomen about the RUQ. He is unconscious and his respirations are 10 breaths/min and shallow, with noisy gurgling sounds.

_____ **11.** What airway technique will you use to open his airway?

   A. Head tilt-neck lift

   B. Jaw thrust

   C. Head tilt-chin lift

   D. None of the above

_____ **12.** After opening the airway, your next priority is to:

   A. provide oxygen at 6 L/min via nonrebreathing mask.

   B. provide oxygen at 15 L/min via nasal cannula.

   C. assist respirations.

   D. suction the airway.

_____ **13.** What method will you use to keep his airway open?

   A. Nasal cannula

   B. Jaw thrust

   C. Oropharyngeal airway

   D. All of the above can be used

_____ **14.** While assisting with respirations, you note gastric distention. In order to prevent or alleviate the distention, you should:

   A. ensure the patient's airway is appropriately positioned.

   B. ventilate the patient at the appropriate rate.

   C. ventilate the patient at the appropriate volume.

   D. all of the above

_____ **15.** The correct ventilation rate for assisting this adult patient is:

   A. 1 breath per 5–6 seconds

   B. 1 breath per 3–5 seconds

   C. 1 breath per 10–12 seconds

   D. There is no need to assist with ventilations for this patient.

_____ **16.** Which is the preferred method of assisting ventilations?

   A. Mouth-to-mask with one-way valve

   B. Two-person BVM device with reservoir and supplemental oxygen

   C. Flow-restricted, oxygen-powered ventilation device

   D. One-person BVM device with oxygen reservoir and supplemental oxygen

_____ **17.** When a person goes _____ minutes without oxygen, brain damage is very likely.

   A. 0-4

   B. 4-6

   C. 6-10

   D. more than 10

_____ **18.** If your partner, while examining a patient, states their lungs are equal and bilateral, you would understand them to mean:

A. both lungs have labored breathing.

B. both lungs are equally bad.

C. they have no lungs.

D. clear and equal lung sounds on both sides.

_____ **19.** What are agonal respirations?

A. Occasional gasping breaths, but adequate to maintain life

B. Occasional gasping breaths, unable to maintain life

C. Painful respirations due to broken ribs

D. Another name for ataxic respirations

_____ **20.** You come upon a patient who is not injured and is breathing on their own with a normal rate and an adequate tidal volume. What would be the advantage of placing them in the recovery position?

A. It's the preferred position of comfort for patients.

B. It helps to protect their cervical spine when injuries are hidden.

C. It helps prevent the aspiration of vomitus.

D. It's easier to load them onto the cot from this position.

Labeling

Label the following diagrams with the correct terms.

1. Upper and Lower Airways

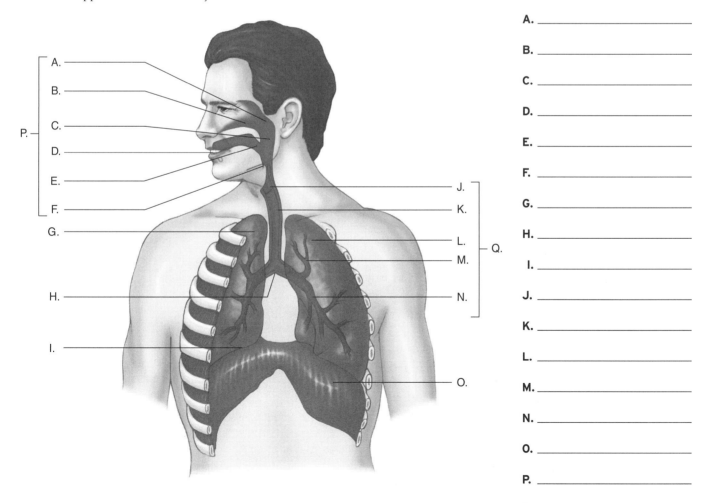

A. _____

B. _____

C. _____

D. _____

E. _____

F. _____

G. _____

H. _____

I. _____

J. _____

K. _____

L. _____

M. _____

N. _____

O. _____

P. _____

Q. _____

2. The Thoracic Cage

A. _____

B. _____

C. _____

D. _____

E. _____

F. _____

G. _____

H. _____

I. _____

J. _____

Fill-in

Read each item carefully, then complete the statement by filling in the missing word(s).

1. Air enters the body through the _____.

2. In exhalation, air pressure in the lungs is _____ than the pressure outside.

3. The air we breathe contains _____ percent oxygen and _____ percent nitrogen.

4. The primary mechanism for triggering breathing is the level of _____ _____ in the blood.

5. During inhalation, the _____ and _____ _____ contract, causing the thorax to enlarge.

6. The drive to breathe is triggered by _____ _____ or _____ _____ _____ levels in arterial blood.

7. Insufficient oxygen in the cells and tissues is called _____.

True/False

If you believe the statement to be more true than false, write the letter "T" in the space provided. If you believe the statement to be more false than true, write the letter "F."

_____ **1.** Nasal airways keep the tongue from blocking the upper airway and facilitate suctioning of the oropharynx.

_____ **2.** Nasal cannulas can deliver a maximum of 44% oxygen at 6 L/min.

_____ **3.** Oral airways should be measured from the tip of the nose to the earlobe.

_____ **4.** Compressed gas cylinders pose no unusual risk.

_____ **5.** The pin-indexing system is used to ensure compatibility between pressure regulators and oxygen flowmeters.

Short Answer

Complete this section with short written answers using the space provided.

1. List the five early signs of hypoxia.

_____

_____

2. What are the normal respiratory rates for adults, children, and infants?

_____

_____

3. How can you avoid gastric distention while performing artificial ventilation?

_____

_____

4. List the six steps for providing one-rescuer artificial ventilation with a BVM device.

_____

_____

5. List six signs of inadequate breathing.

_____

_____

6. What are accessory muscles? Name three.

_____

_____

7. When should medical control be consulted before inserting a nasal airway?

_____

_____

8. List the three steps in nasal airway insertion.

_____

_____

9. What is the best suction tip for suctioning the pharynx, and why?

_____

_____

10. What is the time limit for each episode of suctioning an adult?

_____

_____

Word Fun

The following crossword puzzle is an activity provided to reinforce correct spelling and understanding of medical terminology associated with emergency care and the EMT-B. Use the clues in the column to complete the puzzle.

Across

3. Requires more than normal effort
5. Dying, gasping respirations
7. Larynx, nose, mouth, throat
9. Not enough O_2
11. Face piece attached to a reservoir
12. Exchange of air between lungs and outside

Down

1. Opening in neck connected to trachea
2. Mechanism that causes retching
4. Limits exposure to body fluids
6. Airway inserted in nostril
8. Airway inserted in mouth
10. Diaphragm relaxes

Ambulance Calls

The following case scenarios provide an opportunity to explore the concerns associated with patient management. Read each scenario, then answer each question in detail.

1. You are dispatched to a crash with multiple patients early one morning near the end of your shift. Your patient was the unrestrained driver of one of the vehicles. She is 38 years old and struck her face against the steering wheel and windshield. There is a large laceration on her nose and several teeth are missing. Though unconscious, she has vomited a large amount of food and blood, which is pooling in her mouth. You note gurgling noises as she attempts to breathe.

   How would you best manage this patient?

   _____

   _____

   _____

2. You are dispatched to a local restaurant for an unconscious woman. As you arrive, you are greeted by a frantic restaurant manager. She tells you that one of her staff members went into the restroom to complete the hourly cleaning routine and found a woman lying on the floor, motionless and apparently not breathing. You and your partner enter the cramped ladies' bathroom, to find an older woman who is apneic, cyanotic, and has a carotid pulse. You attempt to ventilate the patient using a BVM device, but are unsuccessful.

   How should you manage this patient?

   _____

   _____

   _____

3. You are outside doing some yard work when you hear one of your neighbors call for help. As you cross the street, you see a husband standing over his wife who is lying on the ground. He tells you that she complained of feeling lightheaded and then she suddenly passed out. He further tells you that he was able to help her to the ground without injury. When you assess her, you hear snoring sounds as she breathes.

   How do you manage this patient?

   _____

   _____

   _____

Skill Drills

Skill Drill 7-1: Positioning the Unconscious Patient

Test your knowledge of skill drills by placing the photos below in the correct order. Number the first step with a "1," the second step with a "2," etc.

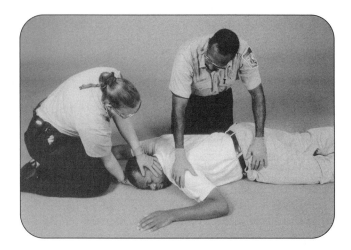

_____

Have your partner place his or her hand on the patient's far shoulder and hip.

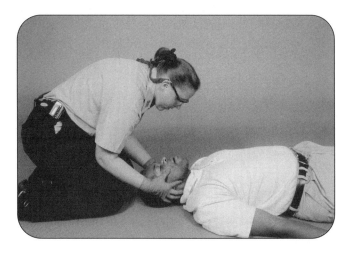

_____

Open and assess the patient's airway and breathing status.

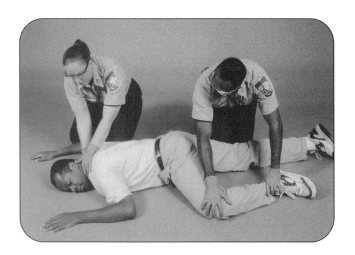

_____

Support the head while your partner straightens the patient's legs.

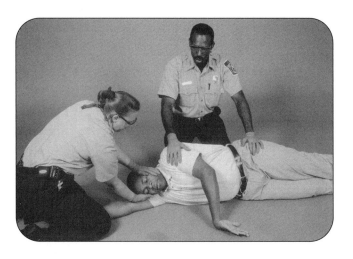

_____

Roll the patient as a unit with the person at the head calling the count to begin the move.

CHAPTER

8 Patient Assessment

Workbook Activities

The following activities have been designed to help you. Your instructor may require you to complete some or all of these activities as a regular part of your EMT-B training program. You are encouraged to complete any activity that your instructor does not assign as a way to enhance your learning in the classroom.

Chapter Review

The following exercises provide an opportunity to refresh your knowledge of this chapter.

Matching

Match each of the terms in the left column to the appropriate definition in the right column.

_____ 1. Triage
_____ 2. Sclera
_____ 3. Subcutaneous emphysema
_____ 4. Paradoxical motion
_____ 5. Conjunctiva
_____ 6. Crepitus
_____ 7. Accessory muscles
_____ 8. Breath sounds
_____ 9. Chief complaint
_____ 10. Frostbite
_____ 11. Jaundice
_____ 12. Orientation

A. indication of air movement in the lungs
B. lining of the eyelid
C. movements in which the skin pulls in around the ribs during inspiration
D. white of the eyes
E. the mental status of a patient
F. the six pain questions
G. yellow skin color due to liver disease or dysfunction
H. a crackling sound
I. examine by touch
J. damage to tissues as the result of exposure to cold
K. the way in which a patient responds to external stimuli
L. secondary muscles of respiration

_____ **13.** OPQRST

_____ **14.** Palpate

_____ **15.** Responsiveness

_____ **16.** Retractions

M. air under the skin

N. motion of a segment of chest wall that is opposite the normal movement during breathing

O. the reason a patient called for help

P. sorting

Multiple Choice

Read each item carefully, then select the best response.

_____ **1.** The scene size-up consists of all of the following, except:

    A. determining mechanism of injury.

    B. requesting additional assistance.

    C. determining level of responsiveness.

    D. PPE/BSI.

_____ **2.** Possible dangers you may observe during the scene size-up include:

    A. oncoming traffic.

    B. unstable surfaces.

    C. downed electrical lines.

    D. all of the above

_____ **3.** With _____, the force of the injury occurs over a broad area, and the skin is usually not broken.

    A. motor vehicle crashes

    B. blunt trauma

    C. penetrating trauma

    D. gunshot wounds

_____ **4.** With _____, the force of the injury occurs at a small point of contact between the skin and the object.

    A. motor vehicle crashes

    B. blunt trauma

    C. penetrating trauma

    D. falls

_____ **5.** You can use the mechanism of injury as a kind of guide to predict the potential for a serious injury by evaluating:

    A. the amount of force applied to the body.

    B. the length of time the force was applied.

    C. the areas of the body that are involved.

    D. all of the above

_____ **6.** In penetrating trauma, the severity of injury depends on all of the following, except:

    A. the geographic location.

    B. the characteristics of the penetrating object.

    C. the amount of force or energy.

    D. the part of the body affected.

_____ **7.** In order to quickly determine the nature of illness, talk with:

    A. the patient.

    B. family members.

    C. bystanders.

    D. all of the above

_____ **8.** When considering the need for additional resources, questions to ask include all of the following, except:

  A. How many patients are there?

  B. Is it raining?

  C. Who contacted EMS?

  D. Does the scene pose a threat to you or your patient's safety?

_____ **9.** The initial assessment includes evaluation of all of the following, except:

  A. mental status.

  B. pupils.

  C. airway.

  D. circulation.

_____ **10.** The best indicator of brain function is the patient's:

  A. pulse rate.

  B. papillary response.

  C. mental status.

  D. respiratory rate and depth.

_____ **11.** An altered mental status may be caused by:

  A. head trauma.

  B. hypoxemia.

  C. hypoglycemia.

  D. all of the above

_____ **12.** All of the following are signs of inadequate breathing except:

  A. tightness in the chest.

  B. two- to three-word dyspnea.

  C. use of accessory muscles.

  D. nasal flaring.

_____ **13.** The _____ of the patient's pulse will give you a general idea of the overall status of the patient's cardiac function.

  A. rate

  B. rhythm

  C. strength

  D. all of the above

_____ **14.** The AED should be used on medical patients who are at least _____ year(s) old and who have been assessed to be unresponsive, apneic, and pulseless.

  A. 1

  B. 8

  C. 9

  D. 10

_____ **15.** In almost all instances, controlling external bleeding is accomplished by:

  A. direct pressure.

  B. elevation.

  C. pressure points.

  D. tourniquet.

_____ **16.** Assessing the _____ is one of the most important and readily accessible ways of evaluating circulation.

  A. pulse

  B. respirations

  C. skin

  D. capillary refill

_____ **17.** Skin color depends on:

    A. pigmentation.

    B. blood oxygen levels.

    C. the amount of blood circulating through the vessels of the skin.

    D. all of the above

_____ **18.** In deeply pigmented skin, you should look for changes in color in areas of the skin that have less pigment, including:

    A. the sclera.

    B. the conjunctiva.

    C. the mucous membranes of the mouth.

    D. all of the above

_____ **19.** Other conditions, not related to the body's circulation, such as _____, may slow capillary refill.

    A. local circulatory compromise

    B. hypothermia

    C. age

    D. all of the above

_____ **20.** While initial treatment is important, it is essential to remember that immediate _____ is one of the keys to the survival of any high-priority patient.

    A. airway control

    B. bleeding control

    C. transport

    D. application of oxygen

_____ **21.** Goals of the focused history and physical exam include:

    A. identifying the patient's chief complaint.

    B. understanding the specific circumstances surrounding the chief complaint.

    C. directing further physical examination.

    D. all of the above

_____ **22.** Understanding the _____ helps you to understand the severity of the patient's problem and provide invaluable information to hospital staff as well.

    A. chief complaint

    B. mechanism of injury

    C. physical exam

    D. focused history

_____ **23.** An integral part of the rapid trauma assessment is evaluation using the mnemonic:

    A. AVPU.

    B. DCAP-BTLS.

    C. OPQRST.

    D. SAMPLE.

_____ **24.** It is particularly important to evaluate the neck before:

    A. log rolling the patient.

    B. examining the chest.

    C. covering it with a cervical collar.

    D. checking for the presence of a carotid pulse.

_____ **25.** To check for motor function, you should ask the patient:

    A. to wiggle his or her fingers or toes.

    B. to identify which extremity you are touching.

    C. if they can feel you touching them.

    D. all of the above

_____ **26.** Baseline vital signs provide useful information about the:

    A. overall functions of the patient's heart.

    B. overall functions of the patient's lungs.

    C. patient's stability.

    D. all of the above

_____ **27.** When performing a detailed exam, check the neck for:

    A. subcutaneous emphysema.

    B. jugular vein distention.

    C. crepitus.

    D. all of the above

_____ **28.** The purpose of the ongoing assessment is to ask and answer the following questions, except:

    A. Is treatment improving the patient's condition?

    B. What is the patient's diagnosis?

    C. Has an already identified problem gotten better? Worse?

    D. What is the nature of any newly identified problems?

_____ **29.** When reevaluating any interventions you started, take a moment to ensure that:

    A. oxygen is still flowing.

    B. backboard straps are still tight.

    C. bleeding has been controlled.

    D. all of the above

_____ **30.** The last step of the scene size-up for the medical patient is:

    A. consider c-spine.

    B. consider additional resources.

    C. initial assessment.

    D. None of the above.

Questions 31-33 are derived from the following scenario: You've been called to a park where an 86-year-old man has fallen and can't get up. He is pale, cool to the touch, and clammy.

_____ **31.** After ensuring the scene is safe, your next step is to:

    A. provide c-spine.

    B. provide oxygen.

    C. BSI.

    D. determine the number of patients.

_____ **32.** After assessing his ABC's, your next step is to:

    A. complete a focused history and physical exam.

    B. complete a rapid trauma assessment.

    C. complete a detailed assessment.

    D. identify patient priority and make a transport decision

_____ **33.** After completing the initial assessment, your next step is to:

    A. evaluate responsiveness.

    B. reconsider the MOI.

    C. rapid trauma assessment.

    D. focused trauma assessment.

_____ **34.** After evaluating the responsiveness of the medical patient during the focused history and physical exam, your next step for the responsive patient is the:

    A. baseline vitals.

    B. history of illness.

    C. sample history.

    D. None of the above

Fill-in

Read each item carefully, then complete the statement by filling in the missing word(s).

1. From a practical point of view, prehospital emergency care is simply a series of _____ about treatment and transport.

2. The best way to reduce your risk of exposure is to follow _____ _____ _____ (BSI) precautions.

3. You cannot help your patient if you become a _____ yourself.

4. You should park your unit in a place that will offer you and your partner the greatest _____ but also rapid access to the patient and your equipment.

5. _____ is a process of identifying the severity of each patient's condition.

6. The _____ _____ is based on your immediate assessment of the environment, the presenting signs and symptoms, mechanism of injury in a trauma patient, and the patient's chief complaint.

7. The first steps in caring for any patient focus on finding and treating the most _____ _____ illnesses and injuries.

8. With an unresponsive patient or one with a decreased level of consciousness, you should immediately assess the _____ of the airway.

9. If a patient seems to have difficulty breathing, you should immediately _____ the airway.

10. Poor perfusion may result in _____ , pale, and clammy skin.

11. Correct identification of high-priority patients is an essential aspect of the _____ _____ and helps to improve patient outcome.

True/False

If you believe the statement to be more true than false, write the letter "T" in the space provided. If you believe the statement to be more false than true, write the letter "F."

_____ 1. Responsiveness is evaluated with the mnemonic DCAP-BTLS.
_____ 2. An ongoing assessment is not necessary for stable patients.
_____ 3. Distinguishing between medical and trauma patients is less important than identifying and treating their problems appropriately.
_____ 4. The apparent absence of a palpable pulse in a responsive patient is not caused by cardiac arrest.
_____ 5. A patient with a poor general impression is considered a priority patient.

Short Answer

Complete this section with short written answers using the space provided.

1. What is the single goal of initial assessment?

_____

_____

2. What is the general impression based on?

_____

_____

3. What do the letters ABC stand for in the assessment process?

_____

_____

4. What four questions are asked when assessing orientation and what purpose do these questions serve?

_____

_____

5. What are the three goals of the focused history and physical exam?

_____

_____

6. List the elements of DCAP-BTLS.

_____

_____

7. List at least five significant mechanisms of injury in an adult.

_____

_____

_____

Word Fun

The following crossword puzzle is an activity provided to reinforce correct spelling and understanding of medical terminology associated with emergency care and the EMT-B. Use the clues in the column to complete the puzzle.

Across

5. Skin pulls in around ribs

8. Involuntary muscle contractions, to protect

9. Pain assessment acronym

11. Acronym for history

12. Below 95 degrees F

13. Distal discomfort or pain

Down

1. Broad area trauma without broken skin

2. Rattling, moist sounds

3. Motion in opposite direction

4. Grating or grinding

6. Formation of clots

7. Times four

10. Coarse breath sounds

Ambulance Calls

The following case scenarios provide an opportunity to explore the concerns associated with patient management. Read each scenario, then answer each question in detail.

1. You are dispatched to a motor vehicle collision where you find a 32-year-old man with extensive trauma to the face and gurgling in his airway. He is responsive only to pain. You also note that the windshield is spider-webbed and there is deformity to the steering wheel. He is not wearing a seat belt.

   How would you best manage this patient? What clues tell you the transport status?

   _____

   _____

   _____

2. You are dispatched to a local residence for "difficulty breathing." You find a man standing in his kitchen, leaning against a counter, and holding a metered-dose inhaler. As you question him, you see that he is working very hard to breathe, hear wheezing, and note that he can only answer you with one- or two-word responses.

   How do you manage this patient?

   _____

   _____

   _____

3. You are dispatched to "man fallen" at a private home. You arrive to find an older man who is semiconscious (response to painful stimuli) and has fallen down a flight of wooden stairs onto a cement basement floor. He has bruising and a small laceration above his left eye.

   How do you manage this patient?

   _____

   _____

   _____

Skill Drills

Skill Drill 8-1: Performing a Rapid Physical Exam

Test your knowledge of this skill drill by placing the photos below in the correct order. Number the first step with a "1," the second step with a "2," etc.

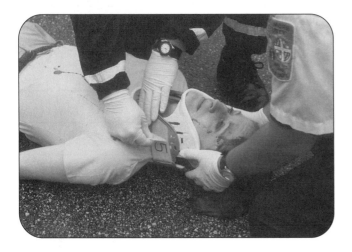

Apply a cervical spinal immobilization device on trauma patients.

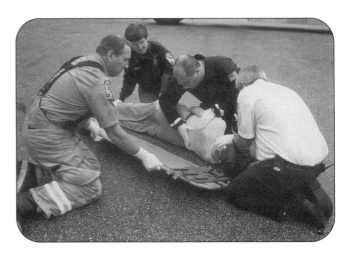

Assess the back. In trauma patients, roll the patient in one motion.

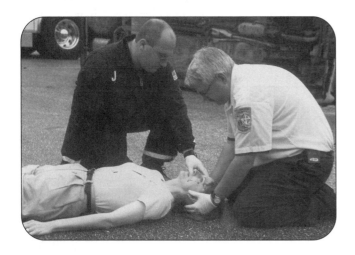

Assess the head. Have your partner maintain in-line stabilization.

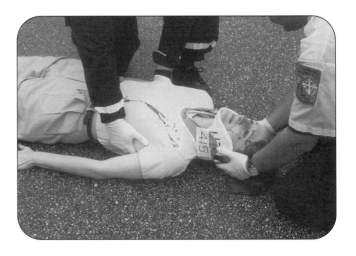

Assess the chest. Listen to breath sounds on both sides of the chest.

(continued)

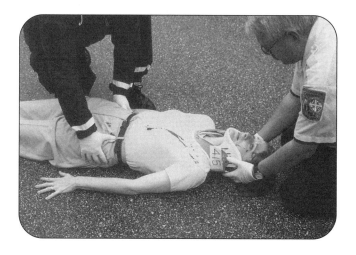

Assess the pelvis. If there is no pain, gently compress the pelvis downward and inward to look for tenderness or instability.

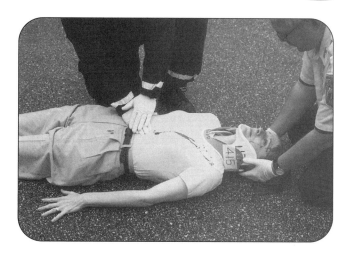

Assess the abdomen.

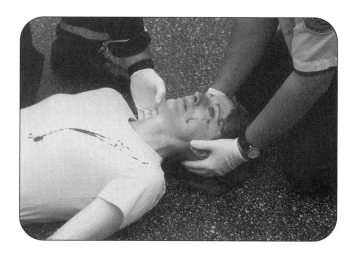

Assess the neck.

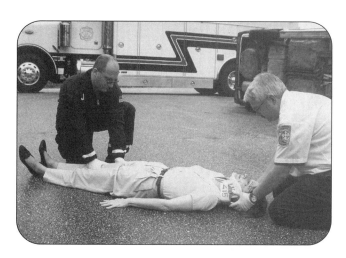

Assess all four extremities. Assess pulse, motor, and sensory function.

Skill Drill 8-3: Performing the Detailed Physical Exam
Test your knowledge of this skill drill by placing the photos below in the correct order. Number the first step with a "1," the second step with a "2," etc.

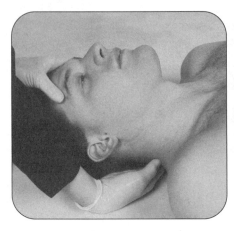

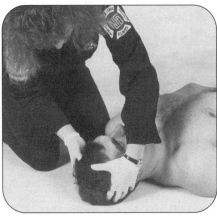

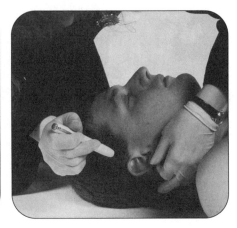

_____

Palpate the front and back of the neck.

_____

Observe and palpate the head.

_____

Check the ears for drainage or blood.

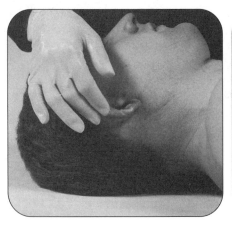

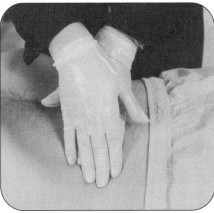

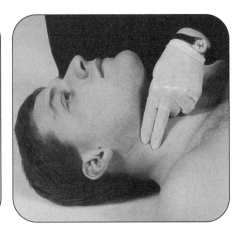

_____

Look behind the ear for Battle's sign.

_____

Gently palpate the abdomen.

_____

Observe for jugular vein distention.

(continued)

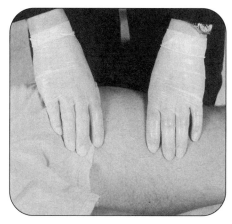

Inspect the extremities; assess distal circulation and motor sensory function.

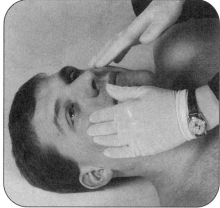

Palpate the maxillae.

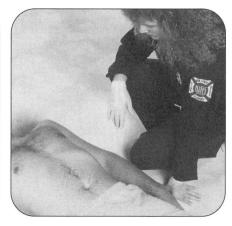

Inspect the chest and observe breathing motion.

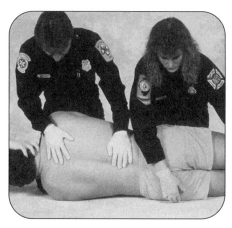

Log roll the patient and inspect the back.

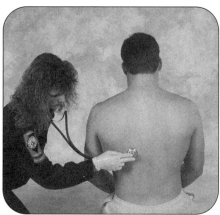

Listen to posterior breath sounds (bases, apices).

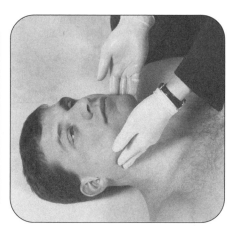

Palpate the mandible.

(continued)

Inspect the area around the eyes and eyelids.

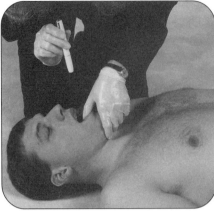

Assess the mouth and nose.

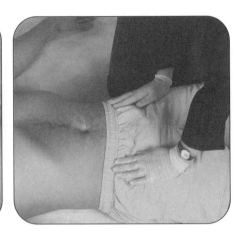

Gently press the iliac crests.

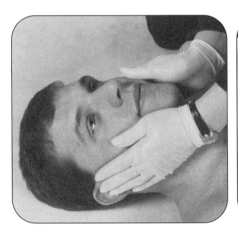

Palpate the zygomas.

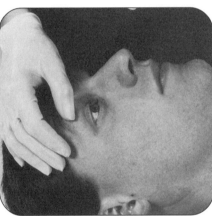

Examine the eyes for redness, contact lenses. Check pupil function.

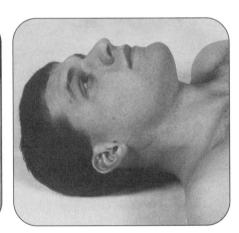

Inspect the neck.

(continued)

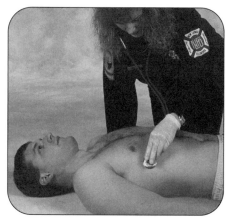

Listen to anterior breath sounds (midaxillary, midclavicular).

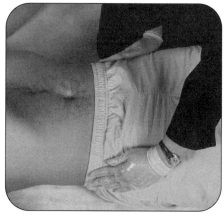

Gently compress the pelvis from the sides.

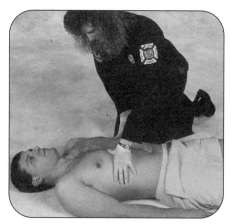

Gently palpate over the ribs.

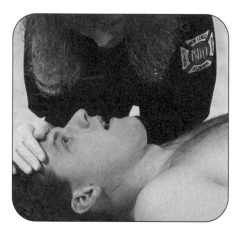

Check for unusual breath odors.

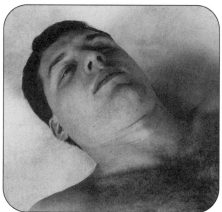

Observe the face.

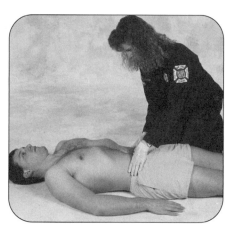

Observe the abdomen and pelvis.

CHAPTER

9 Communications and Documentation

Workbook Activities

The following activities have been designed to help you. Your instructor may require you to complete some or all of these activities as a regular part of your EMT-B training program. You are encouraged to complete any activity that your instructor does not assign as a way to enhance your learning in the classroom.

Chapter Review

The following exercises provide an opportunity to refresh your knowledge of this chapter.

Matching

Match each of the terms in the left column to the appropriate definition in the right column.

_____ 1. Base station

_____ 2. Mobile radio

_____ 3. Portable radio

_____ 4. Repeater

_____ 5. Telemetry

_____ 6. UHF

_____ 7. VHF

_____ 8. Cellular telephone

_____ 9. Dedicated line

_____ 10. MED channels

_____ 11. Scanner

_____ 12. Channel

_____ 13. Rapport

A. "hot line"

B. a trusting relationship built with your patient

C. communication through an interconnected series of repeater stations

D. assigned frequency used to carry voice and/or data communications

E. radio receiver that searches across several frequencies until the message is completed

F. VHF and UHF channels designated exclusively for EMS use

G. vehicle-mounted device that operates at a lower frequency than a base station

H. a process in which electronic signals are converted into coded, audible signals

I. radio frequencies between 30 and 300 MHz

J. hand-carried or hand-held devices that operate at 1 to 5 watts

K. special base station radio that receives messages and signals on one frequency and then automatically retransmits them on a second frequency

L. radio frequencies between 300 and 3,000 MHz

M. radio hardware containing a transmitter and receiver that is located in a fixed location

Multiple Choice

Read each item carefully, then select the best response.

_____ **1.** The base station may be used:

A. in a single place by an operator speaking into a microphone that is connected directly to the equipment.

B. remotely through telephone lines.

C. by radio from a communication center.

D. all of the above

_____ **2.** The transmission range of a(n) _____ is more limited than that of mobile or base station radios.

A. portable radio

B. 800 MHz radio

C. cellular phone

D. UHF radio

_____ **3.** Base stations:

A. usually have more power than mobile or portable radios.

B. have higher, more efficient antenna systems.

C. allow for communication with field units at much greater distances.

D. all of the above

_____ **4.** _____ are helpful when you are away from the ambulance and need to communicate with dispatch, another unit, or medical control.

A. Base stations

B. Portable radios

C. Mobile radios

D. Cellular phones

_____ **5.** Digital signals are also used in some kinds of paging and tone alerting systems because they transmit _____ and allow more choices and flexibility.

A. numerically

B. faster

C. alphanumerically

D. encoded messages

_____ **6.** As with all repeater-based systems, a cellular telephone is useless if the equipment:

A. fails.

B. loses power.

C. is damaged by severe weather or other circumstances.

D. all of the above

_____ **7.** In the simplex mode, all of the following are true, except:

A. when one party transmits, the other must wait to reply.

B. you must push a button to talk.

C. it is called a "pair of frequencies."

D. radio transmissions can occur in either direction, but not simultaneously in both.

_____ **8.** Principle EMS-related responsibilities of the FCC include:

    A. monitoring radio operations.

    B. establishing limitations for transmitter power output.

    C. allocating specific radio frequencies for use by EMS providers.

    D. all of the above

_____ **9.** Information given to the responding unit(s) should include all of the following, except:

    A. the number of patients.

    B. the time the unit will arrive.

    C. the exact location of the incident.

    D. responses by other public safety agencies.

_____ **10.** You must consult with medical control to:

    A. notify the hospital of an incoming patient.

    B. request advice or orders from medical control.

    C. advise the hospital of special situations.

    D. all of the above

_____ **11.** The patient report commonly includes all of the following, except:

    A. a list of the patient's medications.

    B. the patient's age and gender.

    C. a brief history of the patient's current problem.

    D. your estimated time of arrival.

_____ **12.** In most areas, medical control is provided by the _____ who work at the receiving hospital.

    A. nurses

    B. physicians

    C. interns

    D. staff

_____ **13.** For _____ reasons, the delivery of sophisticated care, such as assisting patients in taking medications, must be done in association with physicians.

    A. logical

    B. ethical

    C. legal

    D. all of the above

_____ **14.** Standard radio operating procedures are designed to:

    A. reduce the number of misunderstood messages.

    B. keep transmissions brief.

    C. develop effective radio discipline.

    D. all of the above

_____ **15.** Be sure that you report all patient information in a(n) _____ manner.

    A. objective

    B. accurate

    C. professional

    D. all of the above

_____ **16.** Medical control guides the treatment of patients in the system through all of the following, except:

    A. hands-on care.

    B. protocols.

    C. direct orders.

    D. post-call review.

_____ **17.** Depending upon how the protocols are written, you may need to call medical control for direct orders to:

    A. administer certain treatments.

    B. transport a patient.

    C. request assistance from other agencies.

    D. immobilize a patient.

_____ **18.** During transport:

    A. you must periodically reassess the patient's overall condition.

    B. it is not necessary to report changes in the patient's condition.

    C. you are required to check vital signs once.

    D. once treatment is provided, it is safe to finish your paperwork, since the patient's condition will remain stable.

_____ **19.** While en route to and from the scene, you should report all of the following to the dispatcher, except:

    A. any special hazards.

    B. traffic delays.

    C. abandoned vehicles in the median.

    D. road construction.

_____ **20.** Situations that might require special preparation on the part of the hospital include:

    A. HazMat situations.

    B. mass-casualty incidents.

    C. rescues in progress.

    D. all of the above

_____ **21.** The _____ officially occurs during your oral report at the hospital, not as a result of your radio report en route.

    A. patient report

    B. transfer of care

    C. termination of services

    D. all of the above

_____ **22.** Effective communication between the EMT-B and health care professionals in the receiving facility is an essential cornerstone of _____ patient care.

    A. efficient

    B. effective

    C. appropriate

    D. all of the above

_____ **23.** Components that must be included in the oral report during transfer of care include:

    A. the patient's name.

    B. any important history.

    C. vital signs assessed.

    D. all of the above

_____ **24.** Your _____ are critically important in gaining the trust of both the patient and family.

    A. gestures

    B. body movements

    C. attitude toward the patient

    D. all of the above

_____ **25.** If the patient is hearing impaired, you should:

    A. stand on the patient's left side.

    B. shout.

    C. speak clearly and distinctly.

    D. use baby talk.

_____ **26.** The functional age relates to the person's:

    A. ability to function in daily activities.

    B. mental state.

    C. activity pattern.

    D. all of the above

_____ **27.** When caring for a visually impaired patient, you should:

    A. use sign language.

    B. touch the patient only when necessary to render care.

    C. try to avoid sudden movements.

    D. never walk them to the ambulance.

_____ **28.** When attempting to communicate with non-English-speaking patients, you should:

    A. use short, simple questions and simple words whenever possible.

    B. always use medical terms.

    C. shout.

    D. position yourself so the patient can read your lips.

_____ **29.** The patient information that is included in the minimum data set includes all of the following, except:

    A. chief complaint.

    B. the time that the EMS unit arrived at the scene.

    C. respirations and effort.

    D. skin color and temperature.

_____ **30.** Functions of the prehospital care report include:

    A. continuity of care.

    B. education.

    C. research.

    D. all of the above

_____ **31.** A good prehospital care report documents:

    A. the care that was provided.

    B. the patient's condition on arrival.

    C. any changes.

    D. all of the above

_____ **32.** When completing the narrative section, be sure to:

    A. describe what you see and what you do.

    B. only include positive findings.

    C. record your conclusions about the incident.

    D. use appropriate radio codes.

_____ **33.** Instances in which you may be required to file special reports with appropriate authorities include:

    A. gunshot wounds.

    B. dog bites.

    C. suspected physical, sexual, or substance abuse.

    D. all of the above

Questions 34-38 are derived from the following scenario: You have just finished an ambulance run where a 45-year-old man had run his SUV into a utility pole. The driver was found slumped over the steering wheel, unconscious. A large electrical wire was lying across the hood of the vehicle. After securing scene safety, you were able to approach the patient and completed a rapid physical assessment, in which you found a 6-inch laceration across his forehead. The patient regained responsiveness, was alert and oriented, and refused care.

_____ **34.** Should an EMT document this call, since the patient refused care?

    A. No – You only need to document when you have actually provided care.

    B. No – This was not a billable run.

    C. Yes – and is best signed by the patient as "Refusal of Care."

    D. Both A and B

_____ **35.** What would be important to document?

    A. That the scene needed to be made safe

    B. That ensuring scene safety delayed care

    C. That you completed a rapid physical assessment

    D. All of the above

_____ **36.** While writing the report, you made an error. How should this be corrected?

    A. Draw a single line through it.

    B. Erase the mistake.

    C. Cover up the mistake with correction fluid.

    D. All of the above

_____ **37.** What are the consequences of falsifying a report?

    A. It may result in the suspension and/or revocation of your license.

    B. It gives other health care providers a false impression of assessment/findings.

    C. It results in poor patient care.

    D. All of the above

_____ **38.** If the patient refuses to sign the refusal form:

    A. Sign it yourself and state: "Patient refused to sign."

    B. You cannot let him leave the scene until he either goes with you or signs the form.

    C. Have a credible witness sign the form testifying they witnessed the patient's refusal of care.

    D. If they refuse care, you don't have to document.

Fill-in

Read each item carefully, then complete the statement by filling in the missing word(s).

1. Written communications, in the form of a written _____ _____ _____, provide you with an opportunity to communicate the patient's story to others who may participate in the patient's care in the future.

2. A two-way radio consists of two units: a _____ and a _____.

3. A _____ _____, also known as a hot line, is always open or under the control of the individuals at each end.

4. With _____, electronic signals are converted into coded, audible signals.

5. Low-power portable radios that communicate through a series of interconnected repeater stations called "cells" are known as _____ _____.

6. _____ are commonly used in EMS operations to alert on- and off-duty personnel.

7. When the first call to 9-1-1 comes in, the dispatcher must try to judge its relative _____ to begin the appropriate EMS response using emergency medical dispatch protocols.

8. The principal reason for radio communication is to facilitate communication between you and _____

_____.

9. You could be successfully sued for _____ if you describe a patient in a way that injures his or her reputation.

10. Regardless of your system's design, your link to _____ _____ is vital to maintain the high quality of care that your patient requires and deserves.

11. To ensure complete understanding, once you receive an order from medical control, you must _____ the order back, word for word, and then receive confirmation.

12. By their very nature, _____ _____ do not require direct communication with medical control.

13. Maintaining _____ _____ with your patient builds trust and lets the patient know that he or she is your first priority.

14. Children can easily see through lies or deception, so you must always be _____ with them.

15. If the patient does not speak any English, find a family member or friend to act as a(n) _____.

16. The national EMS community has identified a _____ _____ _____ that should enable communication and comparison of EMS runs between agencies, regions, and states.

17. _____ adult patients have the right to refuse treatment.

True/False

If you believe the statement to be more true than false, write the letter "T" in the space provided. If you believe the statement to be more false than true, write the letter "F."

_____ **1.** The two-way radio is actually at least two units: a transmitter and a receiver.

_____ **2.** Base stations typically have more power and much higher and more efficient antenna systems than mobile or portable radios.

_____ **3.** A cellular telephone is just another kind of portable radio that is available for EMS use.

_____ **4.** The transmission range of a mobile radio is more limited than that of a portable radio.

_____ **5.** A dedicated line, a special telephone line used for specific point-to-point communications, is always open or under the control of the individuals at each end.

_____ **6.** The written report is a vital part of providing emergency medical care and ensuring the continuity of patient care.

_____ **7.** EMS systems that use repeaters are unable to get good signals from portable radios.

_____ **8.** Small changes in your location will not significantly affect the quality of your transmission.

_____ **9.** Your reporting responsibilities end when you arrive at the hospital.

_____ **10.** Patients deserve to know that you can provide medical care and that you are concerned about their well-being.

Short Answer

Complete this section with short written answers using the space provided.

1. List the five principal FCC responsibilities related to EMS.

_____

_____

2. List five guidelines for effective radio communications.

_____

_____

3. List the six functions of a prehospital care report.

_____

_____

4. Describe the two types of written report forms generally in use in EMS systems.

_____

_____

Word Fun

The following crossword puzzle is an activity provided to reinforce correct spelling and understanding of medical terminology associated with emergency care and the EMT-B. Use the clues in the column to complete the puzzle.

Across

2. Frequencies between 300 and 3,000 MHz

4. Electronic signals converted into coded audible signals

7. Point-to-point

8. Agency with jurisdiction over radios

12. Radio receiver that searches

13. Transmit in both ways, but not at the same time

14. Hardware in a fixed location

Down

1. Transmit and receive simultaneously

3. Outline specific directions, protocols

5. Frequencies for exclusive EMS usage

6. Frequencies between 30 and 300 MHz

9. Trusting relationship

10. Receives on one, transmits on another

11. Assigned frequency

Ambulance Calls

The following case scenarios provide an opportunity to explore the concerns associated with patient management. Read each scenario, then answer each question in detail.

1. You are in the dispatch office filling in for the EMS dispatcher who needed to use the restroom. The phone rings and you answer it to find a hysterical female screaming about a child falling in an old well. The only information she is providing is that he is 5 years old and is not making any noise. The address is on the computer display. How would you best manage this situation and what additional help would you call for?

   _____

   _____

   _____

2. You respond to an "unknown medical problem" in an area commonly populated by Hispanic Americans. You arrive to find several individuals speaking to a middle-aged man. They seem to be concerned about him, and motion you toward the patient. You attempt to gain information about the situation, but your patient does not speak English and you do not speak Spanish. The patient has no outward appearance of any problems. What should you do?

   _____

   _____

   _____

3. You are dispatched to the parking lot of a grocery store for a "confused child." You arrive to find a young boy who is developmentally disabled. He cannot communicate who he is or where he lives. He is frightened but appears otherwise unharmed. What should you do?

   _____

   _____

   _____

CHAPTER

10 General Pharmacology

Workbook Activities

The following activities have been designed to help you. Your instructor may require you to complete some or all of these activities as a regular part of your EMT-B training program. You are encouraged to complete any activity that your instructor does not assign as a way to enhance your learning in the classroom.

Chapter Review

The following exercises provide an opportunity to refresh your knowledge of this chapter.

Matching

Match each of the terms in the left column to the appropriate definition in the right column.

_____ **1.** Absorption   **A.** lotions, creams, ointments

_____ **2.** Contraindication   **B.** effect that a drug is expected to have

_____ **3.** Side effect   **C.** the study of the properties and effects of drugs and medications

_____ **4.** Adsorption   **D.** amount of medication given

_____ **5.** Dose   **E.** gelatin shells filled with powdered or liquid medication

_____ **6.** Indication   **F.** any action of a drug other than the desired one

_____ **7.** Action   **G.** to bind or stick to a surface

_____ **8.** Pharmacology   **H.** therapeutic use for a particular medication

_____ **9.** Capsules   **I.** process by which medications travel through body tissues

_____ **10.** Topical medications   **J.** situation in which a drug should not be given

Multiple Choice

Read each item carefully, then select the best response.

_____ **1.** The proper dose of a medication depends on all of the following, except:

   A. the patient's age.

   B. the patient's size.

   C. generic substitutions.

   D. the desired action.

_____ **2.** Nitroglycerin relieves the squeezing or crushing pain associated with angina by:

   A. dilating the arteries to increase the oxygen supply to the heart muscle.

   B. causing the heart to contract harder and increase cardiac output.

   C. causing the heart to beat faster to supply more oxygen to the heart.

   D. all of the above

_____ **3.** The brand name that a manufacturer gives to a medication is called the _____ name.

   A. trade

   B. generic

   C. chemical

   D. prescription

_____ **4.** The fastest way to deliver a chemical substance is by the _____ route.

   A. intravenous

   B. oral

   C. sublingual

   D. intramuscular

_____ **5.** The form the manufacturer chooses for a medication ensures:

   A. the proper route of the medication.

   B. the timing of its release into the bloodstream.

   C. its effects on target organs or body systems.

   D. all of the above

_____ **6.** Solutions may be given:

   A. orally.

   B. intramuscularly.

   C. rectally.

   D. all of the above

_____ **7.** In the prehospital setting, _____ is the preferred method of giving oxygen to patients who have sufficient tidal volume, and can provide up to 95% inspired oxygen.

   A. a nasal cannula

   B. a nonrebreathing mask

   C. a bag-valve mask

   D. any of the above

_____ **8.** Characteristics of epinephrine include:

   A. dilating passages in the lungs.

   B. constricting blood vessels.

   C. increasing the heart rate and blood pressure.

   D. all of the above

_____ **9.** Epinephrine acts as a specific antidote to:

    A. adrenaline.

    B. histamine.

    C. asthma.

    D. bronchitis.

_____ **10.** Nitroglycerin relieves pain because its purpose is to increase blood flow by relieving the spasms or causing the arteries to:

    A. dilate.

    B. constrict.

    C. thicken.

    D. contract.

_____ **11.** Nitroglycerin affects the body in the following ways (select all that apply):

    A. It decreases blood pressure.

    B. It relaxes veins throughout the body.

    C. It often causes a mild headache after administration.

    D. It increases blood return to the heart.

Questions 12-16 are derived from the following scenario: You are called to a home of a known 34-year-old man with diabetes. When you arrive, you find the patient supine and unconscious on the living room floor, snoring.

_____ **12.** Medications most EMT-Bs carry in the ambulance that would pertain to this call include:

    A. insulin, oxygen, and oral glucose.

    B. nitroglycerin, oxygen, and oral glucose.

    C. activated charcoal, oxygen, and oral glucose.

    D. none of the above

_____ **13.** Oral glucose:

    A. is a suspension.

    B. should be given to this patient.

    C. is placed between a patient's cheek and gum.

    D. is not carried by EMT-Bs.

_____ **14.** The government publication listing all drugs in the United States is called the:

    A. United States Pharmacopoeia.

    B. The Department of Transportation Reference Guide.

    C. US Pharmacology.

    D. Nursing Drug Reference.

_____ **15.** Oral glucose is _____ for this patient.

    A. indicated

    B. contraindicated

    C. not normally given

    D. none of the above

_____ **16.** Oxygen:

    A. is a drug.

    B. should be applied to this patient.

    C. does not burn.

    D. all the above

Fill-in

Read each item carefully, then complete the statement by filling in the missing word.

1. _____ is a simple sugar that is readily absorbed by the bloodstream.

2. _____ is the main hormone that controls the body's fight-or-flight response.

3. Nitroglycerin is usually taken _____.

4. In all but the _____ _____ route, the medication is absorbed into the bloodstream through

   various body tissues.

5. When given by mouth, _____ may be absorbed from the stomach fairly quickly because the medication

   is already dissolved.

6. A _____ is a chemical substance that is used to treat or prevent disease or relieve pain.

True/False

If you believe the statement to be more true than false, write the letter "T" in the space provided. If you believe the statement to be more false than true, write the letter "F."

_____ 1. Oxygen is a flammable substance.
_____ 2. Glucose may be administered to an unconscious patient in order to save his or her life.
_____ 3. Epinephrine is a hormone produced by the body to aid in digestion.
_____ 4. Nitroglycerin decreases blood pressure.
_____ 5. Sublingual medications are rapidly absorbed into the digestive tract.
_____ 6. Vital signs should be taken before and after a medication is given.
_____ 7. Even though medications can react with each other, this is not a potentially harmful condition for the patient.
_____ 8. Nitroglycerin should only be administered when the patient's systolic blood pressure is below 100 mm Hg.

Short Answer

Complete this section with short written answers using the space provided.

1. List seven routes of medication administration.

   _____

   _____

2. Describe the general steps of administering medication.

   _____

   _____

3. Describe the action of activated charcoal and the steps of administration that are specific to this medication.

   _____

   _____

4. List three characteristics of epinephrine.

_____

_____

5. How is an epinephrine auto-injector activated?

_____

_____

6. List five effects of nitroglycerin.

_____

_____

7. Explain why metered-dose inhalers are often used with a spacer.

_____

_____

Word Fun

The following crossword puzzle is a good way to reinforce correct spelling and understanding of medical terminology associated with emergency care and the EMT-B. Use the clues in the column to complete the puzzle.

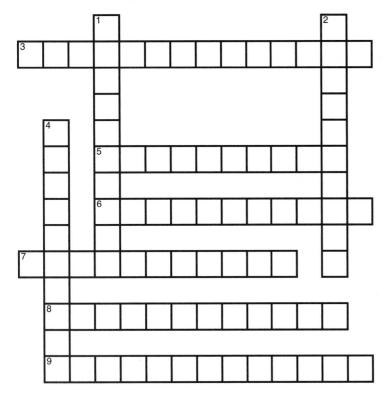

Across

3. Through the skin
5. Process for medication to travel
6. Into the vein
7. Raises heart rate and blood pressure
8. Into the bone
9. Dilates arteries in angina patients

Down

1. Therapeutic use for medication
2. Under the tongue
4. Binding to or sticking to a surface

Ambulance Calls

The following case scenarios provide an opportunity to explore the concerns associated with patient management. Read each scenario, then answer each question in detail.

1. You are dispatched to "difficulty breathing" at one of your town's many parks. As you near the park entrance, you see a crowd of people who frantically wave for you. You arrive to find a city employee who was apparently mowing the park grounds when he accidentally mowed over a yellow jacket nest. He was wearing coveralls, but he was repeatedly stung around his neck and face. He appears to be somewhat confused; you can hear stridor with each inspiration and his blood pressure is 80/40. Your local protocols allow EMT-Bs to carry EpiPens.

How do you best manage this patient?

_____

_____

_____

2. You are dispatched to an "unknown medical problem" at 1212 Main Street. You were called after police were summoned to subdue a combative male shopper. Police officers were able to calm him down, but felt that something was "not right" about him. You arrive to find a calm, but confused man who is sweaty and pale. He has no complaints, but keeps repeating, "I have to get home now." You notice a medical ID bracelet indicating that this patient is an insulin-dependent diabetic.

How do you best manage this patient?

_____

_____

_____

3. You are dispatched to the residence of a 68-year-old man who is complaining of a "crushing" chest pain radiating down his left arm for the past hour. He is pale, cool, diaphoretic, and is very nauseated. He tells you he had a heart attack several years ago and takes nitroglycerin as needed. He took two tablets prior to your arrival and reports no relief.

How would you best manage this patient?

_____

_____

_____

CHAPTER

11 Respiratory Emergencies

Workbook Activities

The following activities have been designed to help you. Your instructor may require you to complete some or all of these activities as a regular part of your EMT-B training program. You are encouraged to complete any activity that your instructor does not assign as a way to enhance your learning in the classroom.

Chapter Review

The following exercises provide an opportunity to refresh your knowledge of this chapter.

Matching

Match each of the terms in the left column to the appropriate definition in the right column.

_____ **1.** Asthma

_____ **2.** Pulmonary edema

_____ **3.** Epiglottitis

_____ **4.** Emphysema

_____ **5.** Pleural effusion

_____ **6.** Pneumothorax

_____ **7.** Dyspnea

_____ **8.** Pneumonia

_____ **9.** Hypoxia

_____ **10.** Bronchitis

_____ **11.** Hyperventilation

_____ **12.** Allergen

_____ **13.** Embolus

A. irritation of the major lung passageways

B. acute spasm of the bronchioles, associated with excessive mucus production and sometimes spasm of the bronchiolar muscles

C. accumulation of air in the pleural space

D. fluid build-up within the alveoli and lung tissue

E. an infectious disease of the lung that damages lung tissue

F. a substance that causes an allergic reaction

G. difficulty breathing

H. bacterial infection that can produce severe swelling

I. a blood clot or other substance in the circulatory system that travels to a blood vessel where it causes blockage

J. disease of the lungs in which the alveoli stretch, lose elasticity, and are destroyed

K. rapid or deep breathing that lowers blood carbon dioxide levels below normal

L. fluid outside of the lung

M. condition in which the body's cells and tissues do not have enough oxygen

Multiple Choice

Read each item carefully, then select the best response.

_____ 1. When treating a patient with dyspnea, you must be prepared to treat:
   A. the symptoms.
   B. the underlying problem.
   C. the patient's anxiety.
   D. all of the above

_____ 2. The oxygen-carbon dioxide exchange takes place in the:
   A. trachea.
   B. bronchial tree.
   C. alveoli.
   D. blood.

_____ 3. Oxygen-carbon dioxide exchange may be hampered if:
   A. the pleural space is filled with air or excess fluid.
   B. the alveoli are damaged.
   C. the air passages are obstructed.
   D. all of the above

_____ 4. If carbon dioxide levels drop too low, the person automatically breathes:
   A. normally.
   B. rapidly and deeply.
   C. slower, less deeply.
   D. fast and shallow.

_____ 5. If the level of carbon dioxide in the arterial blood rises above normal, the patient breathes:
   A. normally.
   B. rapidly and deeply.
   C. slower, less deeply.
   D. fast and shallow.

_____ 6. The level of carbon dioxide in the arterial blood can rise due to:
   A. emphysema.
   B. chronic bronchitis.
   C. cardiovascular disease.
   D. all of the above

_____ 7. The second stimulus that develops in patients with normally high levels of carbon dioxide responds to:
   A. increased oxygen levels.
   B. decreased oxygen levels.
   C. increased carbon dioxide levels.
   D. decreased carbon dioxide levels.

_____ 8. _____ is a sign of hypoxia to the brain.
   A. Altered mental status
   B. Decreased heart rate
   C. Decreased respiratory rate
   D. Delayed capillary refill time

_____ 9. An obstruction to the exchange of gases between the alveoli and the capillaries may result from:
   A. epiglottitis.
   B. pneumonia.
   C. colds.
   D. all of the above

_____ **10.** Pulmonary edema can develop quickly after a major:

A. heart attack.

B. episode of syncope.

C. brain injury.

D. all of the above

_____ **11.** In addition to a major heart attack, pulmonary edema may also be produced by:

A. inhaling large amounts of smoke.

B. traumatic injuries to the chest.

C. inhaling toxic chemical fumes.

D. all of the above

_____ **12.** _____ is a loss of the elastic material around the air spaces as a result of chronic stretching of the alveoli when bronchitic airways obstruct easy expulsion of gases.

A. Emphysema

B. Bronchitis

C. Pneumonia

D. Diphtheria

_____ **13.** Most patients with COPD will:

A. chronically produce sputum.

B. have a chronic cough.

C. have difficulty expelling air from their lungs.

D. all of the above

_____ **14.** The patient with COPD usually presents with:

A. an increased blood pressure.

B. a green or yellow productive cough.

C. a decreased heart rate.

D. all of the above

_____ **15.** A pneumothorax caused by a medical condition without any injury is known as:

A. a tension pneumothorax.

B. a subcutaneous pneumothorax.

C. spontaneous.

D. none of the above

_____ **16.** Asthma produces a characteristic _____ as patients attempt to exhale through partially obstructed air passages.

A. rhonchi

B. stridor

C. wheezing

D. rattle

_____ **17.** An allergic response to certain foods or some other allergen may produce an acute:

A. bronchodilation.

B. asthma attack.

C. vasoconstriction.

D. insulin release.

_____ **18.** Treatment for anaphylaxis and acute asthma attacks include:

A. epinephrine.

B. high-flow oxygen.

C. antihistamines.

D. all of the above

_____ **19.** A collection of fluid outside the lungs on one or both sides of the chest is called a:

    A. spontaneous pneumothorax.

    B. subcutaneous emphysema.

    C. pleural effusion.

    D. tension pneumothorax.

_____ **20.** Always consider _____ in patients who were eating just before becoming short of breath:

    A. upper airway obstruction

    B. anaphylaxis

    C. lower airway obstruction

    D. bronchoconstriction

_____ **21.** _____ is defined as overbreathing to the point that the level of arterial carbon dioxide falls below normal.

    A. Reactive airway syndrome

    B. Hyperventilation

    C. Tachypnea

    D. Pleural effusion

_____ **22.** Slowing of respirations after administration of oxygen to a COPD patient does not necessarily mean that the patient no longer needs the oxygen; he or she may need:

    A. insulin.

    B. even more oxygen.

    C. mouth-to-mouth resuscitation.

    D. none of the above

Questions 23-27 are derived from the following scenario: You respond to a home of a 78-year-old man having difficulty breathing. He is sitting at the kitchen table in a classic tripod position, wearing a nasal cannula. He is cyanotic, smoking, and has his shirt unbuttoned. His respirations are 30 breaths/min and shallow, his heart rate is 110 beats/min, and his blood pressure is 136/88 mm Hg.

_____ **23.** Your first thought as an EMT-B should be to:

    A. apply a nonrebreathing mask at 15L/min.

    B. call for back-up.

    C. consider BSI precautions.

    D. put the cigarette out.

_____ **24.** His brain stem senses the level of _____ in the arterial blood, causing the rapid respirations.

    A. carbon dioxide

    B. oxygen

    C. insulin

    D. none of the above

_____ **25.** Proper management of this patient should include:

    A. supplemental oxygen.

    B. chest compressions.

    C. suctioning.

    D. all the above.

_____ **26.** Which of the following is <u>NOT</u> a sign or symptom of his inadequate breathing?

    A. His cyanosis

    B. His shirt was unbuttoned

    C. He was in a tripod position

    D. His heart rate was over 100 beats/min (tachycardia)

_____ **27.** What should you do during the ongoing assessment?

   A. Assess vital signs every 5 minutes.

   B. Repeat the initial and focused assessment.

   C. Reassess interventions performed.

   D. All of the above

_____ **28.** Questions to ask during the focused history and physical examination include:

   A. What has the patient already done for the breathing problem?

   B. Does the patient use a prescribed inhaler?

   C. Does the patient have any allergies?

   D. All of the above

_____ **29.** Generic names for popular inhaled medications include:

   A. ventolin.

   B. metaprel.

   C. terbutaline.

   D. all of the above

_____ **30.** Contraindications to helping a patient self-administer any MDI medication include:

   A. not obtaining permission from medical control.

   B. noticing that the inhaler is not prescribed for this patient.

   C. noticing that the patient has already met the maximum prescribed dose.

   D. all of the above

_____ **31.** Possible side effects of over-the-counter cold medications may include:

   A. agitation.

   B. increased heart rate.

   C. increased blood pressure.

   D. all of the above

_____ **32.** A prolonged asthma attack that is unrelieved by epinephrine may progress into a condition known as:

   A. pleural effusion.

   B. status epilepticus.

   C. status asthmaticus.

   D. reactive airway disease.

Labeling

Label the following diagrams with the correct terms.

Obstruction, scarring, and dilation of the alveolar sac

A. _____

B. _____

C. _____

D. _____

E. _____

F. _____

G. _____

H. _____

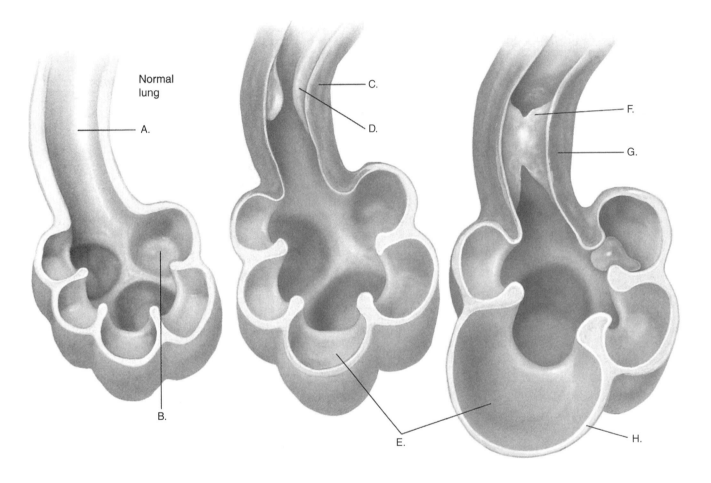

Fill-in

Read each item carefully, then complete the statement by filling in the missing word(s).

1. The level of _____ _____ bathing the brain stem stimulates respiration.

2. The level of _____ in the blood is a secondary stimulus for respiration.

3. _____ passes from the blood through capillaries to tissue cells.

4. Carbon dioxide and oxygen are exchanged in the _____.

5. Air enters the body through the _____.

6. Abnormal breathing is indicated by a rate slower than _____ breaths/min or faster than

_____ breaths/min.

7. During respiration, oxygen is provided to the blood, and _____ _____ is removed from it.

8. Evaluating the adequacy of the pulse can give you an indication of the patient's _____ _____.

9. The last step in the initial assessment is to make a _____ _____.

10. When asking questions about the present illness during the focused history and physical exam, use

_____ and _____ to guide you in your questioning.

True/False

If you believe the statement to be more true than false, write the letter "T" in the space provided. If you believe the statement to be more false than true, write the letter "F."

_____ **1.** Chronic bronchitis is characterized by spasm and narrowing of the bronchioles due to exposure to allergens.

_____ **2.** With pneumothorax, the lung collapses because the negative vacuum pressure in the pleural space is lost.

_____ **3.** Anaphylactic reactions occur only in patients with a previous history of asthma or allergies.

_____ **4.** Decreased breath sounds in asthma occur because fluid in the pleural space has moved the lung away from the chest wall.

_____ **5.** Pulmonary emboli are difficult to diagnose.

_____ **6.** A patient with aspirin poisoning may hyperventilate in response to acidosis.

_____ **7.** The distinction between hyperventilation and hyperventilation syndrome is straightforward and should guide the EMT-B's treatment choices.

_____ **8.** COPD most often results from cigarette smoking.

_____ **9.** Asthma and COPD are characterized by long inspiratory times.

_____ **10.** SARS is a serious, potentially life-threatening viral infection that usually starts with flu-like symptoms and usually progresses to pneumonia and respiratory failure.

_____ **11.** When assessing a patient, the general impression will help you decide whether the patient's condition is stable or unstable.

_____ **12.** Skin color, capillary refill, level of consciousness, and pain measurement are key in evaluating the respiratory patient.

Short Answer

Complete this section with short written answers using the space provided.

1. List five characteristics of normal breathing.

_____

_____

2. List the five most common mechanisms occurring in lung disorders.

_____

_____

3. Under what conditions should you not assist a patient with a metered-dose inhaler?

_____

_____

4. Describe chronic bronchitis.

_____

_____

5. List five signs of inadequate breathing.

_____

_____

6. Explain carbon dioxide retention.

_____

_____

7. When ventilating a patient, how would you determine if your ventilations are adequate?

_____

_____

Word Fun emtb.com vocab explorer

The following crossword puzzle is an activity provided to reinforce correct spelling and understanding of medical terminology associated with emergency care and the EMT-B. Use the clues provided to complete the puzzle.

Across

1. High-pitched, whistling sound
4. Crackling, rattling sounds
6. Difficulty breathing
12. Barking cough
13. Inflammation of leaf-shaped airway cover
14. Muscle spasm in small airways

Down

2. Traveling clot
3. Harsh, high-pitched sound
5. Irritation of major airways
7. Substance causing a reaction
8. Air in pleural space
9. Coarse sounds from mucus in airways
10. Slow process of chronic disruption of airways
11. Low oxygen

Ambulance Calls

The following case scenarios will give you an opportunity to explore the concerns associated with patient management. Read each scenario, then answer each question in detail.

1. You are called to the home of a young boy who is reportedly experiencing difficulty swallowing. You arrive to find concerned parents who tell you that their son seems to 'be sick.' He can't swallow, has a high fever and refuses to lie down. As you enter the child's bedroom, you find him standing with arms outstretched onto the footboard of the bed, drooling and with a very frightened look on his face.

   How do you manage this patient?

   _____

   _____

   _____

2. You are dispatched to a 36-year-old woman complaining of shortness of breath. You arrive to find a slightly overweight female patient who tells you she 'can't catch her breath.' She is a smoker whose only medication is birth control pills.

   How would you best manage this patient?

   _____

   _____

   _____

3. You are called to the home of a 73-year-old man complaining of severe dyspnea. The patient has a history of COPD and is on home oxygen at 2 L/min via nasal cannula. His family tells you he has a long history of breathing problems and emphysema. He is cyanotic around his lips and his respirations are 36 breaths/min and shallow.

   How would you best manage this patient?

   _____

   _____

   _____

Skill Drills

Skill Drill 11-1: Assisting a Patient With a Metered-Dose Inhaler
Test your knowledge of this skill drill by filling in the correct words in the photo captions.

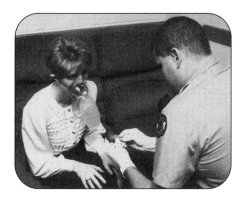

1. Ensure inhaler is at room temperature or _____.

2. Remove oxygen mask. Hand inhaler to patient. Instruct about breathing and _____ _____.

3. Instruct patient to press inhaler and inhale. Instruct about _____ _____.

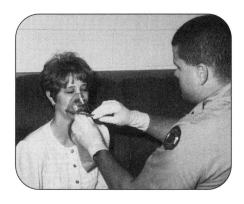

3. Reapply _____. After a few _____, have patient repeat _____ if order/protocol allows.

Assessment Review

Answer the following questions pertaining to the assessment of the types of emergencies discussed in this chapter.

1. You have been assessing a 17-year-old female in respiratory distress, and you have just obtained her baseline vital signs. Your next step is to:

A. make a transport decision.

B. consider a detailed physical examination.

C. contact medical control.

D. make interventions.

2. You have determined that she is hyperventilating. Your emergency care would include:
  A. having her breathe into a small paper sack.
  B. providing oxygen.
  C. have her run in place until the hyperventilation subsides.
  D. none of the above

3. You have been called to a patient who resides in a long-term care facility and who is having difficulty breathing. After assessing and treating life threats to the patient's airway, breathing, and circulation, your next step is to:
  A. make a transport decision.
  B. obtain a "SAMPLE" history.
  C. obtain an "OPQRST" history.
  D. obtain baseline vital signs.

4. During the ongoing assessment, vital signs should be taken every _____ minutes for the unstable patient.
  A. 3
  B. 5
  C. 10
  D. 15

5. During the ongoing assessment, vital signs should be taken every _____ minutes for the stable patient.
  A. 3
  B. 5
  C. 10
  D. 15

Emergency Care Summary

Complete the statements pertaining to emergency care for the types of emergencies discussed in this chapter by filling in the missing word(s).

NOTE: While the steps below are widely accepted, be sure to consult and follow your local protocol.

Respiratory Distress	Asthma	Infection of Upper or Lower Airway
Administer oxygen by placing a _____ mask on the patient and supplying oxygen at a rate of _____ to _____ L/min. For any patient in respiratory distress, use positioning, airway adjuncts (_____ or _____ airway), or positive pressure ventilation as indicated.	Administer oxygen. Allow patient to sit in upright position. Suction large amounts of mucus. Help patient self-administer a _____ inhaler: 1. Obtain order from medical control. 2. Check _____ and whether patient has taken other doses. 3. Ensure inhaler is at room temperature or warmer. 4. Shake inhaler vigorously several times. 5. Remove oxygen mask. Instruct patient to _____ deeply. 6. Instruct patient to press inhaler and inhale. Instruct patient to hold breath as long as is comfortable. 7. Reapply oxygen.	Administer _____ oxygen if available. Do not attempt to suction airway or place an oropharyngeal airway. Transport promptly with patient in position of comfort.

Acute Pulmonary Edema	Chronic Obstructive Pulmonary Disease	Spontaneous Pneumothorax
Administer 100% oxygen and suction any secretions from the airway as necessary. Place in position of comfort and provide _____ _____ as needed. Transport promptly.	Provide _____ oxygen via nonrebreathing mask at _____ /min. If patient is prescribed an inhaler, administer it according to local protocol. Document time and effect on patient with each use. Place in the position of comfort and provide prompt transport.	Provide _____ oxygen and place in position of comfort. Transport promptly. Support airway, breathing, and circulation as necessary.

Pleural Effusions	Obstruction of the Upper Airway	Pulmonary Embolism	Hyperventilation
Provide _____ oxygen at 15 L/min and place in position of comfort. Support airway, breathing, and circulation as necessary. Transport promptly.	For partial or complete foreign body airway obstructions, clear by following _____ guidelines, apply full-flow oxygen at 15 L/min as necessary, and transport promptly.	Clear airway and provide full-flow oxygen at 15 L/min. Place in position of comfort and provide prompt transport. Provide ventilatory support as necessary and be prepared for _____ _____.	Provide full-flow oxygen at 15 L/min and coach _____ slower in a calm manner. Complete an initial assessment and focused history and physical exam. Transport promptly for evaluation.

CHAPTER

12 Cardiovascular Emergencies

Workbook Activities

The following activities have been designed to help you. Your instructor may require you to complete some or all of these activities as a regular part of your EMT-B training program. You are encouraged to complete any activity that your instructor does not assign as a way to enhance your learning in the classroom.

Chapter Review

The following exercises provide an opportunity to refresh your knowledge of this chapter.

Matching

Match each of the terms in the left column to the appropriate definition in the right column.

_____ 1. Atria

_____ 2. Coronary arteries

_____ 3. Atrioventricular (AV) node

_____ 4. Myocardium

_____ 5. Sinus node

_____ 6. Venae cavae

_____ 7. Ventricles

_____ 8. Aorta

_____ 9. Atherosclerosis

_____ 10. Arrhythmia

_____ 11. Ischemia

_____ 12. Infarction

_____ 13. Tachycardia

_____ 14. Asystole

_____ 15. Bradycardia

A. absence of all heart electrical activity

B. calcium and cholesterol buildup inside blood vessels

C. blood vessels that supply blood to the myocardium

D. abnormal heart rhythm

E. unusually slow heart rhythm, less than 60 beats/min

F. lack of oxygen

G. heart muscle

H. lower chambers of the heart

I. tissue death

J. rapid heart rhythm, more than 100 beats/min

K. carry oxygen-poor blood back to the heart

L. upper chambers of the heart

M. body's main artery

N. electrical impulses begin here

O. electrical impulses slow here to allow blood to move from the atria to the ventricles

Multiple Choice

Read each item carefully, then select the best response.

_____ 1. We can help to reduce the number of deaths attributed to cardiovascular disease with:
   A. better public awareness.
   B. early access.
   C. public access defibrillation.
   D. all of the above

_____ 2. The aorta receives its blood supply from the:
   A. right atria.
   B. left atria.
   C. right ventricle.
   D. left ventricle.

_____ 3. Blood enters into the right atrium from the body through the:
   A. vena cava.
   B. aorta.
   C. pulmonary artery.
   D. pulmonary vein.

_____ 4. The only veins in the body to carry oxygenated blood are the:
   A. external jugular veins.
   B. pulmonary veins.
   C. subclavian veins.
   D. inferior vena cava.

_____ 5. Normal electrical impulses originate in the sinus node, just above the:
   A. atria.
   B. ventricles.
   C. AV junction.
   D. Bundle of His.

_____ 6. Dilation of the coronary arteries will _____ blood flow.
   A. shut off
   B. increase
   C. decrease
   D. regulate

_____ 7. The _____ are tiny blood vessels about one cell thick.
   A. arterioles
   B. venules
   C. capillaries
   D. ventricles

_____ 8. _____ carry oxygen to the body's tissues and then remove carbon dioxide.
   A. Red blood cells
   B. White blood cells
   C. Platelets
   D. Veins

_____ 9. _____ is the maximum pressure exerted by the left ventricle as it contracts.
   A. Cardiac output
   B. Diastolic blood pressure
   C. Systolic blood pressure
   D. Stroke volume

_____ **10.** Atherosclerosis can lead to a complete _____ of a coronary artery.

   A. occlusion

   B. disintegration

   C. dilation

   D. contraction

_____ **11.** The lumen of an artery may be partially or completely blocked by the blood-clotting system due to a _____ that exposes the inside of the atherosclerotic wall.

   A. tear

   B. crack

   C. clot

   D. rupture

_____ **12.** Tissues downstream from a blood clot will suffer from lack of oxygen. If blood flow is resumed in a short time, the _____ tissues will recover.

   A. dead

   B. ischemic

   C. necrosed

   D. dry

_____ **13.** Risk factors for myocardial infarction include all of the following except:

   A. male gender.

   B. high blood pressure.

   C. stress.

   D. increased activity level.

_____ **14.** When, for a brief period of time, heart tissues do not get enough oxygen, the pain is called:

   A. AMI.

   B. angina.

   C. ischemia.

   D. CAD.

_____ **15.** Angina pain may be felt in the:

   A. arms.

   B. midback.

   C. epigastrium.

   D. all of the above

_____ **16.** Angina may be associated with:

   A. shortness of breath.

   B. nausea.

   C. sweating.

   D. all of the above

_____ **17.** Because oxygen supply to the heart is diminished with angina, the _____ can be compromised and the person is at risk for significant cardiac rhythm problems.

   A. circulation

   B. cardiac output

   C. electrical system

   D. vasculature

_____ **18.** About _____ minutes after blood flow is cut off, some heart muscle cells begin to die.

   A. 10

   B. 20

   C. 30

   D. 40

_____ **19.** An acute myocardial infarction is more likely to occur in the larger, thick-walled left ventricle, which needs more _____ than in the right ventricle.

    A. oxygen and glucose

    B. force to pump

    C. blood and oxygen

    D. electrical activity

_____ **20.** The pain of AMI differs from the pain of angina because:

    A. it may be caused by exertion.

    B. it is usually relieved by rest.

    C. it does not resolve in a few minutes.

    D. all of the above

_____ **21.** Consequences of AMI may include:

    A. cardiogenic shock.

    B. congestive heart failure.

    C. sudden death.

    D. all of the above

_____ **22.** Sudden death is usually the result of _____, in which the heart fails to generate an effective blood flow.

    A. AMI

    B. atherosclerosis

    C. PVCs

    D. cardiac arrest

_____ **23.** Disorganized, ineffective quivering of the ventricles is known as:

    A. ventricular fibrillation.

    B. asystole.

    C. ventricular stand still.

    D. ventricular tachycardia.

_____ **24.** Causes of congestive heart failure include all of the following except:

    A. chronic hypotension.

    B. heart valve damage.

    C. a myocardial infarction.

    D. longstanding high blood pressure.

_____ **25.** Signs and symptoms of shock include all of the following except:

    A. elevated heart rate.

    B. pale, clammy skin.

    C. air hunger.

    D. elevated blood pressure.

_____ **26.** In patients with CHF, changes in heart function occur, including:

    A. a decrease in heart rate.

    B. enlargement of the left ventricle.

    C. enlargement of the right ventricle.

    D. a decrease in blood pressure.

_____ **27.** Physical findings of AMI include skin that is _____ because of poor cardiac output and the loss of perfusion.

    A. pink

    B. white

    C. gray

    D. red

_____ **28.** All patient assessments begin by determining whether or not the patient:

    A. is breathing.

    B. can talk.

    C. is responsive.

    D. has a pulse.

_____ **29.** To assess chest pain, use the mnemonic:

    A. AVPU.

    B. OPQRST.

    C. SAMPLE.

    D. CHART.

_____ **30.** When using the mnemonic OPQRST, the "P" stands for:

    A. parasthesia.

    B. pain.

    C. provocation.

    D. predisposing factors.

_____ **31.** Nitroglycerin may be in the form of a:

    A. skin patch.

    B. spray.

    C. pill.

    D. all of the above

_____ **32.** When administering nitroglycerin to a patient, you should check the patient's _____ within 5 minutes after each dose.

    A. level of consciousness

    B. breathing

    C. pulse

    D. blood pressure

_____ **33.** In general, a maximum of _____ dose(s) of nitroglycerin are given for any one episode of chest pain.

    A. one

    B. two

    C. three

    D. four

_____ **34.** _____ are inserted when the electrical control system of the heart is so damaged that it cannot function properly.

    A. Stents

    B. Pacemakers

    C. Balloon angioplasties

    D. Defibrillation

_____ **35.** When the battery wears out in a pacemaker, the patient may experience:

    A. syncope.

    B. dizziness.

    C. weakness.

    D. all of the above

_____ **36.** The computer inside the AED is specifically programmed to recognize rhythms that require defibrillation to correct, most commonly:

    A. asystole.

    B. ventricular tachycardia.

    C. ventricular fibrillation.

    D. supraventricular tachycardia.

_____ **37.** You should apply the AED only to unresponsive patients with no:

   A. significant medical problems.

   B. cardiac history.

   C. pulse.

   D. brain activity.

_____ **38.** _____ usually refers to a state of cardiac arrest despite an organized electrical complex.

   A. Asystole

   B. Pulseless electrical activity

   C. Ventricular fibrillation

   D. Ventricular tachycardia

_____ **39.** The links in the chain of survival include all of the following except:

   A. early access and CPR.

   B. early ACLS.

   C. early administration of nitroglycerin.

   D. early defibrillation.

_____ **40.** An AED may fail to function properly due to:

   A. the batteries not working.

   B. improper maintenance.

   C. operator error.

   D. all of the above

Questions 41-45 are derived from the following scenario: At 5:00 in the morning, you respond to the home of a 76-year-old man complaining of chest pain. Upon arrival, the patient states that he has been sleeping in the recliner all night due to indigestion, when the pain woke him up. He also tells you he has taken two nitroglycerin tablets.

_____ **41.** Your first priority is to:

   A. apply an AED.

   B. provide high-flow oxygen.

   C. evaluate the need to administer a third nitroglycerin tablet.

   D. size up the scene.

_____ **42.** His respirations are 16 breaths/min, pulse is 98 beats/min, his blood pressure is 92/76 mm Hg, and he is still complaining of chest pain. What actions should you take to intervene?

   A. Provide high-flow oxygen.

   B. Administer a third nitroglycerin tablet.

   C. Apply an AED.

   D. All of the above

_____ **43.** Without warning, the patient loses consciousness, stops breathing, and has no pulse. After verifying pulselessness and apnea, you should:

   A. begin CPR.

   B. provide two respirations with a BVM device or pocket mask.

   C. call for ALS backup.

   D. apply the AED that is at hand.

_____ **44.** After applying an AED to this patient, the AED states; "No shock advised." What is your next step of action?

   A. Load and transport the patient.

   B. Push to reanalyze.

   C. Perform CPR for 2 minutes, then have the AED reanalyze.

   D. None of the above

_____ **45.** Your patient is now conscious and you are en route to the hospital. You are six blocks away when the patient stops breathing and no longer has a pulse. You should:

A. continue to the hospital.

B. continue to the hospital and analyze the rhythm.

C. stop the vehicle and analyze the rhythm.

D. none of the above

Labeling

Label the following diagrams with the correct terms.

1. The Right and Left Sides of the Heart
Where arrows appear, indicate the substance and its origin and destination.

Part A:

A. _____

B. _____

C. _____

D. _____

E. _____

F. _____

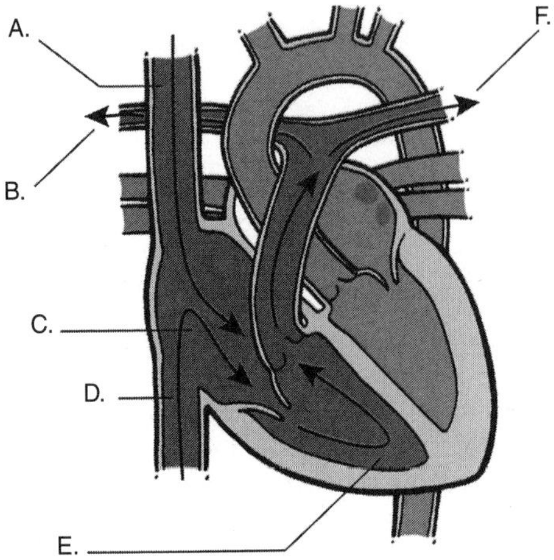

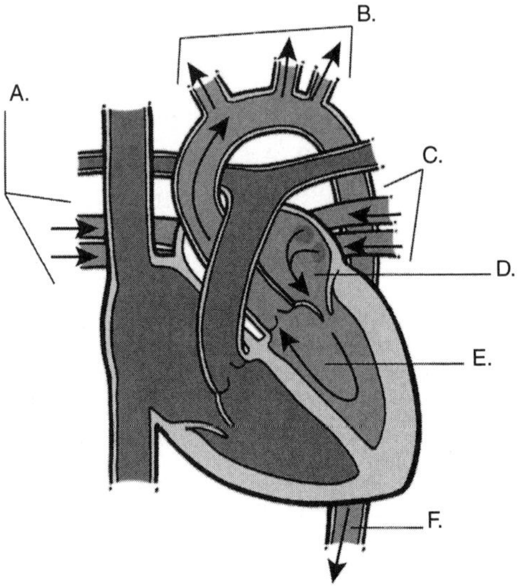

Part B:

A. _____

B. _____

C. _____

D. _____

E. _____

F. _____

2. Electrical Conduction

A. _____

B. _____

C. _____

D. _____

E. _____

F. _____

G. _____

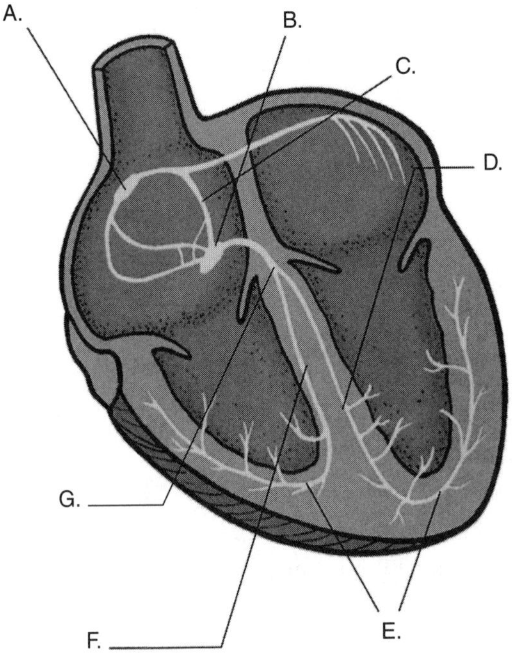

3. Pulse Points

State the name of the artery that is being assessed at each pulse point below:

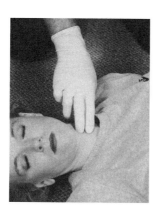

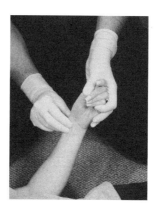

_____

_____

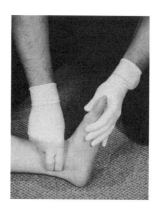

_____

_____

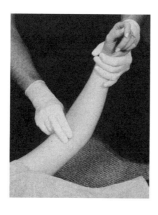

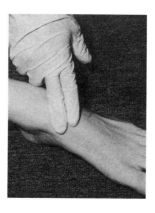

_____

_____

Fill-in

Read each item carefully, then complete the statement by filling in the missing word.

1. The heart is divided down the middle by a wall called the _____.

2. The _____ is the body's main artery.

3. The _____ ventricle pumps blood in through the pulmonary circulation.

4. Electrical impulses spread from the _____ node to the ventricles.

5. Blood supply to the heart is increased by _____ of the coronary arteries.

6. _____ _____ cells remove carbon dioxide from the body's tissues.

7. _____ blood pressure reflects the pressure on the walls of the arteries when the ventricle is at rest.

8. The heart has _____ chambers.

9. The _____ side of the heart is more muscular because it must pump blood into the aorta and all the other arteries of the body.

True/False

If you believe the statement to be more true than false, write the letter "T" in the space provided. If you believe the statement to be more false than true, write the letter "F."

_____ 1. The right side of the heart pumps oxygen-rich blood to the body.

_____ 2. In the normal heart, the need for increased blood flow to the myocardium is easily met by an increase in heart rate.

_____ 3. Atherosclerosis results in narrowing of the lumen of coronary arteries.

_____ 4. Infarction is a temporary interruption of the blood supply to the tissues.

_____ 5. Angina can result from a spasm of the artery.

_____ 6. The pain of angina and the pain of AMI are easily distinguishable.

_____ 7. Nitroglycerin works in most patients within 5 minutes to relieve the pain of AMI.

_____ 8. If an AED malfunctions during use, you must report that problem to the manufacturer and the Department of Human Resources.

_____ 9. Angina occurs when the heart's need for oxygen exceeds its supply.

_____ 10. White blood cells are the most numerous and help the blood to clot.

_____ 11. Cardiac arrest in younger children is less common than in older children and is usually caused by a breathing problem.

_____ 12. An AED with special pediatric pads is indicated for use on pediatric medical patients between the ages of 1 and 8 years who have been assessed to be unresponsive, not breathing, and pulseless.

Short Answer

Complete this section with short written answers in the space provided.

1. Name and describe the two basic types of defibrillators.

_____

_____

2. What are the three most common errors of AED use?

_____

_____

3. If ALS is not responding to the scene, what are the three points at which transport should be initiated for a cardiac arrest patient?

_____

_____

4. List six safety considerations for operating an AED.

_____

_____

5. What is the procedure for assisting a patient with nitroglycerin administration?

_____

_____

6. List three ways in which AMI pain differs from angina pain.

_____

_____

7. List three serious consequences of AMI.

_____

_____

8. Name at least five signs and symptoms associated with AMI.

_____

_____

9. Describe the technique for AED pad placement.

_____

_____

Word Fun

The following crossword puzzle is an activity provided to reinforce correct spelling and understanding of medical terminology associated with emergency care and the EMT-B. Use the clues in the column to complete the puzzle.

Across

1. Less than 60 beats/min
5. Greater than 100 beats/min
6. Lower chamber
7. Lack of oxygen to tissues
8. Widening
9. Absence of electrical activity

Down

2. Irregular heart rhythm
3. To shock the heart
4. Blockage
9. Main artery of body

Ambulance Calls

The following case scenarios provide an opportunity to explore the concerns associated with patient management. Read each scenario, then answer each question in detail.

1. You are dispatched to the residence of a 58-year-old man complaining of chest pain. He states that it feels like "somebody is standing on my chest." He sat down when it started and took a nitroglycerin tablet. He is still a little nauseated and sweaty, but feels better. He is very anxious.

   How would you best manage this patient?

   _____

   _____

   _____

2. You are dispatched to the home of a 45-year-old man experiencing chest pain. He told his wife that he was fine, but she decided to call 9-1-1. When you arrive, you find your patient sitting in the living room looking anxious. He is sweaty, pale and admits the pain is worse than just a few moments ago. He tells you that he is a very athletic person, so the pain must just be stress-related and it will go away after he relaxes for a while. He tells you he doesn't want to be taken to the hospital.

   How would you best manage this patient?

   _____

   _____

   _____

3. You are dispatched to the home of a 60-year-old woman complaining of sudden weakness. She tells you that she usually has enough energy to perform daily tasks around the house, but today she's suddenly very tired, has some pain in her jaw and has some nausea. She denies any history of recent illness including cough, cold or fever. She is otherwise healthy and does not take any medications.

How would you best manage this patient?

_____

_____

_____

Skill Drills

Skill Drill 12-2: AED and CPR

Test your knowledge of this skill drill by placing the photos below in the correct order. Number the first step with a "1," the second step with a "2," etc.

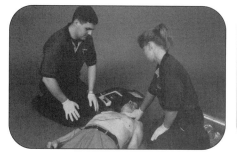

_____

If no pulse advised, check for a pulse.

If pulse is present, check breathing.

If breathing adequately, give oxygen and transport. If not breathing adequately, provide artificial ventilation and transport.

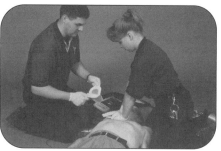

_____

If pulseless, begin CPR.

Prepare the AED pads.

Turn on the AED.

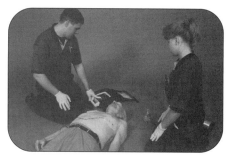

_____

Apply AED pads.

Stop CPR.

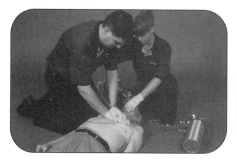

_____

If no pulse, perform 2 minutes of CPR.

Gather additional information about the arrest event.

After 2 minutes of CPR, reassess the pulse and reanalyze the cardiac rhythm.

If necessary, repeat the cycle of one shock and 2 minutes of CPR until ALS arrives.

Transport and contact medical control as needed.

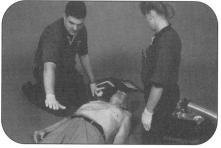

_____

Verbally and visually clear the patient.

Push the Analyze button if there is one.

Wait for the AED to analyze rhythm.

If no shock advised, perform CPR for 2 minutes and then reassess.

If shock advised, recheck that all are clear and push the Shock button.

Immediately perform CPR for 2 minutes after the shock has been delivered.

_____

Stop CPR if in progress.

Assess responsiveness.

Check breathing and pulse.

If unresponsive and not breathing adequately, give two ventilations and check pulse.

Assessment Review

Answer the following questions pertaining to the assessment of the types of emergencies discussed in this chapter.

_____ **1.** What type of additional resources may be required for someone with chest pain?

    A. Lift assistance

    B. Advanced life support

    C. Police

    D. All the above

_____ **2.** When taking a SAMPLE history of a conscious person with chest pain, what specific questions should the EMT-B ask of their patient?

    A. Whether he or she has had a heart attack before

    B. How long did he or she stay in the hospital?

    C. Did his or her physician inform him or her of risk factors associated with heart disease?

    D. All of the above

_____ **3.** What steps should be taken to complete a focused history and physical exam on an unconscious patient with a suspected cardiac problem?

    A. Perform a rapid physical exam.

    B. Obtain vital signs.

    C. Obtain history from family or bystanders.

    D. All of the above

_____ **4.** What questions should you ask during the focused history and physical exam of your pulseless, apneic patient?

    A. Have you had a heart attack before?

    B. Have you had heart problems in the past?

    C. Did you take any aspirin today?

    D. None of the above

_____ **5.** What types of BSI should be taken for someone in cardiac arrest?

    A. Gloves

    B. Eye protection

    C. Face mask/shield

    D. All the above

Emergency Care Summary

Complete the statements pertaining to emergency care for the types of emergencies discussed in this chapter by filling in the missing words.

NOTE: While the steps below are widely accepted, be sure to consult and follow your local protocol.

Chest Pain	Cardiac Arrest

Chest Pain

Depending on local protocol, prepare to administer baby aspirin and assist with prescribed nitroglycerin. Check condition of medication(s) and expiration date(s).

Aspirin

Administer according to protocols.

Nitroglycerin

1. Obtain permission from medical control.
2. Take patient's blood pressure. Continue only if _____ pressure greater than _____ mm Hg.
3. Check that you have the right medication, right patient, and right delivery route.
4. Question patient about last _____ and effects. Ensure patient understands route of administration. Prepare to have the patient lie down to prevent _____.
5. Ask patient to lift his or her tongue. Place tablet _____ tongue or spray under tongue if medication is in spray form. Have patient keep mouth closed until dissolved/absorbed.
6. Recheck blood pressure within _____ minutes. Record medication and time of administration. If chest pain persists and systolic blood pressure is greater than _____ mm Hg, repeat the dose every 5 minutes as authorized by medical control.

Reevaluate transport decision. Do not delay transport to assist with nitroglycerin.

Cardiac Arrest

Defibrillation

For unwitnessed adult cardiac arrest:

1. Stop CPR if it is in progress. Assess responsiveness. If the patient is responsive, do not apply the AED.
2. If unresponsive, pulseless, and apneic, perform five cycles (approximately 2 minutes) of CPR.
3. Prepare to use the AED. Power on the AED unit.
4. Remove any clothing from the patient's chest area if not already done. Apply the pads to the chest: one to the right of the sternum just below the clavicle, the other on the lower left chest wall (top of pad 2" to 3" below the armpit).
5. Stop CPR. Make sure that on one is touching the patient and loudly call "clear".
6. Push the analyze button, if there is one, and wait for the AED unit to analyze.
7. If shock is not indicated, perform CPR for 2 ninutes and then reassess pulse and analyze cardiac rhythm. If a shock is advised, make sure that no one is touching the patient. When the patient and the area around the patient are clear, push the shock button. Resume CPR, starting with chest compressions, immediately after the shock has been delivered.
8. After 2 minutes of CPR, check for a pulse and analyze the cardiac rhythm. If the patient has a pulse, check the patient's breathing. If the patient is breathing adequately, give oxygen via nonrebreathing mask and transport. If the patient is not breathing adequately, perform artificial ventilation (10 to 12 breaths/min) with 100% oxygen and transport.
9. If the patient has no pulse, perform 2 minutes of CPR, starting with chest compressions.
10. After 2 minutes of CPR, check for a pulse and analyze the cardiac rhythm (as applicable).
11. If necessary, repeat the cycle of administrating one shock and performing 2 minutes of CPR until ALS arrives.
12. Transport the patient and contact medical control as needed. Note: If the patient's cardiac arrest was not witnessed, especially if the call-to-arrival interval is greater than 5 minutes, perform five cycles (approximately 2 minutes) of CPR and then apply the AED. Follow local protocols regarding witnessed versus unwitnessed cardiac arrest and the use of the AED.

CHAPTER

13 Neurologic Emergencies

Workbook Activities

The following activities have been designed to help you. Your instructor may require you to complete some or all of these activities as a regular part of your EMT-B training program. You are encouraged to complete any activity that your instructor does not assign as a way to enhance your learning in the classroom.

Chapter Review

The following exercises provide an opportunity to refresh your knowledge of this chapter.

Matching

Match each of the terms in the left column to the appropriate definition in the right column.

_____ **1.** Brain stem

_____ **2.** Foramen magnum

_____ **3.** Spinal nerves

_____ **4.** Cerebrum

_____ **5.** Cranial nerves

_____ **6.** Cerebellum

_____ **7.** Embolism

_____ **8.** Aneurysm

_____ **9.** Status epilepticus

_____ **10.** Hypoglycemia

_____ **11.** Aphasia

_____ **12.** Berry aneurysm

_____ **13.** Dysarthria

A. slurred, hard to understand speech

B. the back part of this area of the brain processes sight

C. weakness in an artery wall

D. controls most basic functions of the body

E. exit at each vertebra and carry messages to and from the body

F. this area of the brain controls muscle and body coordination

G. hole in the base of the skull

H. low blood glucose

I. innervate eyes, ears, face

J. clot that forms elsewhere and travels to the site of damage

K. weakness in a blood vessel that resembles a tiny balloon

L. inability to produce or understand speech

M. seizures that recur every few minutes

Multiple Choice

Read each item carefully, then select the best response.

_____ **1.** Seizures may occur as a result of:

    A. metabolic problems.

    B. brain tumor.

    C. a recent or old head injury.

    D. all of the above

_____ **2.** The _____ controls the most basic functions of the body, such as breathing, blood pressure, swallowing, and pupil constriction.

    A. brain stem

    B. cerebellum

    C. cerebrum

    D. spinal cord

_____ **3.** At each vertebra in the neck and back, _____ nerves, called spinal nerves, branch out from the spinal cord and carry signals to and from the body.

    A. two

    B. three

    C. four

    D. five

_____ **4.** Brain disorders include all of the following except:

    A. coma.

    B. infection.

    C. hypoglycemia.

    D. tumor.

_____ **5.** When blood flow to a particular part of the brain is cut off by a blockage inside a blood vessel, the result is:

    A. a hemorrhagic stroke.

    B. atherosclerosis.

    C. an ischemic stroke.

    D. a cerebral embolism.

_____ **6.** Patients who are at the highest risk of hemorrhagic stroke are those who have:

    A. untreated hypertension.

    B. an aneurysm.

    C. a berry aneurysm.

    D. atherosclerosis.

_____ **7.** Patients with a subarachnoid hemorrhage typically complain of a sudden severe:

    A. bout of dizziness.

    B. headache.

    C. altered mental status.

    D. thirst.

_____ **8.** The plaque that builds up in atherosclerosis obstructs blood flow, and interferes with the vessel's ability to:

    A. constrict.

    B. dilate.

    C. diffuse.

    D. exchange gases.

_____ **9.** A TIA, or mini-stroke, is the name given to a stroke when symptoms go away on their own in less than:

    A. half an hour.

    B. 1 hour.

    C. 12 hours.

    D. 24 hours.

_____ **10.** Seizures characterized by unconsciousness and a generalized severe twitching of all the body's muscles that lasts several minutes or longer is called a:

    A. grand mal seizure.

    B. petit mal seizure.

    C. focal motor seizure.

    D. febrile seizure.

_____ **11.** Metabolic seizures may be due to:

    A. epilepsy.

    B. a brain tumor.

    C. high fevers.

    D. hypoglycemia.

_____ **12.** When assessing a patient with a history of seizure activity, it is important to:

    A. determine whether this episode differs from any previous ones.

    B. recognize the postictal state.

    C. look for other problems associated with the seizure.

    D. all of the above

_____ **13.** Signs and symptoms of possible seizure activity include:

    A. altered mental status.

    B. incontinence.

    C. rapid and deep respirations.

    D. all of the above

_____ **14.** Common causes of altered mental status include all of the following, except:

    A. body temperature abnormalities.

    B. hypoxemia.

    C. unequal pupils.

    D. hypoglycemia.

_____ **15.** The principle difference between a patient who has had a stroke and a patient with hypoglycemia almost always has to do with the:

    A. papillary response.

    B. mental status.

    C. blood pressure.

    D. capillary refill time.

_____ **16.** Consider the possibility of _____ in a patient who has had a seizure.

    A. brain injury

    B. hyperglycemia

    C. hypoglycemia

    D. hypertension

_____ **17.** Individuals with chronic alcoholism can have abnormalities in liver function and in their blood-clotting and immune systems, which can predispose them to:

    A. intracranial bleeding.

    B. brain and bloodstream infections.

    C. hypoglycemia.

    D. all of the above

_____ **18.** Low oxygen levels in the bloodstream will affect the entire brain, causing:

   A. anxiety.

   B. restlessness.

   C. confusion.

   D. all of the above

_____ **19.** Patients with _____ may have trouble understanding speech but can speak clearly.

   A. aphasia

   B. receptive aphasia

   C. expressive aphasia

   D. dysarthria

_____ **20.** High blood pressure in stroke patients should not be treated in the field because:

   A. the brain is raising the blood pressure in an attempt to force more oxygen into its injured parts.

   B. quite often, blood pressure will return to normal or may drop significantly on its own.

   C. many times it is a response to bleeding in the brain.

   D. all of the above

_____ **21.** The following conditions may simulate a stroke except:

   A. hyperglycemia.

   B. a postictal state.

   C. hypoglycemia.

   D. subdural bleeding.

_____ **22.** When assessing a patient with a possible CVA, you should check the _____ first.

   A. pulse

   B. airway

   C. pupils

   D. blood pressure

_____ **23.** A patient with a GCS of 12 has:

   A. no dysfunction.

   B. mild dysfunction.

   C. moderate to severe dysfunction.

   D. severe dysfunction.

_____ **24.** Assess the mental status using the mnemonic:

   A. OPQRST.

   B. SAMPLE.

   C. AVPU.

   D. PEARRL.

Questions 25-29 are derived from the following scenario: You are called to a long-term care facility, and find a 92-year-old woman supine in her bed. Though she is able to track your movement with her eyes, she is unresponsive to you and the nurse standing by. The nurse explains this is abnormal behavior for her, as she is normally talkative and very friendly.

_____ **25.** How would you best determine the patient's level of consciousness?

   A. By using AVPU

   B. By using the Cincinnati Stroke Scale

   C. By using the Glasgow Coma Scale

   D. All of the above

_____ **26.** A score of _____ from the Glasgow Coma Scale would indicate severe dysfunction.

   A. 15 or less

   B. 14-15

   C. 11-13

   D. 10 or less

_____ **27.** Your patient may have a hemorrhagic or ischemic stroke. What statement is true regarding the two conditions?

   A. Hemorrhagic stroke occurs as a result of bleeding inside the brain.

   B. Ischemic stroke occurs when a part of the brain is cut off by a blockage inside a blood vessel.

   C. Neither A or B is true.

   D. Both A and B are true.

_____ **28.** If the receiving facility told you the cause of her stroke was due to a build up of calcium and cholesterol, forming a plaque inside the walls of her blood vessels, you would know that this patient had:

   A. atherosclerosis.

   B. a TIA.

   C. arteriolosclerosis.

   D. a CVA.

_____ **29.** If this patient is having a seizure, what could be the cause?

   A. Hypoglycemia

   B. Hypoxia

   C. Drug overdose

   D. All the above

Labeling

Label the following diagrams with the correct terms.

1. Brain

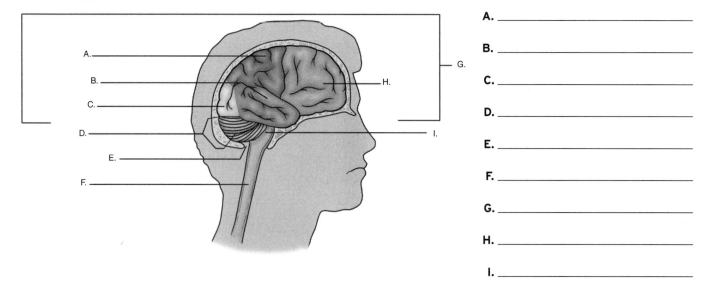

A. _____

B. _____

C. _____

D. _____

E. _____

F. _____

G. _____

H. _____

I. _____

2. Spinal Cord

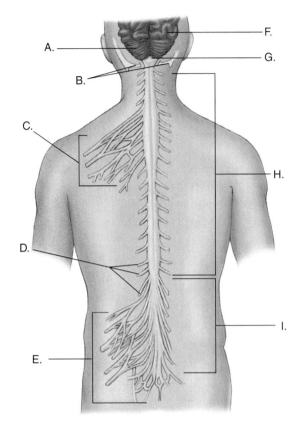

A. _____

B. _____

C. _____

D. _____

E. _____

F. _____

G. _____

H. _____

I. _____

Fill-in

Read each item carefully, then complete the statement by filling in the missing word(s).

1. There are _____ cranial nerves.

2. Playing the piano is coordinated by the _____.

3. The front part of the cerebrum controls _____ and _____.

4. The cranial nerves run to the _____.

5. The brain is divided into _____ major parts.

6. All messages traveling to and from the brain travel along _____.

7. Each hemisphere of the cerebrum controls activities on the _____ side of the body and the _____ side of the face.

8. The _____ is the largest part of the brain.

9. _____ is a loss of bowel and bladder control, and can be due to a generalized seizure.

10. The _____ is the body's computer.

11. Onset of _____ bleeding is usually very rapid after injury.

12. Weakness on one side of the body is known as _____.

13. No matter what the cause, you should consider _____ _____ _____ to be an emergency that

requires immediate attention, even when it appears that the cause may simply be alcohol intoxication or a minor

car crash or fall.

True/False

If you believe the statement to be more true than false, write the letter "T" in the space provided. If you believe the
statement to be more false than true, write the letter "F."

_____ **1.** The postictal state following a seizure commonly lasts only about 3 to 5 minutes.

_____ **2.** Metabolic seizures result from an area of abnormality in the brain.

_____ **3.** Febrile seizures result from sudden high fevers and are generally well tolerated by children.

_____ **4.** Hemiparesis is the inability to speak or understand speech.

_____ **5.** The dura covers the brain.

_____ **6.** Right-sided facial droop is most likely an indication of a problem in the right cerebral hemisphere.

Short Answer

Complete this section with short written answers using the space provided.

1. List and describe the three key tests for assessing stroke.

_____

_____

2. Why is prompt transport of stroke patients critical?

_____

_____

3. Describe the characteristics of a postictal state.

_____

_____

4. What is the difference between infarcted and ischemic cells?

_____

_____

5. List three conditions that may simulate stroke.

_____

_____

Word Fun emtb vocab explorer

The following crossword puzzle is an activity provided to reinforce correct spelling and understanding of medical terminology associated with emergency care and the EMT-B. Use the clues in the column to complete the puzzle.

Across

6. Cholesterol and calcium build-up

7. One-sided weakness

8. Loss of brain function

Down

1. Related to high temperatures, particularly with children

2. Inability to pronounce clearly

3. Low blood glucose level

4. Clotting of vessel

5. Inability to speak

Ambulance Calls

The following case scenarios will give you an opportunity to explore the concerns associated with patient management. Read each scenario, then answer each question in detail.

1. You are dispatched to a private residence for a "confused man." You arrive to find an older man sitting in a recliner. As you begin your assessment, you notice that he has right-sided weakness and doesn't seem to understand your questions. He is alone in the home, and it appears that no one lives with him in the residence. How do you best manage this patient?

_____

_____

_____

2. You are dispatched to a 36-year-old man who had seizure activity at least an hour ago. The patient is incontinent, cold, clammy, and unresponsive. His friends tell you that the "shaking" stopped and he has not woke up. They thought he might just be tired until they discovered they could not wake him. He has no history of seizure activity. He has diabetes for which he takes medication. How would you best manage this patient?

_____

_____

_____

3. You are dispatched to a local business for "woman with severe headache." The 55-year-old patient states that she has had headaches in the past, but this headache is the worst she's ever had in her life. She feels like the room is spinning around; she's seeing "double" and she feels sick to her stomach. She has a history of hypertension. She tells you that she stopped taking her blood pressure medicine about 6-8 months ago because she could no longer afford it.

How do you best manage this patient?

_____

_____

_____

Assessment Review

Answer the following questions pertaining to the assessment of the types of emergencies discussed in this chapter.

You are called to the home of a 52-year-old man who was found unconscious. You have been greeted at the door by an obviously distraught wife who leads you to a bathroom in the master bedroom. As you peer into the room, you see the unresponsive man sitting on the toilet with his pants around his ankles, and your partner finds a small handgun lying on the bedroom floor.

_____   **1.** How should you respond?

   A. Determine if he has been shot.

   B. Ask the wife who else is in the home.

   C. Pick-up and secure the gun.

   D. Drop your equipment and leave the home.

_____   **2.** After it has been determined this was not a shooting, you begin to assess the patient. He regains consciousness and states he must have fainted while having a bowel movement. What type of exam should you perform?

   A. Rapid physical exam

   B. Focused physical exam

   C. Detailed physical exam

   D. No exam is necessary for this patient.

_____   **3.** How should you intervene for this patient?

   A. Provide oxygen.

   B. Determine the patient's glucose level.

   C. Determine the patient's Glasgow Coma Scale score.

   D. All of the above

_____   **4.** As you intervene for this patient, his muscles begin to twitch and his eyes roll back into his head. You respond by:

   A. laying him on the floor.

   B. calling for ALS.

   C. obtaining a SAMPLE history.

   D. all the above

_____   **5.** He continues to twitch for several minutes. How do you respond?

   A. This is normal, and you need to wait it out.

   B. He is reacting to low blood sugar levels, and needs glucose.

   C. You suspect status epilepticus, and call for ALS.

   D. You suspect a CVA, and provide rapid transport once he is done.

Emergency Care Summary

Complete the statements pertaining to emergency care for the types of emergencies discussed in this chapter by filling in the missing words.

NOTE: While the steps below are widely accepted, be sure to consult and follow your local protocol.

Stroke	Seizure	Altered Mental Status

Stroke

Cincinnati Stroke Scale

1. Ask patient to show _____ or smile to determine facial droop.
2. Ask patient to close eyes and hold both arms out with palms up to measure _____ _____.
3. Ask patient to say, "The sky is blue in _____," to monitor speech.

Seizure

Glasgow Coma Scale

Eye Opening

_____	4
Responsive to speech	3
Responsive to pain	2
None	1

Best Verbal Response

Oriented conversation	5
Confused conversation	4
_____ _____	3
Incomprehensible sounds	2
None	1

Best Motor Response

Obeys commands	6
Localizes pain	5
_____ _____ _____	4
Abnormal flexion	3
Abornormal extension	2
None	1

Add the total points selected from all three categories to determine the patient's

_____ _____ _____

_____.

Altered Mental Status

Use the _____ _____ _____ and/or the _____ _____ _____ to aid in your assessment.

CHAPTER

14 The Acute Abdomen

Workbook Activities

The following activities have been designed to help you. Your instructor may require you to complete some or all of these activities as a regular part of your EMT-B training program. You are encouraged to complete any activity that your instructor does not assign as a way to enhance your learning in the classroom.

Chapter Review

The following exercises provide an opportunity to refresh your knowledge of this chapter.

Matching

Match each of the terms in the left column to the appropriate definition in the right column.

_____ 1. Aneurysm

_____ 2. Colic

_____ 3. Retroperitoneal

_____ 4. Ulcer

_____ 5. Hernia

_____ 6. Ileus

_____ 7. Guarding

_____ 8. Anorexia

_____ 9. Emesis

_____ 10. Referred pain

_____ 11. PID

_____ 12. Cystitis

_____ 13. Strangulation

_____ 14. Peritonitis

_____ 15. Peritoneum

A. paralysis of the bowel

B. pain felt in an area of the body other than the actual source

C. protective, involuntary abdominal muscle contractions

D. acute, intermittent cramping abdominal pain

E. behind the peritoneum

F. vomiting

G. common cause of acute abdomen in women

H. a membrane lining the abdomen

I. swelling or enlargement of a weakened arterial wall

J. loss of hunger or appetite

K. protrusion of a loop of an organ or tissue through an abnormal body opening

L. obstruction of blood circulation resulting from compression or entrapment of organ tissue

M. abrasion of the stomach or small intestine

N. inflammation of the bladder

O. inflammation of the peritoneum

Multiple Choice

Read each item carefully, then select the best response.

_____ 1. Peritonitis, with associated fluid loss, may lead to _____ shock.
  A. hemorrhagic
  B. septic
  C. hypovolemic
  D. metabolic

_____ 2. Distention of the abdomen is gauged by:
  A. visualization.
  B. auscultation.
  C. palpation.
  D. the patient's complaint of pain around the umbilicus.

_____ 3. A hernia that returns to its proper body cavity is said to be:
  A. reducible.
  B. extractable.
  C. incarcerated.
  D. replaceable.

_____ 4. Sensory nerves from the spinal cord to the skin and muscles are part of the:
  A. somatic nervous system.
  B. peripheral nervous system.
  C. autonomic nervous system.
  D. sympathetic nervous system.

_____ 5. When an organ of the abdomen is enlarged, rough palpation may cause _____ of the organ.
  A. distention
  B. nausea
  C. swelling
  D. rupture

_____ 6. Severe back pain may be associated with which condition?
  A. abdominal aortic aneurysm
  B. PID
  C. appendicitis
  D. mittelschmerz

_____ 7. The _____ are found in the retroperitoneal space.
  A. stomach and gallbladder
  B. kidneys, genitourinary structures, and large vessels
  C. liver and pancreas
  D. adrenal glands and uterus

_____ 8. A(n) _____ may occur as a result of a surgical wound that has failed to heal properly.
  A. ectopic pregnancy
  B. strangulation
  C. hernia
  D. ulcer

_____ 9. The peritoneal membrane that can perceive the sensations of pain, pressure, and cold is the:
  A. meningeal.
  B. parietal.
  C. retroperitoneal.
  D. visceral.

_____ **10.** Common disease(s) that produce(s) signs of an acute abdomen include:

    A. diverticulitis.

    B. cholecystitis.

    C. acute appendicitis.

    D. all of the above

_____ **11.** A patient with peritonitis may present with rapid, shallow breaths due to:

    A. hypovolemia.

    B. ileus.

    C. pain.

    D. inflammation.

Questions 12-16 are derived from the following scenario: You have been dispatched to the home of a 21-year-old woman with severe abdominal pain.

_____ **12.** A possible cause for her pain may be:

    A. peritonitis.

    B. appendicitis.

    C. mittelschmerz.

    D. all of the above

_____ **13.** As you attempt to palpate her abdomen, she pushes your hand away. This is called:

    A. referred pain.

    B. guarding.

    C. protecting.

    D. all of the above

_____ **14.** She informs you that she missed her last menstrual cycle. You may begin to suspect:

    A. mittelschmerz.

    B. cystitis.

    C. ectopic pregnancy.

    D. PID.

_____ **15.** You should transport her:

    A. in the position of comfort.

    B. supine.

    C. left lateral recumbent.

    D. in the recovery position.

_____ **16.** Her respirations are 18 breaths/min, pulse is 78 beats/min, and her blood pressure is 132/80 mm Hg. How often should you reassess her vital signs?

    A. Every 5 minutes

    B. Every 10 minutes

    C. Every 15 minutes

    D. Every 20 minutes

Labeling
Label the following diagrams with the correct terms.

1. Solid Organs

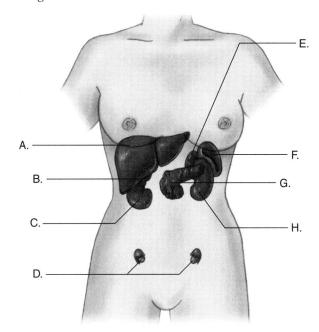

A. _____

B. _____

C. _____

D. _____

E. _____

F. _____

G. _____

H. _____

2. Hollow Organs

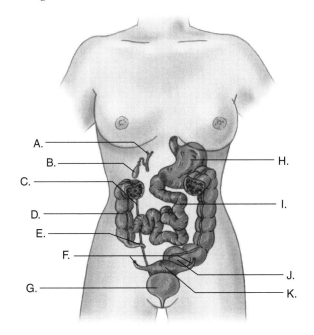

A. _____

B. _____

C. _____

D. _____

E. _____

F. _____

G. _____

H. _____

I. _____

J. _____

K. _____

3. Retroperitoneal Organs

A. _____

B. _____

C. _____

D. _____

E. _____

F. _____

G. _____

H. _____

I. _____

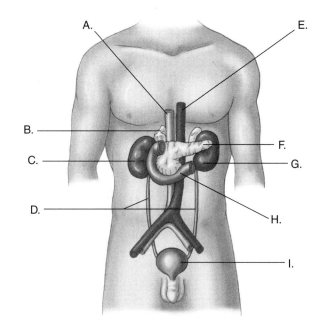

True/False

If you believe the statement to be more true than false, write the letter "T" in the space provided. If you believe the statement to be more false than true, write the letter "F."

_____ **1.** Referred pain is a result of connection between ligaments in the abdominal and chest cavities.

_____ **2.** Abdominal pain in women is usually related to the menstrual cycle and is rarely serious.

_____ **3.** A leading aorta, being retroperitoneal, will not cause peritonitis.

_____ **4.** It is important to accurately diagnose the cause of acute abdominal pain in order to properly treat the patient.

_____ **5.** The parietal peritoneum lines the walls of the abdominal cavity.

_____ **6.** The patient with peritonitis usually reports relief of pain when lying left lateral recumbent with the knees pulled in.

_____ **7.** When palpating the abdomen, always start with the quadrant where the patient complains of the most severe pain.

_____ **8.** Massive hemorrhaging is associated with rupture of an abdominal aortic aneurysm.

_____ **9.** Pneumonia may cause abdominal pain.

Short Answer

Complete this section with short written answers using the space provided.

1. Explain the phenomenon of referred pain.

_____

_____

2. Should an EMT-B attempt to diagnose the cause of abdominal pain? Why or why not?

_____

_____

3. Why does abdominal distention accompany ileus?

_____

_____

4. What two conditions may result in hypovolemic shock in the patient with an acute abdomen?

_____

_____

5. List the general EMT-B emergency care for patients with acute abdomen.

_____

_____

Word Fun

The following crossword puzzle is an activity provided to reinforce correct spelling and understanding of medical terminology associated with emergency care and the EMT-B. Use the clues in the column to complete the puzzle.

Across

3. Felt elsewhere in the body

8. Abrasion of stomach or small intestine

10. Complete obstruction

11. Paralysis of the bowels

12. Lining of the abdominal cavity

Down

1. Protrusion of a loop of an organ

2. Involuntary muscle contractions for protection

4. Vomiting

5. Sudden onset of pain below diaphragm

6. Lack of appetite

7. Results in the weakening of an arterial wall

9. Acute cramping abdominal pain

Ambulance Calls

The following case scenarios provide an opportunity to explore the concerns associated with patient management. Read each scenario, then answer each question in detail.

1. You are called to the local high school nurse's office for a 16-year-old female complaining of severe, abrupt abdominal pain. She is pale, cool, and clammy with absent radial pulses. The school nurse tells you that the patient found out 4 weeks ago that she is pregnant.

How would you best manage this patient?

_____

_____

_____

2. You are dispatched to a long-term care facility for a geriatric man with abdominal pain. Upon your arrival, one of the staff members tells you that this patient has been bedridden and taking pain medications for the last several weeks, and has recently had problems passing a normal bowel movement. He now has a distended, tender abdomen with nausea, vomiting and tachycardia.

How do you best manage this patient?

_____

_____

_____

3. You are dispatched to the home of another responder for "severe back pain." You arrive to find your coworker writhing in pain on the floor. He tells you that he tried to urinate (unsuccessfully) when he immediately experienced a sharp, cramping sensation in his right side. As you are talking to him he throws up. After vomiting, he tells you the pain is worse and is now spreading to his groin.

How do you best manage this patient?

_____

_____

_____

Assessment Review

Answer the following questions pertaining to the assessment of the types of emergencies discussed in this chapter. Questions 1-5 are derived from the following scenario:

You respond the home of a 6-year-old boy complaining of severe back pain. As you approach the boy, he is in the fetal position crying, and is in obvious pain. As you ask questions of the child, he remains silent and his dad answers your questions.

_____ **1.** What will you determine during the initial assessment?

    A. The priority of care

    B. The child's level of consciousness

    C. Finding and treating any life threats

    D. All of the above

_____ **2.** Your partner takes Dad to a separate room to obtain a SAMPLE history. Which question(s) should she ask?

   A. "Do you know what medications he is currently taking?"

   B. "Where is his mother?"

   C. "Sir, we need to know, did you or your wife hurt him?"

   D. All of the above

_____ **3.** As you take a SAMPLE history of the boy, he admits to you his Dad kicked him in the stomach a few days ago and it has been hurting since. How should you proceed?

   A. Confront the father to see if this is true.

   B. Call for police back-up.

   C. Take the child without the father's knowledge.

   D. All of the above

_____ **4.** Upon examination, you note that his lower right quadrant is distended. The likely cause of the distention is:

   A. strangulation.

   B. cholecystitis.

   C. ileus.

   D. none of the above

_____ **5.** During your ongoing assessment, you should:

   A. repeat the initial assessment.

   B. repeat the focused assessment.

   C. reassure the patient.

   D. all of the above

Emergency Care Summary

Complete the statements pertaining to emergency care for the types of emergencies discussed in this chapter by filling in the missing words.

   NOTE: While the steps below are widely accepted, be sure to consult and follow your local protocol.

Acute Abdomen

General Management

1. Explain to the patient what you are about to do.
2. Place the patient in a _____ position with legs drawn up and _____ at the knees unless _____ is suspected.
3. Determine whether the patient is restless or quiet, whether motion causes pain, or whether any characteristic position, _____, or obvious _____ is present.
4. _____ the four _____ of the abdomen.
5. Determine whether the patient can relax the _____ _____ on command.
6. Determine whether the abdomen is tender when _____.

CHAPTER

15 Diabetic Emergencies

Workbook Activities

The following activities have been designed to help you. Your instructor may require you to complete some or all of these activities as a regular part of your EMT-B training program. You are encouraged to complete any activity that your instructor does not assign as a way to enhance your learning in the classroom.

Chapter Review

The following exercises provide an opportunity to refresh your knowledge of this chapter.

Matching

Match each of the terms in the left column to the appropriate definition in the right column.

_____ 1. Hormone

_____ 2. Polydipsia

_____ 3. Type I diabetes

_____ 4. Acidosis

_____ 5. Insulin

_____ 6. Diabetic coma

_____ 7. Polyuria

_____ 8. Type II diabetes

_____ 9. Polyphagia

_____ 10. Insulin shock

_____ 11. Glucose

_____ 12. Kussmaul respirations

_____ 13. Hyperglycemia

_____ 14. Diabetes

A. altered level of consciousness caused by insufficient glucose in the blood

B. diabetes that usually starts in childhood; requires insulin

C. excessive eating

D. deep, rapid breathing

E. excessive urination

F. excessive thirst persisting for a long period of time

G. diabetes with onset later in life; may be controlled by diet and oral medication

H. chemical produced by a gland that regulates body organs

I. literal meaning: "A passer through; a siphon"

J. extremely high blood glucose level

K. pathologic condition resulting from the accumulation of acids in the body

L. hormone that enables glucose to enter the cells

M. primary fuel, along with oxygen, for cellular metabolism

N. state of unconsciousness resulting from several problems, including ketoacidosis, dehydration, and hyperglycemia

Multiple Choice

Read each item carefully, then select the best response.

_____ **1.** Patients with which type of diabetes are more likely to have metabolic problems and organ damage?

   A. Type I

   B. Type II

   C. Sugar diabetes

   D. HHNC

_____ **2.** Normal blood glucose levels range from _____ mg/dL.

   A. 80 to 120

   B. 90 to 140

   C. 70 to 110

   D. 60 to 100

_____ **3.** Patients with diabetes mellitus and a lack of insulin excrete excess glucose through their:

   A. lymphatic system.

   B. sweat.

   C. respiratory efforts.

   D. urine.

_____ **4.** Diabetes is a metabolic disorder in which the hormone _____ is missing or ineffective.

   A. estrogen

   B. adrenaline

   C. insulin

   D. epinephrine

_____ **5.** Complications of diabetes include:

   A. paralysis.

   B. cardiovascular disease.

   C. brittle bones.

   D. hepatitis.

_____ **6.** The accumulation of ketones and fatty acids in blood tissue can lead to a dangerous condition in diabetic patients known as:

   A. diabetic ketoacidosis.

   B. insulin shock.

   C. HHNC.

   D. hypoglycemia.

_____ **7.** The term for excessive eating as a result of cellular "hunger" is:

   A. polyuria.

   B. polydipsia.

   C. polyphagia.

   D. polyphony.

_____ **8.** Insulin is produced by the:

   A. adrenal glands.

   B. hypothalamus.

   C. spleen.

   D. pancreas.

_____ **9.** Factors that may contribute to diabetic coma include:
    A. infection.
    B. alcohol consumption.
    C. insufficient insulin.
    D. all of the above

_____ **10.** The only organ that does not require insulin to allow glucose to enter its cells is the:
    A. liver.
    B. brain.
    C. pancreas.
    D. heart.

_____ **11.** The sweet or fruity odor on the breath of a diabetic patient is caused by _____ in the blood.
    A. acetone
    B. ketones
    C. alcohol
    D. insulin

_____ **12.** The term for excessive thirst is:
    A. polyuria.
    B. polydipsia.
    C. polyphagia.
    D. polyphony.

_____ **13.** Oral diabetic medications include:
    A. Micronase.
    B. Glucotrol.
    C. Diabinase.
    D. all of the above

_____ **14.** _____ is one of the basic sugars in the body.
    A. Dextrose
    B. Sucrose
    C. Fructose
    D. Syrup

_____ **15.** _____ is the hormone that is normally produced by the pancreas that enables glucose to enter the cells
    A. Insulin
    B. adrenaline
    C. Estrogen
    D. Epinephrine

_____ **16.** The term for excessive urination is:
    A. polyuria.
    B. polydipsia.
    C. polyphagia.
    D. polyphony.

_____ **17.** When fat is used as an immediate energy source, _____ and fatty acids are formed as waste products.
    A. dextrose
    B. sucrose
    C. ketones
    D. bicarbonate

_____ **18.** When using a glucometer, the patient tests his or her glucose level in a drop of:
    A. urine.
    B. blood.
    C. saliva.
    D. cerebrospinal fluid.

_____ **19.** The onset of hypoglycemia can occur within:
   A. seconds.
   B. minutes.
   C. hours.
   D. days.

_____ **20.** Without _____, or with very low levels, brain cells rapidly suffer permanent damage.
   A. epinephrine
   B. ketones
   C. bicarbonate
   D. glucose

_____ **21.** _____ is/are a potentially life-threatening complication of insulin shock.
   A. Kussmaul respirations
   B. Hypotension
   C. Seizures
   D. Polydipsia

_____ **22.** Blood glucose levels are measured in:
   A. micrograms per deciliter.
   B. milligrams per deciliter.
   C. milliliters per decigram.
   D. microliters per decigram.

_____ **23.** Diabetic coma may develop as a result of:
   A. too little insulin.
   B. too much insulin.
   C. overhydration.
   D. metabolic alkalosis.

_____ **24.** Always suspect hypoglycemia in any patient with:
   A. Kussmaul respirations.
   B. an altered mental status.
   C. nausea and vomiting.
   D. all of the above

_____ **25.** The most important step in caring for the unresponsive diabetic patient is to:
   A. give oral glucose immediately.
   B. perform a focused assessment.
   C. open the airway.
   D. obtain a SAMPLE history.

_____ **26.** Determination of diabetic coma or insulin shock should:
   A. be made before transport of the patient.
   B. be made before administration of oral glucose.
   C. be determined by a urine glucose test.
   D. be based upon your knowledge of the signs and symptoms of each condition.

_____ **27.** An unresponsive patient involved in a motor vehicle crash with a diabetic identification bracelet should be suspected to be _____ until proven otherwise.
   A. hypoglycemic
   B. hyperglycemic
   C. intoxicated
   D. in shock

_____ **28.** Contraindications for the use of oral glucose include:
   A. unconsciousness.
   B. known alcoholic.
   C. insulin shock.
   D. all of the above

_____ **29.** When reassessing the diabetic patient after administration of oral glucose, watch for:

    A. airway problems.

    B. seizures.

    C. sudden loss of consciousness.

    D. all of the above

_____ **30.** Signs and symptoms associated with hypoglycemia include:

    A. warm, dry skin.

    B. rapid, weak pulse.

    C. Kussmaul respirations.

    D. anxious or combative behavior.

_____ **31.** The patient in insulin shock is experiencing:

    A. hyperglycemia.

    B. hypoglycemia.

    C. diabetic ketoacidosis.

    D. a low production of insulin.

_____ **32.** Signs of dehydration include:

    A. good skin turgor.

    B. elevated blood pressure.

    C. sunken eyes.

    D. all of the above

_____ **33.** Diabetic patients who complain of "not feeling so well" should:

    A. have their glucose level checked.

    B. have a rapid trauma assessment completed.

    C. be rapidly transported to the closest medical facility.

    D. immediately be given oral glucose.

_____ **34.** Causes of insulin shock include:

    A. taking too much insulin.

    B. vigorous exercise without sufficient glucose intake.

    C. nausea, vomiting, anorexia.

    D. all of the above

_____ **35.** Insulin shock can develop more often and more severely in children than in adults due to their:

    A. high activity level and failure to maintain a strict schedule of eating.

    B. genetic makeup.

    C. smaller body size.

    D. all of the above

_____ **36.** Because diabetic coma is a complex metabolic condition that usually develops over time and involves all the tissues of the body, correcting this condition may:

    A. be accomplished quickly through the use of oral glucose.

    B. require rapid infusion of IV fluid to prevent permanent brain damage.

    C. take many hours in a hospital setting.

    D. include a reduction in the amount of insulin normally taken by the patient.

_____ **37.** A patient in insulin shock or a diabetic coma may appear to be:

    A. having a heart attack.

    B. perfectly normal.

    C. intoxicated.

    D. having a stroke.

Questions 38-42 are derived from the following scenario: A 54-year-old golfer collapsed on the 17th green at the golf course. His friend said he wasn't feeling well after the 8th hole, but insisted on walking and finishing out the game. He is pale, cool and diaphoretic, and answers incoherently to your questions.

_____ **38.** During your rapid physical exam, you discover a medical alert necklace around his neck that reads "Type 2 Diabetic". This tells you that:

    A. he developed diabetes later in life.

    B. his body produces inadequate amounts of insulin.

    C. he is likely to be taking non-insulin type of oral medications.

    D. all of the above

_____ **39.** His blood glucose level is 65 mg/dL. You:

    A. do not suspect hypoglycemia and begin to think cardiac in nature.

    B. suspect hyperglycemia and proceed to give oral glucose.

    C. suspect hypoglycemia and proceed to give oral glucose.

    D. suspect hypoglycemia but oral glucose is contraindicated for him.

_____ **40.** The patient loses consciousness and a second blood glucose level reads 48 mg/dL. You should:

    A. call for, or transport to, ALS.

    B. ensure a patent airway.

    C. provide high-flow oxygen.

    D. all of the above

_____ **41.** Because of his blood glucose level and rapid respirations, you suspect:

    A. insulin shock.

    B. diabetic coma.

    C. renal failure.

    D. mittelschmerz.

_____ **42.** Since the patient is unconscious and his blood glucose level is 48 mg/dL, how should the glucose be delivered?

    A. Between the cheek and gum

    B. Placed on the back of the tongue

    C. Placed on the tip of the tongue

    D. None of the above

Fill-in

Read each item carefully, then complete the statement by filling in the missing word(s).

1. The full name of diabetes is _____ _____.

2. Diabetes is considered to be a(n) _____ problem, in which the body becomes allergic to its own tissues and literally destroys them.

3. Diabetes is defined as a lack of or _____ action of insulin.

4. Too much blood glucose by itself does not always cause _____ _____, but on some occasions, it can lead to it.

5. A patient in insulin shock needs _____ immediately and a patient in a diabetic coma needs _____ and IV fluid therapy.

True/False

If you believe the statement to be more true than false, write the letter "T" in the space provided. If you believe the statement to be more false than true, write the letter "F."

_____ 1. When patients use fat for energy, the fat waste products increase the amount of acid in the blood and tissue.

_____ 2. The level of consciousness can be affected if a patient has not exercised enough.

_____ 3. If blood glucose levels remain low, a patient may lose consciousness or have permanent brain damage.

_____ 4. Signs and symptoms can develop quickly in children because their level of activity can exhaust their glucose levels.

_____ 5. Diabetic emergencies can occur when a patient's blood glucose level gets too high or when it drops too low.

_____ 6. Diabetic patients may require insulin to control their blood glucose.

_____ 7. Glucose is a hormone that enables insulin to enter the cells of the body.

_____ 8. Insulin is one of the basic sugars essential for cell metabolism in humans.

_____ 9. Diabetes can cause kidney failure, blindness, and damage to blood vessels.

_____ 10. Most children with diabetes are insulin dependent.

_____ 11. Many adults with diabetes can control their blood glucose levels with diet alone.

Short Answer

Complete this section with short written answers using the space provided.

1. What is insulin and what is its role in metabolism?

_____

_____

2. What are two trade names for oral glucose?

_____

_____

3. When should you not give oral glucose to a patient experiencing a suspected diabetic emergency?

_____

_____

4. List the trade names of three oral medications used by diabetics.

_____

_____

5. What are the three problems associated with the development of diabetic coma?

_____

_____

6. List the physical signs of diabetic coma.

_____

_____

7. If a diabetic patient was "fine" 2 hours ago and now is unconscious and unresponsive, which diabetes-related condition would you suspect and why?

_____

_____

8. Why should oral glucose be given to any diabetic patient with an altered level of consciousness?

_____

_____

Word Fun

The following crossword puzzle is an activity provided to reinforce correct spelling and understanding of medical terminology associated with emergency care and the EMT-B. Use the clues in the column to complete the puzzle.

Across

1. Unconscious state with high glucose levels
3. Excessive eating
6. Disease that usually starts early in life, generally requires daily injections
8. Excessive thirst
10. Chemical substance produced by a gland
11. May result from hypoglycemia

Down

2. Disease that starts later in life, usually controlled by diet and oral medications
4. Opposite of alkalosis
5. Hormone produced in pancreas
7. Excessive urination
9. One of the basic sugars

Ambulance Calls

The following case scenarios provide an opportunity to explore the concerns associated with patient management. Read each scenario, then answer each question in detail.

1. You are called to a local residence where you find a 22-year-old woman supine in bed, unresponsive to your attempts to rouse her. She is cold and clammy with gurgling respirations. Her mother tells you that her only history is diabetes, which she has had since she was a small child. What steps would you take in managing this patient?

_____

_____

_____

2. You are requested by local law enforcement to assist with "a man acting strangely." You arrive to find a man in the back of the patrol car, singing and rocking back and forth. The officer tells you he thinks this man is "off his rocker." Apparently, the man was walking down Main Street when he decided to remove all of his clothing. He continued to walk around downtown, naked, until restrained by law enforcement officers.

How do you best manage this patient?

_____

_____

_____

3. You are dispatched to assist with a diabetic patient well known in your department for being noncompliant with his medications and diet. You have responded numerous times to his residence, all for instances of low blood sugar. Family members greet you at the door and say, "It's Jon again. Just give him some sugar like you usually do." You walk into the patient's bedroom to discover him unconscious with snoring respirations.

How do you best manage this patient?

_____

_____

_____

Skill Drills

Skill Drill 15-1: Administering Glucose
Test your knowledge of this skill drill by filling in the correct words in the photo captions.

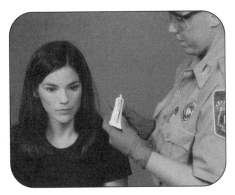  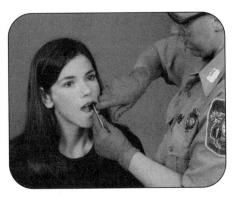

1. Make sure that the tube of glucose is intact and has not _____ .

2. Squeeze a generous amount of oral glucose onto the _____ _____ of a _____ _____ or tongue depressor.

3. Open the patient's _____ . Place the tongue depressor on the _____ _____ between the cheek and the gum with the _____ _____ next to the cheek. Repeat until the entire tube has been used.

Comparison Table

Complete the following table on the characteristics of diabetic emergencies.

	Hyperglycemia	Hypoglycemia
History Food intake Insulin dosage Onset Skin Infection		
Gastrointestinal Tract Thirst Hunger Vomiting		
Respiratory System Breathing Odor of breath		
Cardiovascular System Blood pressure Pulse		
Nervous System Consciousness		
Urine Sugar Acetone		
Treatment Response		

Assessment Review

Answer the following questions pertaining to the assessment of the types of emergencies discussed in this chapter. Questions 1-5 are derived from the following scenario:

While driving back to the station, you and your partner find an unconscious person lying on the grass in front of a home.

_____ **1.** After completing scene size-up, your fist step is to:

A. take appropriate BSI precautions.

B. form a general impression of the patient.

C. apply oxygen.

D. none of the above

_____ **2.** The patient appears to be a woman in her mid 30s, and she is unresponsive to verbal or painful stimulus. Because there is no one around, you are unable to complete a focused history. Which physical exam should you perform first?

A. Focused physical exam

B. Rapid physical exam

C. Detailed physical exam

D. Blood glucose level

_____ **3.** Her respirations are 28 breaths/min, her pulse is 110 beats/min, and her blood pressure is 94/52 mm Hg. How should you intervene for her?

A. Give her oral glucose.

B. Give her insulin.

C. Provide high-flow oxygen.

D. All of the above

_____ **4.** Your protocols do not allow you to measure blood glucose levels, and you are unsure as to the nature of her illness. You should:

A. provide oral glucose anyway.

B. provide insulin found in her purse.

C. provide nitroglycerin found in her purse.

D. none of the above

_____ **5.** When relaying information to medical control, you should inform them of:

A. the patient's condition.

B. any changes of consciousness.

C. any difficulty the patient may experience in breathing.

D. all of the above

Emergency Care Summary

Complete the statements pertaining to emergency care for the types of emergencies discussed in this chapter by filling in the missing words.

NOTE: While the steps below are widely accepted, be sure to consult and follow your local protocol.

Diabetic Emergencies

Administering Glucose

1. Examine the tube to ensure that it is not open or broken. Check the _____ _____.

2. Squeeze a _____ amount onto the bottom _____ of a bite stick or tongue depressor.

3. Open the patient's mouth. Place the tongue depressor on the _____ _____ between the cheek and gum, with the gel side next to the _____.

CHAPTER

16 Allergic Reactions and Envenomations

Workbook Activities

The following activities have been designed to help you. Your instructor may require you to complete some or all of these activities as a regular part of your EMT-B training program. You are encouraged to complete any activity that your instructor does not assign as a way to enhance your learning in the classroom.

Chapter Review

The following exercises provide an opportunity to refresh your knowledge of this chapter.

Matching

Match each of the terms in the left column to the appropriate definition in the right column.

_____ 1. Allergic reaction

_____ 2. Leukotrienes

_____ 3. Wheezing

_____ 4. Urticaria

_____ 5. Stridor

_____ 6. Allergen

_____ 7. Wheal

_____ 8. Toxin

A. substance made by body; released in anaphylaxis

B. harsh, high-pitched inspiratory sound, usually resulting from upper airway obstruction

C. raised, swollen area on skin resulting from an insect bite or allergic reaction

D. an exaggerated immune response to any substance

E. multiple raised areas on the skin that itch or burn

F. a poison or harmful substance

G. substance that causes an allergic reaction

H. high-pitched, whistling breath sound usually resulting from blockage of the airway and typically heard on expiration

Multiple Choice

Read each item carefully, then select the best response.

_____ **1.** Steps for assisting a patient with administration of an EpiPen include:

    A. taking body substance isolation precautions.

    B. placing the tip of the auto-injector against the medial part of the patient's thigh.

    C. recapping the injector before placing it in the trash.

    D. all of the above

_____ **2.** Allergens may include:

    A. food.

    B. animal bites.

    C. semen.

    D. all of the above

_____ **3.** Anaphylaxis is not always life threatening, but it typically involves:

    A. multiple organ systems.

    B. wheezing.

    C. urticaria.

    D. wheals.

_____ **4.** Signs and symptoms of insect stings or bites include:

    A. swelling.

    B. wheals.

    C. localized heat.

    D. all of the above

_____ **5.** Prolonged respiratory difficulty can cause _____, shock, and even death.

    A. tachypnea

    B. pulmonary edema

    C. tachycardia

    D. airway obstruction

_____ **6.** Speed is essential because more than two-thirds of patients who die of anaphylaxis do so within the first:

    A. 10 minutes.

    B. 30 minutes.

    C. hour.

    D. 3 hours.

_____ **7.** Questions to ask when obtaining a history from a patient appearing to have an allergic reaction include:

    A. whether the patient has a history of allergies.

    B. what the patient was exposed to.

    C. how the patient was exposed.

    D. all of the above

_____ **8.** The dosage of epinephrine in an adult EpiPen is:

    A. 0.10 mg.

    B. 0.15 mg.

    C. 0.30 mg.

    D. 0.50 mg.

_____ **9.** Epinephrine, whether made by the body or by a drug manufacturer, works rapidly to:

    A. raise the pulse rate and blood pressure.

    B. inhibit an allergic reaction.

    C. dilate the bronchioles.

    D. all of the above

_____ **10.** Because the stinger of the honeybee is barbed and remains in the wound, it can continue to inject venom for up to:

  A. 1 minute.

  B. 15 minutes.

  C. 20 minutes.

  D. several hours.

_____ **11.** You should not use tweezers or forceps to remove an embedded stinger because:

  A. squeezing may cause the stinger to inject more venom into the wound.

  B. the stinger may break off in the wound.

  C. the tweezers are not sterile and may cause infection.

  D. removing the stinger may cause bleeding.

_____ **12.** Your assessment of the patient experiencing an allergic reaction should include evaluations of the:

  A. respiratory system.

  B. circulatory system.

  C. skin.

  D. all of the above

_____ **13.** Eating certain foods, such as shellfish or nuts, may result in a relatively _____ reaction that still can be quite severe.

  A. mild

  B. fast

  C. slow

  D. rapid

_____ **14.** In dealing with allergy-related emergencies, you must be aware of the possibility of acute _____ and cardiovascular collapse.

  A. hypotension

  B. tachypnea

  C. airway obstruction

  D. shock

_____ **15.** Wheezing occurs because excessive _____ and mucus are secreted into the bronchial passages.

  A. fluid

  B. carbon dioxide

  C. blood

  D. all of the above

Questions 16-20 are derived from the following scenario: You have been called to a park where a local church is holding a pot-luck dinner. As you exit your ambulance, a woman is holding her 7-year-old son who is wheezing and having difficulty breathing. She informs you that he had inadvertently eaten a brownie with nuts, and he is allergic to nuts.

_____ **16.** You lift his shirt and find small, raised areas that he is trying to itch. They are likely to be:

  A. leukotrienes.

  B. histamines.

  C. urticaria.

  D. none of the above

_____ **17.** Why is he wheezing?

  A. He has had an envenomation.

  B. His bronchioles are constricting.

  C. His bronchioles are dilating.

  D. His uvula has swollen.

_____ **18.** His mother has an EpiPen that contains the appropriate dose of epinephrine for a child. What dose would that be?

   A. 0.8 mg

   B. 0.5 mg

   C. 0.4 mg

   D. 0.15 mg

_____ **19.** When assisting with an auto-injector, how long should you hold the pen against the thigh?

   A. 3 minutes

   B. 5 minutes

   C. 10 minutes

   D. none of the above

_____ **20.** After removing the auto-injector from his thigh, you should:

   A. record the time.

   B. record the dose.

   C. reassess his vital signs.

   D. all of the above

Fill-in

Read each item carefully, then complete the statement by filling in the missing word(s).

1. Wheezing occurs because excessive fluid and mucus are secreted into the _____ _____.

2. Small areas of generalized itching or burning that appear as multiple, small, raised areas on the skin are called

_____.

3. The stinger of the honeybee is _____, so the bee cannot withdraw it.

4. A reaction involving the entire body is called _____.

5. The presence of _____ or respiratory distress indicates that the patient is having a severe enough allergic

reaction to lead to death.

6. Epinephrine inhibits the allergic reaction and dilates the _____.

7. Your ability to recognize and manage the many signs and symptoms of allergic reactions may be the only thing

standing between a patient and _____ _____.

True/False

If you believe the statement to be more true than false, write the letter "T" in the space provided. If you believe the statement to be more false than true, write the letter "F."

_____ **1.** Allergic reactions can occur in response to almost any substance.

_____ **2.** An allergic reaction occurs when the body has an immune response to a substance.

_____ **3.** Wheezing is a high-pitched breath sound usually resulting from blockage of the airway and heard on expiration.

_____ **4.** For a patient appearing to have an allergic reaction, give 100% oxygen via nasal cannula.

Short Answer

Complete this section with short written answers using the space provided.

1. Common side effects of epinephrine include:

_____

_____

2. What are five stimuli that most often cause allergic reactions?

_____

_____

3. What are the steps for administering or assisting with administration of an epinephrine auto-injector?

_____

_____

4. What are the common respiratory and circulatory signs or symptoms of an allergic reaction?

_____

_____

Word Fun

The following crossword puzzle is an activity provided to reinforce correct spelling and understanding of medical terminology associated with emergency care and the EMT-B. Use the clues in the column to complete the puzzle.

Across

5. Small spots; itching, raised areas

6. Poison or harmful substance

8. Adrenaline

Down

1. Chemical substance made by the body, leads to anaphylaxis

2. Severe allergic reaction

3. Harsh, high-pitched airway sounds

4. Responsible for allergy symptoms

7. Raised, swollen area from a sting

Ambulance Calls

The following case scenarios provide an opportunity to explore the concerns associated with patient management. Read each scenario, then answer each question in detail.

1. You are dispatched to care for a wanted suspect who was fleeing from police and was eventually subdued by a canine officer. The man has several puncture wounds on his left arm and no gross bleeding is present. How do you best manage this patient?

_____

_____

_____

2. You are dispatched to assist a 12-year-old child who was climbing a tree and apparently disturbed a wasp nest. When you arrive, the child is lying under the tree and the nest is on the ground next to her. How do you best manage this patient?

_____

_____

_____

3. You are dispatched to a local seafood restaurant for a person who is having difficulty breathing. Upon arrival, you find a 22-year-old woman with facial edema, cyanosis around the lips, audible wheezing, and urticaria on her face and upper body. Her boyfriend tells you she ate shrimp and she is allergic to them. He also tells you she has some medicine in her purse and hands you an EpiPen prescribed to her. How would you best manage this patient?

_____

_____

_____

Skill Drills

Skill Drill 16-1: Using an Auto-injector

Test your knowledge of this skill drill by filling in the correct words in the photo captions.

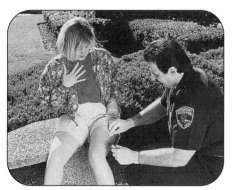  

1. Remove the auto-injector's _____ _____, and quickly wipe the thigh with _____.

2. Place the tip of the auto-injector against the _____ part of the thigh.

3. Push the _____ firmly against the _____, and hold it in place until all the medication is injected.

Skill Drill 16-2: Using an AnaKit
Test your knowledge of this skill drill by placing the following photos in the correct order. Number the first step with a "1," the second step with a "2," etc.

If available, apply a cold pack to the sting site.

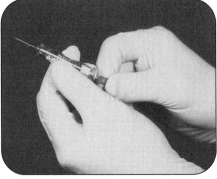

Turn the plunger one-quarter turn.

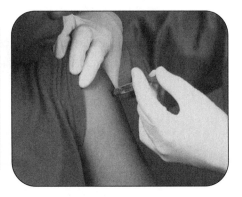

Hold the syringe steady, and push the plunger until it stops.

Quickly insert the needle into the muscle.

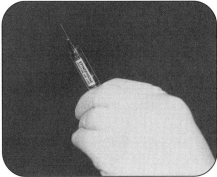

Hold the syringe upright, and carefully use the plunger to remove air.

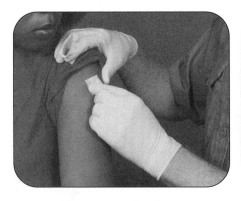

Prepare the injection site with antiseptic, and remove the needle cover.

Have the patient chew and swallow the Chlo-Amine antihistamine tablets provided in the kit.

Assessment Review

Answer the following questions pertaining to the assessment of the types of emergencies discussed in this chapter.

_____ **1.** As you begin to assess a patient suspected of anaphylactic shock, what should be done first?
  A. Assist with an EpiPen.
  B. Take a set of vital signs.
  C. Provide high-flow oxygen.
  D. Obtain a pulse oximetry reading.

_____ **2.** After assisting with an EpiPen:
  A. apply high-flow oxygen if it has not been done already.
  B. take a set of vital signs.
  C. place the used EpiPen in a biohazard container.
  D. all of the above

_____ **3.** When using the auto-injector:
  A. remove the safety cap.
  B. wipe the thigh with antiseptic.
  C. push the autoinjector firmly against the thigh for about 10 seconds.
  D. all of the above

_____ **4.** What does the letter "M" stand for in SAMPLE history?
  A. Money—how will they pay for this?
  B. Maintenance—have they been compliant with their medications?
  C. Medications—what medications are they currently taking?
  D. None of the above

_____ **5.** What does the letter "P" stand for in SAMPLE history?
  A. Pain—rate your pain from 1-10
  B. Probable cause—what caused this to happen?
  C. Priority—is this a high or low priority patient?
  D. Past pertinent history

Emergency Care Summary

Complete the statements pertaining to emergency care for the types of emergencies discussed in this chapter by filling in the missing words.

NOTE: While the steps below are widely accepted, be sure to consult and follow your local protocol.

Allergic Reactions

Using an Auto-injector
1. Remove the auto-injector's safety cap, and quickly wipe the thigh with antiseptic.
2. Place the tip of the auto-injector against the _____ part of the thigh.
3. Push the auto-injector firmly against the thigh, and hold it in place until all the medication is injected (about _____ seconds).

Using an Anakit
1. Prepare the injection site with antiseptic, and remove the needle cover.

2. Hold the syringe _____, and carefully use the plunger to remove the air.
3. Turn the plunger _____ turn.
4. Quickly insert the needle into the _____.
5. Hold the syringe steady, and push the plunger until it stops.
6. Have the patient chew and swallow the Chlo-Amine _____ tablets provided in the kit.
7. If available, apply a _____ _____ to the sting site.

CHAPTER

17 Substance Abuse and Poisoning

Workbook Activities

The following activities have been designed to help you. Your instructor may require you to complete some or all of these activities as a regular part of your EMT-B training program. You are encouraged to complete any activity that your instructor does not assign as a way to enhance your learning in the classroom.

Chapter Review

The following exercises provide an opportunity to refresh your knowledge of this chapter.

Matching
Match each of the terms in the left column to the appropriate definition in the right column.

_____ 1. Poison

_____ 2. Substance abuse

_____ 3. Antidote

_____ 4. Tolerance

_____ 5. Cholinergic

_____ 6. Ingestion

_____ 7. Hematemesis

_____ 8. Stimulant

_____ 9. Opioid

_____ 10. Sedative-hypnotic

_____ 11. Anticholinergic

A. Xanax, Librium, Valium

B. drug or agent with actions similar to morphine

C. Atropine, Benadryl, some cyclic antidepressants

D. need for increasing amounts of a drug to obtain the same effect

E. agent that produces an excited state

F. any substance whose chemical action can damage body structures or impair body functions

G. a substance that will counteract the effects of a particular poison

H. the misuse of any substance to produce a desired effect

I. taking a substance by mouth

J. overstimulates body functions controlled by parasympathetic nerves

K. vomiting blood

Multiple Choice
Read each item carefully, then select the best response.

_____ **1.** Activated charcoal is in the form of a(n):

   A. elixir.

   B. suspension.

   C. syrup.

   D. emulsion.

_____ **2.** The presence of burning or blistering of the mucous membranes suggests:

   A. ingestion of depressants.

   B. ingestion of poison.

   C. overdose of heroin.

   D. the patient may be a heavy smoker.

_____ **3.** Treatment for ingestion of poisonous plants includes:

   A. assessing the ABCs.

   B. taking the plant to the emergency department.

   C. prompt transport.

   D. all of the above

_____ **4.** The most important consideration in caring for a patient who has been exposed to an organophosphate insecticide or some other cholinergic agent is to:

   A. maintain the airway.

   B. apply high-flow oxygen.

   C. avoid exposure yourself.

   D. initiate CPR.

_____ **5.** Objects that may provide clues to the nature of the poison include:

   A. a needle or syringe.

   B. scattered pills.

   C. chemicals.

   D. all of the above

_____ **6.** The most worrisome avenue of poisoning is:

   A. ingestion.

   B. inhalation.

   C. injection.

   D. absorption.

_____ **7.** The major side effect of ingesting activated charcoal is:

   A. depressed respirations.

   B. overproduction of stomach acid.

   C. black stools.

   D. increased blood pressure.

_____ **8.** Alcohol is a powerful CNS depressant. It:

   A. sharpens the sense of awareness.

   B. slows reflexes.

   C. increases reaction time.

   D. all of the above

_____ **9.** Frequently abused synthetic opioids include:
  A. heroin.
  B. morphine.
  C. Demerol.
  D. all of the above

_____ **10.** Treatment of patients who have overdosed with sedative-hypnotics and have respiratory depression is to:
  A. provide airway clearance.
  B. provide ventilatory assistance.
  C. provide prompt transport.
  D. all of the above

_____ **11.** Anticholinergic medications have properties that block the _____ nerves.
  A. parasympathetic
  B. sympathetic
  C. adrenergic
  D. parasympatholytic

_____ **12.** _____ crack produces the most rapid means of absorption and therefore the most potent effect.
  A. Injected
  B. Absorbed
  C. Smoked
  D. Ingested

_____ **13.** "Nerve gases" overstimulate normal body functions that are controlled by parasympathetic nerves causing:
  A. increased salivation.
  B. increased heart rate.
  C. increased urination.
  D. all of the above

_____ **14.** Signs and symptoms of staphylococcal food poisoning include:
  A. difficulty in speaking.
  B. nausea, vomiting, diarrhea.
  C. blurred vision.
  D. all of the above

_____ **15.** Inhalant effects range from mild drowsiness to coma, but unlike most other sedative-hypnotics, these agents may often cause:
  A. seizures.
  B. vomiting.
  C. swelling of the tongue.
  D. all of the above

_____ **16.** Cocaine may be taken via:
  A. inhalation.
  B. injection.
  C. absorption.
  D. all of the above

_____ **17.** Abusable substances include:
  A. vitamins.
  B. nasal decongestants.
  C. food.
  D. all of the above

_____ **18.** A person who has been using marijuana rarely needs transport to the hospital. Exceptions may include someone who is:

A. hallucinating.

B. very anxious.

C. paranoid.

D. all of the above

_____ **19.** Sympathomimetics are CNS stimulants that frequently cause:

A. hypotension.

B. tachycardia.

C. pinpoint pupils.

D. all of the above

_____ **20.** Carbon monoxide:

A. is odorless.

B. produces severe hypoxia.

C. does not damage or irritate the lungs.

D. all of the above

_____ **21.** Chlorine:

A. is odorless.

B. does not damage or irritate the lungs.

C. causes pulmonary edema.

D. all of the above

_____ **22.** Localized signs and symptoms of absorbed poisoning include:

A. a history of exposure.

B. burns, irritation of the skin.

C. dyspnea.

D. all of the above

_____ **23.** Poisoning by injection is almost always the result of:

A. repetitive bee stings.

B. pit viper envenomation.

C. deliberate drug overdose.

D. homicide.

_____ **24.** When dealing with substances such as phosphorous and elemental sodium, you should:

A. brush the chemical off the patient.

B. remove contaminated clothing.

C. apply a dry dressing to the burn area.

D. all of the above

_____ **25.** Injected poisons are impossible to dilute or remove, as they are usually _____ or cause intense local tissue destruction.

A. absorbed quickly into the body

B. bound to hemoglobin

C. large compounds

D. combined with the cerebrospinal fluid

_____ **26.** Medical problems that may cause the patient to present as intoxicated include:

A. head trauma.

B. toxic reactions.

C. uncontrolled diabetes.

D. all of the above

_____ **27.** Signs and symptoms of alcohol withdrawal include:

    A. agitation and restlessness.

    B. fever, sweating.

    C. seizures.

    D. all of the above

_____ **28.** Treatments for inhaled poisons include:

    A. moving the patient into fresh air.

    B. apply an SCBA to the patient.

    C. covering the patient to prevent spread of the poison.

    D. all of the above

_____ **29.** Signs and symptoms of chlorine exposure include:

    A. cough.

    B. chest pain.

    C. wheezing.

    D. all of the above

_____ **30.** Ingested poisons include:

    A. contaminated food.

    B. household cleaners.

    C. plants.

    D. all of the above

_____ **31.** Ingestion of an opiate, sedative, or barbituate can cause depression of the CNS and:

    A. paralysis of the extremities.

    B. dilation of the pupils.

    C. carpopedal spasms.

    D. slow breathing.

_____ **32.** Inhaled poisons include:

    A. chlorine.

    B. venom.

    C. dieffenbachia.

    D. all of the above

_____ **33.** The most important treatment for poisoning is _____ and/or physically removing the poisonous agent.

    A. administering a specific antidote

    B. high-flow oxygen

    C. diluting

    D. syrup of ipecac

Questions 34-38 are derived from the following scenario: You have responded to home of a 26-year-old woman who had reportedly taken a large amount of pills in an attempt to commit suicide. As you enter the living room, you see her sleeping in her chair, with several empty alcohol containers. She is breathing heavily.

_____ **34.** You are able to arouse her consciousness for a short period of time. Which course of action takes priority?

    A. Apply 15 L/min oxygen via nasal cannula.

    B. Apply 12 L/min oxygen via nonrebreathing mask.

    C. Have her take activated charcoal while she is conscious.

    D. Call for ALS.

_____ **35.** You have determined to give her activated charcoal. How much should you give her?

    A. A half glass

    B. 12.5 to 25 mL

    C. 25 to 50 mL

    D. None of the above

_____ **36.** What would be the desired goal of giving her activated charcoal?

    A. To vomit the drugs and alcohol

    B. To bind the toxin and prevent absorption

    C. To teach her a lesson

    D. None of the above

_____ **37.** If she does not want to take the activated charcoal, you should:

    A. Restrain her, pinch her nose, and make her drink it.

    B. Have her sign a patient refusal form.

    C. Persuade her.

    D. All of the above

_____ **38.** Side effects of ingesting activated charcoal include:

    A. vomiting.

    B. black stools.

    C. nausea.

    D. all of the above

Fill-in

Read each item carefully, then complete the statement by filling in the missing word(s).

1. When dealing with exposure to chemicals, treatment focuses on support: assessing and maintaining the patient's

_____.

2. The most commonly abused drug in the United States is _____.

3. Activated charcoal works by _____, or sticking to, many commonly ingested poisons, preventing the toxin

from being absorbed into the body.

4. If the patient has a chemical agent in the eyes, you should irrigate them quickly and thoroughly, at least

_____ for acid substances and _____ for alkalis.

5. Opioid analgesics are CNS depressants and can cause severe _____ _____.

6. Severe acute alcohol ingestion may cause _____.

7. Your primary responsibility to the patient who has been poisoned is to _____ that a poisoning occurred.

8. The usual dosage for activated charcoal for an adult or child is _____ of activated charcoal per

_____ of body weight.

9. As you irrigate the eyes, make sure that the fluid runs from the bridge of the nose _____.

10. Approximately 80% of all poisoning is by _____, including plants, contaminated food, and most drugs.

11. Patients experiencing alcohol withdrawal may develop _____ _____ if they no longer have their daily

source of alcohol.

12. Phosphorus and elemental sodium _____ when they come in contact with water.

13. Increasing tolerance of a substance can lead to _____.

14. _____ may develop from sweating, fluid loss, insufficient fluid intake, or vomiting associated with DTs.

True/False

If you believe the statement to be more true than false, write the letter "T" in the space provided. If you believe the statement to be more false than true, write the letter "F."

_____ **1.** The usual adult dose of activated charcoal is 25 to 50 g.

_____ **2.** The general treatment of a poisoned patient is to induce vomiting.

_____ **3.** Activated charcoal is a standard of care in all ingestions.

_____ **4.** Inhaled chlorine produces profound hypoxia without lung irritation.

_____ **5.** Shaking activated charcoal decreases its effectiveness.

_____ **6.** Opioid overdose typically presents with pinpoint pupils.

_____ **7.** Cholinergics are chemicals such as nerve gases, organophosphate insecticides, or certain wild mushrooms.

_____ **8.** Alcohol is a stimulant.

_____ **9.** Demerol, Dilaudid, and Vicodin are all examples of opioids.

_____ **10.** Cocaine is one of the most addicting substances known.

Short Answer

Complete this section with short written answers using the space provided.

1. How does activated charcoal work to counteract ingested poison?

_____

_____

2. What are four routes of contact for poisoning?

_____

_____

3. List the typical signs and symptoms of an overdose of sympathomimetics.

_____

_____

4. What are the two main types of food poisoning?

_____

_____

5. What differentiates the presentation of acetaminophen poisoning from that of other substances? What does this mean to the prehospital caregiver?

_____

_____

6. What condition do the mnemonics DUMBELS and SLUDGE pertain to, and what do they mean?

_____

_____

7. In addition to alcohol and marijuana, what are the seven categories of drugs seen in overdoses/poisoning?

_____

_____

8. What five questions should you ask a possible poisoning victim?

_____

_____

9. Why should phosphorous or elemental sodium poisoning victims not be irrigated?

_____

_____

Word Fun

The following crossword puzzle is an activity provided to reinforce correct spelling and understanding of medical terminology associated with emergency care and the EMT-B. Use the clues in the column to complete the puzzle.

Across

1. Induces sleep
4. Overwhelming obsession or physical need
5. Substance whose chemical action causes damage
7. Poison or harmful substance; produced by animals, plants
8. Swallowing
9. Need for increasing amounts of a drug
11. Decreases activity and excitement
12. Vomiting blood

Down

2. Similar to morphine
3. Produces false perceptions
4. Used to neutralize or counteract a poison
6. Misuse of any substance to produce a desired effect
10. Material in emesis

Ambulance Calls

The following case scenarios provide an opportunity to explore the concerns associated with patient management. Read each scenario, then answer each question in detail.

1. You are dispatched to a private residence for "accidental ingestion." You arrive to find a 3-year-old, who parents tell you "got into some rat poison." The child is alert, crying and responding appropriately to parents and environmental stimulus.

   How do you best manage this patient?

   _____

   _____

   _____

2. You are dispatched to the sidewalk in front of a small business for "an intoxicated male." You arrive to find a 60-year-old man sitting on the curb, holding a bottle inside a paper bag. He is not fully alert, and can only tell you that his name is Andy. He allows you to take his blood pressure, and as you roll up his sleeve you notice needle marks along his veins.

   How do you best manage this patient?

   _____

   _____

   _____

3. You are called to a possible suicide attempt. You arrive on the scene to find police and a neighbor in the home of a 25-year-old woman who is unresponsive, supine on her bed. The neighbor tells you that the patient recently broke up with her boyfriend and has been very distraught. There is an empty pill bottle on the nightstand. When you look at the label you see that the prescription was filled yesterday and there were 30 tablets dispensed. There is also an empty liquor bottle on the floor.

   How would you best manage this patient?

   _____

   _____

   _____

Assessment Review

Answer the following questions pertaining to the assessment of the types of emergencies discussed in this chapter.

_____ 1. When would it be more important to provide activated charcoal before applying oxygen?
   A. It wouldn't. Oxygen is always the top priority.
   B. When there's a life threat
   C. When oxygen is contraindicated
   D. None of the above

_____ **2.** What life threat can occur from applying a nonrebreathing mask after a patient has consumed activated charcoal?

    A. Vomiting into the nonrebreathing mask

    B. Aspirating the vomitus

    C. Choking on the vomitus

    D. All of the above

_____ **3.** The "O" in OPQRST stands for:

    A. OPQRST is for trauma only.

    B. On a scale of 1 to 10, rate your pain.

    C. Onset

    D. None of the above

_____ **4.** Why is "T" important in OPQRST?

    A. You will learn what Triggered the event.

    B. You will learn what Time the event started.

    C. You will obtain the patient's Temperature.

    D. You will find out what the patient has Taken.

_____ **5.** When would you not give activated charcoal?

    A. If the patient drank gasoline

    B. If the patient overdosed on aspirin

    C. If the patient overdosed on antidepressants

    D. None of the above

Emergency Care Summary

Complete the statements pertaining to emergency care for the types of emergencies discussed in this chapter by filling in the missing words.

    NOTE: While the steps below are widely accepted, be sure to consult and follow your local protocol.

Substance Abuse and Poisoning

General Management

1. Have trained rescuers remove patient from any poisonous environment.

2. Establish and maintain airway, _____ as needed. Provide _____ _____.

3. Obtain _____ history and _____ _____. Ascertain which drug(s) have been taken.

4. Request _____ when available.

5. Take all containers, bottles, and labels of poisons to the receiving hospital.

For patients who have taken _____, provide calm, prompt transport.

For _____ agents, it is critical to take sufficient body substance isolation precautions.

For patients who have _____ poisoning, notify regional poison control center for assistance in identifying the plant.

For patients who have _____ poisoning, if there are more than two patients with the same illness, transport the food suspected to be responsible for the poisoning.

Administer _____ _____ for poisonous ingestions according to local protocol. Follow these steps:

1. Do not give if the patient exhibits _____ _____ _____, has ingested acids or alkalis, or is unable to _____.

2. Obtain order from medical direction or follow protocol.

3. Shake container.

4. Place in cup with straw and have patient drink _____ to 25 g (for infants and children) or 25 to _____ g (for adults).

CHAPTER

18 Environmental Emergencies

Workbook Activities

The following activities have been designed to help you. Your instructor may require you to complete some or all of these activities as a regular part of your EMT-B training program. You are encouraged to complete any activity that your instructor does not assign as a way to enhance your learning in the classroom.

Chapter Review

The following exercises provide an opportunity to refresh your knowledge of this chapter.

Matching

Match each of the terms in the left column to the appropriate definition in the right column.

_____ 1. Conduction

_____ 2. Air embolism

_____ 3. Evaporation

_____ 4. Hyperthermia

_____ 5. Diving reflex

_____ 6. Core temperature

_____ 7. Convection

_____ 8. Laryngospasm

_____ 9. Electrolytes

_____ 10. Radiation

_____ 11. Hypothermia

_____ 12. Ambient temperature

_____ 13. Heat cramps

_____ 14. Drowning

A. slowing of heart rate caused by sudden immersion in cold water

B. salts and other chemicals dissolved in body fluids

C. severe constriction of the larynx and vocal cords

D. death from suffocation after submersion in water

E. heat loss resulting from standing in a cold room

F. core temperature greater than 101° F

G. condition when the entire body temperature falls

H. condition caused by air bubbles in the blood vessels

I. heat loss that occurs from helicopter rotor blade downwash

J. heat loss resulting from sitting on snow

K. heat loss resulting from sweating

L. painful muscle spasms that occur after vigorous exercise

M. temperature of the surrounding environment

N. temperature of the central part of the body

Multiple Choice
Read each item carefully, then select the best response.

_____ 1. _____ causes body heat to be lost, as warm air in the lungs is exhaled into the atmosphere and cooler air is inhaled.
  A. Convection
  B. Conduction
  C. Radiation
  D. Respiration

_____ 2. Evaporation, the conversion of any liquid to a gas, is a process that requires:
  A. energy.
  B. circulating air.
  C. a warmer ambient temperature.
  D. all of the above

_____ 3. The rate and amount of heat loss by the body can be modified by:
  A. increasing heat production.
  B. moving to an area where heat loss is decreased.
  C. wearing insulated clothing.
  D. all of the above

_____ 4. The characteristic appearance of blue lips and/or fingertips seen in hypothermia is the result of:
  A. lack of oxygen in venous blood.
  B. frostbite.
  C. blood vessels constricting.
  D. bruising.

_____ 5. Signs and symptoms of severe systemic hypothermia include all of the following, except:
  A. weak pulse.
  B. coma.
  C. confusion.
  D. very slow respirations.

_____ 6. Hypothermia is more common among:
  A. older individuals.
  B. infants and children.
  C. those who are already ill.
  D. all of the above

_____ 7. To assess a patient's general temperature, pull back your glove and place the back of your hand on the patient's:
  A. abdomen, underneath the clothing.
  B. forehead.
  C. forearm, on the inside of the wrist.
  D. neck, at the area where you check the carotid pulse.

_____ 8. Never assume that a(n) _____, pulseless patient is dead.
  A. apneic
  B. cyanotic
  C. cold
  D. hyperthermic

_____ **9.** Management of hypothermia in the field consists of all of the following except:

    A. stabilizing vital functions.

    B. removing wet clothing.

    C. preventing further heat loss.

    D. massaging the cold extremities.

_____ **10.** When exposed parts of the body become very cold but not frozen, the condition is called:

    A. frostnip.

    B. chilblains.

    C. immersion foot.

    D. all of the above

_____ **11.** When the body is exposed to more heat energy than it loses, _____ results.

    A. hyperthermia

    B. heat cramps

    C. heat exhaustion

    D. heatstroke

_____ **12.** Contributing factors to the development of heat illnesses include:

    A. high air temperature.

    B. vigorous exercise.

    C. high humidity.

    D. all of the above

_____ **13.** Keeping yourself hydrated while on duty is very important. Drink at least _____ of water per day, and more when exertion or heat is involved.

    A. 8 glasses

    B. 1 liter

    C. 2 liters

    D. 3 liters

_____ **14.** The following statements concerning heat cramps are true except:

    A. they only occur when it is hot outdoors.

    B. they may be seen in well-conditioned athletes.

    C. the exact cause of heat cramps is not well understood.

    D. dehydration may play a role in the development of heat cramps.

_____ **15.** Signs and symptoms of heat exhaustion and associated hypovolemia include:

    A. cold, clammy skin with ashen pallor.

    B. dizziness, weakness, or faintness.

    C. normal vital signs.

    D. all of the above

_____ **16.** Be prepared to transport the patient to the hospital for aggressive treatment of hyperthermia if:

    A. the symptoms do not clear up promptly.

    B. the level of consciousness improves.

    C. the temperature drops.

    D. all of the above

_____ **17.** Often, the first sign of heatstroke is:

    A. a change in behavior.

    B. an increase in pulse rate.

    C. an increase in respirations.

    D. hot, dry, flushed skin.

_____ **18.** The least common but most serious illness caused by heat exposure, occurring when the body is subjected to more heat than it can handle and normal mechanisms for getting rid of the excess heat are overwhelmed, is:

A. hyperthermia.

B. heat cramps.

C. heat exhaustion.

D. heatstroke.

_____ **19.** _____ is the body's attempt at self-preservation by preventing water from entering the lungs.

A. Bronchoconstriction

B. Laryngospasm

C. Esophageal spasms

D. Swelling in the oropharynx

_____ **20.** Treatment of drowning/near drowning begins with:

A. opening the airway.

B. ventilation with 100% oxygen via BVM device.

C. suctioning the lungs to remove the water.

D. rescue and removal from the water.

_____ **21.** After removing a near drowning patient from the water, it may be difficult to find a pulse because of:

A. dilation of peripheral blood vessels.

B. body temperature at the core.

C. low cardiac output.

D. all of the above

_____ **22.** If the near drowning victim has evidence of upper airway obstruction by foreign matter, attempt to clear it by:

A. removing the obstruction manually.

B. suction.

C. using abdominal thrusts.

D. all of the above

_____ **23.** You should never give up on resuscitating a cold-water drowning victim because:

A. when the patient is submerged in water colder than body temperature, heat is maintained in the body.

B. the resulting hypothermia can protect vital organs from the lack of oxygen.

C. the resulting hypothermia raises the metabolic rate.

D. all of the above

_____ **24.** The three phases of a dive, in the order they occur, are:

A. ascent, descent, bottom.

B. descent, bottom, ascent.

C. orientation, bottom, ascent.

D. descent, orientation, ascent.

_____ **25.** Areas usually affected by descent problems include:

A. the lungs.

B. the skin.

C. the joints.

D. vision.

_____ **26.** Potential problems associated with rupture of the lungs include:

A. air emboli.

B. pneumomediastinum.

C. pneumothorax.

D. all of the above

_____ **27.** The organs most severely affected by air embolism are the:

    A. brain and spinal cord.

    B. brain and heart.

    C. heart and lungs.

    D. brain and lungs.

_____ **28.** Black widow spiders may be found in:

    A. New Hampshire.

    B. woodpiles.

    C. Georgia.

    D. all of the above

_____ **29.** Coral snakes may be found in:

    A. Florida.

    B. Kansas.

    C. New Jersey.

    D. all of the above

_____ **30.** Rocky Mountain spotted fever and Lyme disease are both spread through the tick's:

    A. saliva.

    B. blood.

    C. hormones.

    D. all of the above

_____ **31.** Signs of envenomation by a pit viper include:

    A. swelling.

    B. severe burning pain at the site of the injury.

    C. ecchymosis.

    D. all of the above

_____ **32.** Removal of a tick should be accomplished by:

    A. suffocating it with gasoline.

    B. burning it with a lighted match to cause it to release its grip.

    C. using fine tweezers to pull it straight out of the skin.

    D. suffocating it with Vaseline.

_____ **33.** Treatment for a black widow spider bite consists of maintaining:

    A. the airway.

    B. breathing.

    C. circulation.

    D. all of the above

_____ **34.** Treatment of a snake bite from a pit viper includes:

    A. calming the patient.

    B. providing BLS as needed if the patient shows no sign of envenomation.

    C. marking the skin with a pen over the swollen area to note whether swelling is spreading.

    D. all of the above

Questions 35-39 are derived from the following scenario: At 2:00 in the afternoon in July, the weather is 105°F and very humid. You have been called for a "man down" at the park. As you arrive, you recognize him as an alcoholic who has been a "frequent-flyer" with your service. It looks like he had been sitting under a tree when he fell over, unconscious.

_____ **35.** As you assess the patient, he has cold, clammy skin and a dry tongue. You suspect that:

    A. he is dehydrated.

    B. he has suffered heat exhaustion.

    C. he is hypothermic.

    D. all of the above

_____ **36.** As you look closer, you note that he is shivering and his respirations are 20 breaths/min. You begin to have a stronger suspicion that he is:

   A. hyperthermic.

   B. hypothermic.

   C. drunk.

   D. none of the above

_____ **37.** The direct transfer of heat from his body to the cold ground is called:

   A. conduction.

   B. convection.

   C. radiation.

   D. evaporation.

_____ **38.** You pull back on your glove and place the back of your hand on his skin at the abdomen, and the skin feels cool. Again, you suspect:

   A. hyperthermia.

   B. hypothermia.

   C. that he is drunk.

   D. none of the above

_____ **39.** How will you treat this patient?

   A. Prevent conduction heat loss.

   B. Prevent convection heat loss.

   C. Remove the patient from the environment.

   D. All of the above

Fill-in

Read each item carefully, then complete the statement by filling in the missing word.

1. Do not attempt to rewarm patients who have _____ _____ _____ hypothermia, because they are prone to developing arrhythmias unless handled very carefully.

2. Most significant diving injuries occur during _____.

3. When treating a patient with frostbite, never attempt _____ if there is any chance that the part may freeze again before the patient reaches the hospital.

4. As with so many hazards, you cannot help others if you do not practice _____.

5. _____, a common effect of hypothermia, is the body's attempt to maintain heat.

6. Whenever a person dives or jumps into very cold water, the _____ _____ may cause immediate bradycardia.

True/False

If you believe the statement to be more true than false, write the letter "T" in the space provided. If you believe the statement to be more false than true, write the letter "F."

_____ **1.** Normal body temperature is 98.6°F (37.0°C).

_____ **2.** To assess the skin temperature in a patient experiencing a generalized cold emergency, you should feel the patient's skin.

_____ **3.** Mild hypothermia occurs when the core temperature drops to 85°F.

_____ **4.** The body's most efficient heat-regulating mechanisms are sweating and dilation of skin blood vessels.

_____ **5.** People who are at greatest risk for heat illnesses are the elderly and children.

_____ **6.** The strongest stimulus for breathing is an elevation of oxygen in the blood.

_____ **7.** Immediate bradycardia after jumping in cold water is called the diving reflex.

_____ **8.** Ice should be promptly applied to any insect sting or snake bite with swelling.

_____ **9.** The most common type of pit viper is the copperhead.

_____ **10.** Cottonmouths are known for aggressive behavior.

_____ **11.** Ticks should be removed by firmly grasping them with tweezers while rotating them counterclockwise.

_____ **12.** The pain of coelenterate stings may respond to flushing with cold water.

Short Answer

Complete this section with short written answers using the space provided.

1. What are three ways to modify heat loss? Give an example of each.

_____

_____

2. What are the steps in treating heatstroke?

_____

_____

3. What is an air embolism and how does it occur?

_____

_____

4. For what diving emergencies are hyperbaric chambers used?

_____

_____

5. How should a frostbitten foot be treated?

_____

_____

6. What are four "Do Nots" in relation to local cold injuries?

_____

_____

7. What treatments for a snake bite assist with slowing and monitoring the spread of venom?

_____

_____

8. What are the two most common poisonous spiders in the United States and how do their bites differ?

_____

_____

Word Fun emtb vocab explorer

The following crossword puzzle is an activity provided to reinforce correct spelling and understanding of medical terminology associated with emergency care and the EMT-B. Use the clues in the column to complete the puzzle.

Across

1. Slowing heart from sudden submersion in cold water
3. Loss of heat by direct contact
5. Less than 60 beats/min
7. Salts and chemicals in body fluids
8. Used for breathing underwater
9. Loss of heat by air movement
10. Temperature greater than 101°F

Down

2. Loss of heat to colder environment
4. Excessive heat problem, may be fatal
6. Air bubbles in blood vessels

Ambulance Calls

The following case scenarios provide an opportunity to explore the concerns associated with patient management. Read each scenario, then answer each question in detail.

1. You are called to the local airport for a 52-year-old man who is the pilot of his own aircraft. He tells you he is having severe abdominal pain and joint pain. History reveals that the patient is returning from a dive trip off the coast. He says he has had "the bends" before and this feels similar.

   How would you best manage this patient?

   _____

   _____

   _____

2. You are dispatched to a long-term care facility for an Alzheimer patient with an "unknown problem." You arrive to find several staff members who greet you at the front door. They explain that the patient wandered out the back exit door of the Alzheimer unit when the nurse was on her other rounds. They aren't sure how this happened as each patient wears a necklace that triggers an alarm if patients leave this specialized wing of the facility. They found him outside in the snow, and he has possibly been outdoors for 45 minutes.

   How do you best manage this patient?

   _____

   _____

   _____

3. You are dispatched to an unconscious female at a local beauty spa locally known for its "body wraps." You arrive to find a conscious, yet confused, 17-year-old female. A friend tells you that the patient has been using over-the-counter 'water pills' and laxative agents, along with strenuous exercise, in order to lose weight for the prom. She tells you that her friend is otherwise healthy, but she complained of feeling dizzy prior to undergoing the spa treatment.

How would you best manage this patient?

_____

_____

_____

Skill Drills

Skill Drill 18-1: Treating for Heat Exhaustion
Test your knowledge of this skill drill by filling in the correct words in the photo captions.

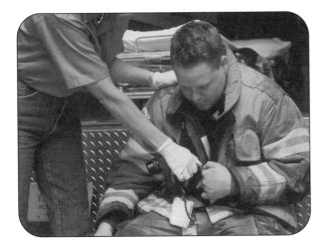

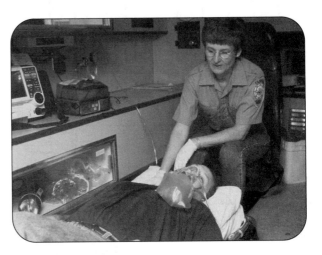

1. Remove _____ _____.

2. Move the patient to a _____ _____.
Give _____.
Place the patient in a _____ position, elevate the legs, and _____ the patient.

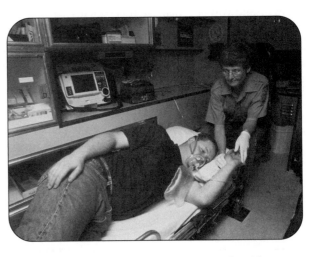

3. If the patient is _____ _____, give water by mouth.

4. If nausea develops, _____ on the side.

Skill Drill 18-2: Stabilizing a Suspected Spinal Injury in the Water

Test your knowledge of this skill drill by placing the photos below in the correct order. Number the first step with a "1," the second step with a "2," etc.

_____

Secure the patient to the backboard.

_____

Turn the patient to a supine position by rotating the entire upper half of the body as a single unit.

_____

Cover the patient with a blanket and apply oxygen if breathing. Begin CPR if breathing and pulse are absent.

_____

Float a buoyant backboard under the patient.

_____

As soon as the patient is turned, begin artificial ventilation using the mouth-to-mouth method or a pocket mask.

_____

Remove the patient from the water.

Assessment Review

Answer the following questions pertaining to the assessment of the types of emergencies discussed in this chapter.

_____ **1.** When assessing the circulation of a patient with cold injuries, how long should you evaluate the pulse?

  A. 15 seconds

  B. 30 seconds

  C. 60 seconds

  D. 90 seconds

_____ **2.** If a patient has a cold skin temperature, he or she likely is:

  A. hypothermic.

  B. hyperthermic.

  C. alcoholic.

  D. none of the above

_____ **3.** If a patient has a hot skin temperature, he or she likely is:

  A. hypothermic.

  B. hyperthermic.

  C. alcoholic.

  D. none of the above

_____ **4.** When treating multiple victims of lightning strikes, who should you concentrate your efforts on first?

  A. Conscious patients

  B. Unconscious patients in respiratory or cardiac arrest

  C. All unconscious patients

  D. None of the above

_____ **5.** What is the best method of inactivating a jellyfish sting?

  A. Urinating

  B. Flushing the site with cold water

  C. Applying vinegar

  D. Applying an ice pack

Emergency Care Summary

Complete the statements pertaining to emergency care for the types of emergencies discussed in this chapter by filling in the missing words.

NOTE: While the steps below are widely accepted, be sure to consult and follow your local protocol.

Cold Injuries

1. Remove wet clothing and keep the patient dry.
2. Prevent heat loss. Move the patient away from any wet or cold surface.
3. _____ all exposed body parts by wrapping them in a blanket or dry, bulky material.
4. Prevent _____ heat loss by erecting a wind barrier around the patient.
5. Remove the patient from the _____ _____ as promptly as possible.

Heat Injuries

1. Remove any excess layers of clothing.
2. Move the patient promptly from the hot environment and out of the sun.
3. Provide _____, if not already done during the initial assessment.
4. Encourage the patient to lie _____ with legs _____. Loosen any tight clothing and _____ the patient.
5. If the patient is alert, encourage him or her to sit up and slowly drink a _____ of water if _____ does not develop.
6. Transport patient in the _____ _____ recumbent position.

Drowning and Diving Injuries

1. Turn the patient to a _____ position by rotating the entire upper half of the body as a single unit.
2. Begin artificial ventilation using the mouth-to-mouth method or a _____ _____.
3. Float a buoyant _____ under the patient.
4. Secure the patient to the backboard.
5. Remove the patient from the water.
6. Cover the patient with a blanket and apply _____ if breathing. Begin CPR if breathing and pulse are absent.

Lightning Injuries

1. Move patient to a sheltered area.
2. Those who are _____ following a lightning strike are much less likely to develop delayed respiratory or _____ _____; most of these victims will survive. Therefore, you should focus your efforts on those who are in respiratory or _____ _____.

Spider Bites

Black Widow Spider
1. Provide _____ for patient in respiratory distress.
2. Transport the patient to the emergency department as soon as possible for treatment of pain and muscle _____.
3. If possible, bring the _____.

Brown Recluse Spider
If bite causes systemic symptoms:
1. Provide BLS.
2. Transport to emergency department.
3. If possible, bring the spider.

Snake Bites

Using an AnaKit
1. Prepare the injection site with antiseptic, and remove the needle cover.
2. Hold the syringe upright, and carefully use the _____ to remove air.
3. Turn the plunger _____ _____ turn.
4. Quickly insert the needle into the _____.
5. Holding the syringe steady, push the plunger until it stops.
6. Have the patient chew and swallow the Chlo-Amine _____ tablets provided in the kit.
7. If available, apply a _____ _____ to the sting site.

Injuries From Marine Animals

1. Limit further _____ of nematocysts by avoiding fresh water, wet sand, showers, or careless manipulation of the tentacles. Keep the patient calm and _____ motion of the affected extremity.
2. Inactivate the nematocysts by applying _____.
3. Remove the remaining tentacles by scraping them off with the edges of a sharp, stiff object such as a _____ _____.
4. Provide prompt transport.

CHAPTER

19 Behavioral Emergencies

Workbook Activities

The following activities have been designed to help you. Your instructor may require you to complete some or all of these activities as a regular part of your EMT-B training program. You are encouraged to complete any activity that your instructor does not assign as a way to enhance your learning in the classroom.

Chapter Review

The following exercises provide an opportunity to refresh your knowledge of this chapter.

Matching

Match each of the terms in the left column to the appropriate definition in the right column.

_____ 1. Behavioral crisis

_____ 2. Psychogenic

_____ 3. Organic brain syndrome

_____ 4. Depression

_____ 5. Functional disorder

_____ 6. Behavior

A. what you can see of a person's response to the environment; his or her actions

B. temporary or permanent dysfunction of the brain caused by a disturbance in brain tissue function

C. any reaction to events that interferes with activities of daily living or is unacceptable to the patient or others

D. a persistent mood of sadness, despair, or discouragement

E. abnormal operation of an organ that cannot be traced to an obvious change in structure or physiology of the organ

F. a symptom or illness caused by mental factors as opposed to physical ones

Multiple Choice
Read each item carefully, then select the best response.

_____ **1.** A psychological or behavioral crisis may be due to:

   A. mind-altering substances.

   B. the emergency situation.

   C. stress.

   D. all of the above

_____ **2.** A normal reaction to a crisis situation would be:

   A. Monday morning blues that last until Friday.

   B. feeling "blue" after the break up of a long-term relationship.

   C. feeling depressed week after week with no discernible cause.

   D. all of the above

_____ **3.** The cause of a behavioral crisis experienced by an unmanageable patient may be:

   A. drug use.

   B. a history of mental illness.

   C. alcohol abuse.

   D. all of the above

_____ **4.** Learning to adapt to a variety of situations in daily life, including stresses and strains, is called:

   A. disruption.

   B. adjustment.

   C. behavior.

   D. functional.

_____ **5.** If the interruption of daily routine tends to recur on a regular basis, the behavior is also considered a _____ problem.

   A. mental health

   B. functional disorder

   C. behavioral

   D. psychogenic

_____ **6.** If an abnormal or disturbing pattern of behavior lasts for at least _____, it is regarded as a matter of concern from a mental health standpoint.

   A. 6 weeks

   B. a month

   C. 6 months

   D. a year

_____ **7.** A person who is no longer able to respond appropriately to the environment may be having what is called a psychological or _____ emergency.

   A. psychiatric

   B. behavioral

   C. functional

   D. adjustment

_____ **8.** Mental disorders may be caused by a:

   A. social disturbance.

   B. chemical disturbance.

   C. biological disturbance.

   D. all of the above

_____ **9.** An altered mental status may arise from:

    A. an oxygen saturation of 98%.

    B. moderate temperatures.

    C. an inadequate blood flow to the brain.

    D. adequate glucose levels in the blood.

_____ **10.** Organic brain syndrome may be caused by:

    A. hypoglycemia.

    B. excessive heat or cold.

    C. lack of oxygen.

    D. all of the above

_____ **11.** An example of a functional disorder would be:

    A. schizophrenia.

    B. organic brain syndrome.

    C. Alzheimer's.

    D. all of the above

_____ **12.** When documenting abnormal behavior, it is important to:

    A. record detailed, subjective findings.

    B. avoid judgmental statements.

    C. avoid quoting the patient's own words.

    D. all of the above

_____ **13.** Safety guidelines for behavioral emergencies include:

    A. assessing the scene.

    B. being prepared to spend extra time.

    C. encouraging purposeful movement.

    D. all of the above

_____ **14.** In evaluating a situation that is considered a behavioral emergency, the first things to consider are:

    A. airway and breathing.

    B. scene safety and patient response.

    C. history of medications.

    D. respiratory and circulatory status.

_____ **15.** Psychogenic circumstances may include:

    A. severe depression.

    B. death of a loved one.

    C. a history of mental illness.

    D. all of the above

_____ **16.** Risk factors for suicide may include:

    A. denial of alcohol use.

    B. recent marriage.

    C. holidays.

    D. all of the above

_____ **17.** Suicidal patients may also be:

    A. homicidal.

    B. hypoxic.

    C. joking.

    D. seeking attention.

_____ **18.** Causes of altered behavior in geriatric patients may include:

    A. constipation.

    B. diabetes.

    C. stroke.

    D. all of the above

_____ **19.** Restraint of a person must be ordered by:

    A. a physician.

    B. a court order.

    C. a law enforcement officer.

    D. all of the above

_____ **20.** When restraining a patient without an appropriate order, legal actions may involve charges of:

    A. abandonment.

    B. negligence.

    C. battery.

    D. breach of duty.

_____ **21.** When restraining a patient face down on a stretcher, it is necessary to constantly reassess the patient's:

    A. level of consciousness.

    B. airway.

    C. emotional status.

    D. pulse rate.

Questions 22-26 are derived from the following scenario: Dean, a man in his 50s, is acting irrationally. His wife states he thinks he is a dictator of a small country, and is wearing nothing but a baseball cap and a belt with a small handgun attached to it.

_____ **22.** What is your best course of action?

    A. Call ALS.

    B. Assess Dean from a distance.

    C. Have his wife take the gun from him.

    D. Call for police back-up.

_____ **23.** The scene is safe. Dean now tells you he is "God," and can do anything he wants to. Is this a psychiatric emergency?

    A. Yes.

    B. No.

    C. This is normal behavior.

    D. None of the above

_____ **24.** How should you approach Dean?

    A. Identify yourself.

    B. Be direct.

    C. Express interest in Dean's story.

    D. All of the above

_____ **25.** Dean becomes agitated and states, "You'll never take me alive." Ordinarily, who can order restraint?

    A. Physician

    B. Court

    C. Law enforcement officer

    D. All of the above

_____ **26.** How many people should be present to restrain Dean?

    A. 2

    B. 4

    C. 6

    D. None of the above

Fill-in

Read each item carefully, then complete the statement by filling in the missing word.

1. _____ is what you can see of a person's response to the environment; his or her actions.

2. A _____ _____ or emergency is any reaction to events that interferes with the activities of daily living or has become unacceptable to the patient, family, or community.

3. Chronic _____, or a persistent feeling of sadness or despair, may be a symptom of a mental or physical disorder.

4. _____ _____ _____ is a temporary or permanent dysfunction of the brain caused by a disturbance in the physical or physiologic functioning of the brain.

5. Any time you encounter an emotionally depressed patient, you must consider the possibility of _____.

True/False

If you believe the statement to be more true than false, write the letter "T" in the space provided. If you believe the statement to be more false than true, write the letter "F."

_____ 1. Depression lasting 2 to 3 weeks after being fired from a job is a normal mental health response.

_____ 2. Low blood glucose or lack of oxygen to the brain may cause behavioral changes, but not to the degree that a psychiatric emergency could exist.

_____ 3. From a mental health standpoint, a pattern of abnormal behavior must last at least 3 months to be a matter of concern.

_____ 4. A disturbed patient should always be transported with restraints.

_____ 5. It is sometimes helpful to allow a patient with a behavioral emergency some time alone to calm down and collect their thoughts.

_____ 6. It is important to avoid looking directly at the patient when dealing with a behavioral crisis.

_____ 7. A patient should never be asked if he or she is considering suicide.

_____ 8. Urinary tract infections can cause behavioral changes in elderly patients.

_____ 9. All individuals with mental health disorders are dangerous, violent, or otherwise unmanageable.

_____ 10. When completing the documentation, it is important to record detailed, subjective findings that support the conclusion of abnormal behavior.

_____ 11. When restraining a patient, at least four people should be present to carry out the restraint.

Short Answer

Complete this section with short written answers using the space provided.

1. What is the distinction between a behavioral crisis and a mental health problem?

_____

_____

2. What three major areas should be considered in evaluating the possible source of a behavioral crisis?

_____

_____

3. What are three factors to consider in determining the level of force required to restrain a patient?

_____

_____

4. List ten safety guidelines for dealing with behavioral emergencies.

_____

_____

5. List ten risk factors for suicide.

_____

_____

Word Fun

The following crossword puzzle is a good way to review correct spelling and meaning of medical terminology associated with emergency care and the EMT-B. Use the clues in the column to complete the puzzle.

Across

3. Reactions interfere with normal activities

5. Illness with psychological symptoms

6. Caused by mental factors; not physical

7. Persistent sadness; despair

Down

1. A change in behavior

2. No known physiologic reason for abnormality

4. Basic doings of a normal person

Ambulance Calls

The following case scenarios provide an opportunity to explore the concerns associated with patient management. Read each scenario, then answer each question in detail.

1. You are dispatched to a non-emergency transport of a female from a local hospital emergency department to a care facility that provides treatment for emotionally disturbed teenagers. She became violent in the ED and was placed in 4-point restraints. As you begin transporting the patient, she begins to cry and asks you to remove the restraints.

   How do you best manage this patient?

   _____

   _____

   _____

2. You are dispatched to a "suicide attempt" at a private residence. When you arrive on scene, you are greeted by a calm, middle-aged man who appears to have been crying. He tells you that he was on the phone with his sister who lives out of state, and that she must have called for the ambulance. Dispatch informed you via cellular phone that this man recently lost his wife of 15 years to breast cancer.

   How do you best manage this patient?

   _____

   _____

   _____

3. You are dispatched to the residence of a 40-year-old woman who is upset over the loss of her mother 5 weeks ago. She tells you that she has no family and has cared for her elderly mother for the past 7 years. She has not eaten for several days and is severely depressed.

   How would you best manage this patient?

   _____

   _____

   _____

Assessment Review

Answer the following questions pertaining to the assessment of the types of emergencies discussed in this chapter. Questions 1-5 are based on risk factors in assessing the level of danger in a behavior call.

_____ 1. When assessing the history, what past behavior(s) do you want to know the patient has exhibited?

   A. Hostile behavior

   B. Overly aggressive behavior

   C. Violent behavior

   D. All of the above

_____ **2.** Physical tension is often a warning signal of impending hostility. What sign might warn you of physical tension?

   A. Posture

   B. Eye movement

   C. Facial expression

   D. Laughter

_____ **3.** What warning signs can be detected from the scene?

   A. Is the patient holding a potentially lethal object?

   B. Is he or she near a potentially lethal object?

   C. Does he or she keep looking at a potentially lethal object?

   D. All of the above

_____ **4.** What kind of speech may be an indicator of emotional distress?

   A. Loud speech

   B. Obscene speech

   C. Erratic and bizarre speech

   D. All of the above

_____ **5.** What type of physical activity may be an indicator of risk to the EMT-B?

   A. Tense muscles

   B. Clenched fists

   C. Pacing or glaring eyes

   D. All of the above

Emergency Care Summary

Complete the statements pertaining to emergency care for the types of emergencies discussed in this chapter by filling in the missing words.

   NOTE: While the steps below are widely accepted, be sure to consult and follow your local protocol.

Behavioral Emergencies

The main task is to diffuse and _____ the situation and _____ _____ the patient. Risk factors to assess the level of danger include:

1. **History**—Has the patient previously exhibited hostile, overly aggressive, or violent behavior?

2. **Posture**—Physical _____ is often a warning signal of impending hostility.

3. **The scene**—Is the patient holding or near potentially lethal objects?

4. **Vocal activity**—What kind of speech is the patient using? Loud, obscene, erratic, and bizarre speech patterns usually indicate _____ _____.

5. **Physical activity**—The patient who has tense muscles, clenched fists, or glaring eyes; is pacing; cannot sit still; or is fiercely protecting personal space requires careful watching. Agitation may predict a quick escalation to _____.

CHAPTER

20 Obstetric and Gynecologic Emergencies

Workbook Activities

The following activities have been designed to help you. Your instructor may require you to complete some or all of these activities as a regular part of your EMT-B training program. You are encouraged to complete any activity that your instructor does not assign as a way to enhance your learning in the classroom.

Chapter Review

The following exercises provide an opportunity to refresh your knowledge of this chapter.

Matching

Match each of the terms in the left column to the appropriate definition in the right column.

_____ 1. Cervix
_____ 2. Perineum
_____ 3. Placenta
_____ 4. Amniotic sac
_____ 5. Fetus
_____ 6. Birth canal
_____ 7. Uterus
_____ 8. Umbilical cord
_____ 9. Vagina
_____ 10. Breech presentation
_____ 11. Limb presentation
_____ 12. Multipara
_____ 13. Nuchal cord
_____ 14. Presentation
_____ 15. Miscarriage

A. an umbilical cord that is wrapped around the infant's neck

B. a fluid-filled, bag-like membrane inside the uterus that grows around the developing fetus

C. the area of skin between the vagina and the anus

D. the neck of the uterus

E. Connects mother and infant

F. the outermost part of a woman's reproductive system

G. the part of the infant that appears first

H. the vagina and lower part of the uterus

I. a woman who has had more than one live birth

J. spontaneous abortion

K. delivery in which the presenting part is a single arm, leg, or foot

L. tissue that develops on the wall of the uterus and is connected to the fetus

M. the hollow organ inside the female pelvis where the fetus grows

N. the developing baby in the uterus

O. delivery in which the buttocks come out first

Multiple Choice
Read each item carefully, then select the best response.

_____ **1.** In the event of a nuchal cord, proper procedure is to:
   A. gently slip the cord over the infant's head or shoulder.
   B. clamp the cord and cut it before delivering the infant.
   C. clamp the cord and cut it, then gently unwind it from around the neck if wrapped around more than once.
   D. all of the above

_____ **2.** If the amniotic fluid is greenish instead of clear or has a foul odor, this is called:
   A. nuchal rigidity.
   B. meconium staining.
   C. placenta previa.
   D. bloody show.

_____ **3.** Once the entire infant is delivered, you should immediately:
   A. wrap it in a towel and place it on one side with head lowered.
   B. be sure the head is covered and keep the neck in a neutral position.
   C. use a sterile gauze pad to wipe the infant's mouth, then suction again.
   D. all of the above

_____ **4.** You may help control bleeding by massaging the _____ after delivery of the placenta.
   A. perineum
   B. fundus
   C. lower back
   D. inner thighs

_____ **5.** The APGAR score should be calculated at _____ minutes after birth.
   A. 1 and 5
   B. 3 and 7
   C. 2 and 10
   D. 4 and 8

_____ **6.** Once the infant is delivered, feel for a brachial pulse or the pulsations in the umbilical cord. The pulse rate should be at least _____ beats/min and if not, begin artificial ventilations.
   A. 60
   B. 80
   C. 100
   D. 120

_____ **7.** When assisting ventilations in a newborn with a BVM device, the rate is _____ breaths/min.
   A. 20 to 30
   B. 30 to 50
   C. 35 to 45
   D. 40 to 60

_____ **8.** When performing CPR on a newborn, you should perform a combined total of _____ ventilations and compressions per minute.
   A. 90
   B. 100
   C. 110
   D. 120

_____ **9.** You cannot successfully deliver a _____ presentation in the field.

    A. limb

    B. breech

    C. vertex

    D. all of the above

_____ **10.** Care for a mother with a prolapsed cord includes:

    A. positioning the mother to keep the weight of the infant off the cord.

    B. high-flow oxygen and rapid transport.

    C. use your hand to physically hold the infant's head off the cord.

    D. all of the above

_____ **11.** When handling a delivery of a drug- or alcohol-addicted mother, your first concern should be for:

    A. the airway of the mother.

    B. your personal safety.

    C. the airway of the infant.

    D. the need for CPR for the infant.

_____ **12.** The stages of labor include:

    A. dilation of the cervix.

    B. expulsion of the baby.

    C. delivery of the placenta.

    D. all of the above

_____ **13.** The first stage of labor begins with the onset of contractions and ends when:

    A. the infant is born.

    B. the cervix is fully dilated.

    C. the water breaks.

    D. the placenta is delivered.

_____ **14.** Signs of the beginning of labor include:

    A. bloody show.

    B. contractions of the uterus.

    C. rupture of the amniotic sac.

    D. all of the above

_____ **15.** The second stage of labor begins when the cervix is fully dilated and ends when:

    A. the infant is born.

    B. the water breaks.

    C. the placenta delivers.

    D. the uterus stops contracting.

_____ **16.** The third stage of labor begins with the birth of the infant and ends with the:

    A. release of milk from the breasts.

    B. cessation of uterine contractions.

    C. delivery of the placenta.

    D. cutting of the umbilical cord.

_____ **17.** The difference between pre-eclampsia and eclampsia is the onset of:

    A. seeing spots.

    B. seizures.

    C. swelling in the hands and feet.

    D. headaches.

_____ **18.** You should consider the possibility of a(n) _____ in women who have missed a menstrual cycle and complain of a sudden stabbing and usually unilateral pain in the lower abdomen.

   A. PID

   B. ectopic pregnancy

   C. miscarriage

   D. placenta abruptio

_____ **19.** Consider delivery of the fetus at the scene when:

   A. delivery can be expected within a few minutes.

   B. when a natural disaster, or other problem, makes it impossible to reach the hospital.

   C. no transportation is available.

   D. all of the above

_____ **20.** When in doubt about the possibility of an imminent delivery:

   A. go for help if you are alone.

   B. insert two fingers into the vagina to feel for the head.

   C. contact medical control for further guidance.

   D. all of the above

_____ **21.** Low blood pressure resulting from compression of the inferior vena cava by the weight of the fetus when the mother is supine is called:

   A. pregnancy-induced hypertension.

   B. placenta previa.

   C. placenta abruptio.

   D. supine hypotensive syndrome.

_____ **22.** _____ is a situation in which the umbilical cord comes out of the vagina before the infant.

   A. Eclampsia

   B. Placenta previa

   C. Placenta abruptio

   D. Prolapsed cord

_____ **23.** Premature separation of the placenta from the wall of the uterus is known as:

   A. eclampsia.

   B. placenta previa.

   C. placenta abruptio.

   D. prolapsed cord.

_____ **24.** _____ is a condition in which the placenta develops over and covers the cervix.

   A. Eclampsia

   B. Placenta previa

   C. Placenta abruptio

   D. Prolapsed cord

_____ **25.** _____ is heralded by the onset of convulsions, or seizures, resulting from severe hypertension in the pregnant woman.

   A. Eclampsia

   B. Placenta previa

   C. Placenta abruptio

   D. Supine hypotensive syndrome

_____ **26.** _____ is a condition of infants who are born to alcoholic mothers; it is characterized by physical and mental retardation and a variety of congenital abnormalities.

   A. Pregnancy-induced hypertension

   B. Ectopic pregnancy

   C. Fetal alcohol syndrome

   D. Supine hypotensive syndrome

Questions 27-31 are derived from the following scenario: You have been dispatched to the side of a highway where a woman is reported to be delivering a baby. As you approach the vehicle, you see her lying down in the back seat.

_____ 27. Which of the following signs/symptoms tell you that the birth is imminent?

    A. Her water has broken.

    B. Her contractions are 30 to 60 minutes apart.

    C. Multigravida 4, Multipara 3

    D. All of the above

_____ 28. As you prepare her for delivery, what position should she be placed in?

    A. Supine, on the back seat

    B. Supine, on the ground

    C. Supine, one foot on the ground and one foot on the seat

    D. Seated

_____ 29. As you perform a visual exam, you note crowning. This means that:

    A. the baby is making a crowing-type of sound.

    B. the baby cannot be visualized.

    C. you can visualize the baby's head.

    D. the father is excited and needs care.

_____ 30. Once the head has been delivered:

    A. suction the nose, and then the mouth.

    B. apply oxygen over the vagina.

    C. suction the mouth, then the nose.

    D. apply a nasal cannula at 3 L/min to the child.

_____ 31. Concerning the delivery of the placenta, which of the following are emergency situations?

    A. More than 30 minutes have elapsed and the placenta has not delivered.

    B. There is more than 500 mL of bleeding before delivery of the placenta.

    C. There is significant bleeding after delivery of the placenta.

    D. All of the above

Tables

Complete the following table for the APGAR scoring system by listing the characteristics for each score.

Area of Activity	Score		
	2	1	0
Appearance			
Pulse			
Grimace or Irritability			
Activity or Muscle Tone			
Respiration			

Labeling

Label the following diagram with the correct terms.

1. Anatomic structures of the pregnant woman

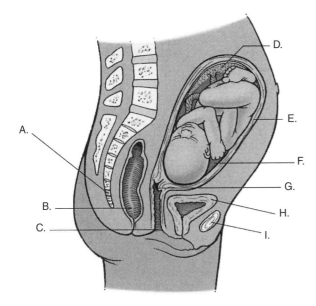

A. _____

B. _____

C. _____

D. _____

E. _____

F. _____

G. _____

H. _____

I. _____

Fill-in

Read each item carefully, then complete the statement by filling in the missing word(s).

1. After delivery, the _____, or afterbirth, separates from the uterus and is delivered.

2. The umbilical cord contains two _____ and one _____.

3. The amniotic sac contains about _____ _____ _____ of amniotic fluid, which helps to insulate and protect the floating fetus as it develops.

4. A full-term pregnancy is from _____ to _____ weeks, counting from the first day of the last menstrual cycle.

5. The pregnancy is divided into three _____ of about 3 months each.

6. There is a high potential of exposure due to _____ _____ released during childbirth.

7. The leading cause of maternal death in the first trimester is internal hemorrhage into the abdomen following rupture of an _____ _____.

8. In serious trauma, the only chance to save the infant is to adequately _____ the mother.

9. During the delivery, be careful that you do not poke your fingers into the infant's eyes or into the two soft spots, called _____, on the head.

True/False

If you believe the statement to be more true than false, write the letter "T" in the space provided. If you believe the statement to be more false than true, write the letter "F."

_____ **1.** The small mucous plug from the cervix that is discharged from the vagina, often at the beginning of labor, is called a bloody show.

_____ **2.** Crowning occurs when the baby's head obstructs the birth canal, preventing normal delivery.

_____ **3.** Labor begins with the rupture of the amniotic sac and ends with the delivery of the baby's head.

_____ **4.** A woman who is having her first baby is called a multigravida.

_____ **5.** Once labor has begun, it can be slowed by holding the patient's legs together.

_____ **6.** Delivery of the buttocks before the baby's head is called a breech delivery.

_____ **7.** After delivery, the baby should be kept at the same level as the mother's vagina until after the cord is cut.

_____ **8.** The placenta and cord should be properly disposed of in a biohazard container after delivery.

_____ **9.** The umbilical cord may be gently pulled to aid in delivery of the placenta.

_____ **10.** A limb presentation occurs when the baby's arm, leg, or foot is emerging from the vagina first.

_____ **11.** Multiple births may have more than one placenta.

Short Answer

Complete this section with short written answers using the space provided.

1. What are some possible causes of vaginal hemorrhage in early and late pregnancy?

_____

_____

2. In what position should pregnant patients who are not delivering be transported and why?

_____

_____

3. List three signs that indicate the beginning of labor.

_____

_____

4. Under what three circumstances should you consider delivering the patient at the scene?

_____

_____

5. Once the baby's head emerges, what actions should be taken to prevent too rapid a delivery?

_____

_____

6. Why is it important to avoid pushing on the fontanels?

_____

_____

7. How can you help decrease perineal tearing?

_____

_____

8. What are the two situations in which an EMT-B may insert his or her fingers into a patient's vagina?

_____

_____

9. What are three fetal effects of maternal drug or alcohol addiction?

_____

_____

Word Fun

The following crossword puzzle is an activity provided to reinforce correct spelling and understanding of medical terminology associated with emergency care and the EMT-B. Use the clues in the column to complete the puzzle.

Across

3. Fluid-filled bag for developing fetus

6. Umbilicus wrapped around baby's neck

8. Develops over and covers the cervix

9. Rating of newborn on 5 factors

Down

1. Buttocks first

2. Previously been pregnant

4. Head showing during labor

5. Convulsions from hypertension in pregnant woman

7. Area of skin between anus and vagina, subject to tearing during birth process

Ambulance Calls

The following case scenarios provide an opportunity to explore the concerns associated with patient management. Read each scenario, then answer each question in detail.

1. You are dispatched to a grocery store for an "unknown medical problem." You arrive to find a 20-year-old woman who is in her 22nd week of pregnancy. She tells you that this is her first pregnancy and she had made an appointment to see her obstetrician later that day because she was not feeling well. She's been experiencing a headache, swelling in her hands and feet, and transient problems with her vision. She states that she suddenly felt lightheaded and had to sit down. She appears unhurt.

   How do you best manage this patient?

   _____

   _____

   _____

2. You are enjoying a quiet lunch at the fire station when you hear the back doorbell ring. As you look out the window, you see a young woman running away from the building. You open the door to call to her and when you look down, you see something bundled in a wet sheet. It's a newborn; she's wet with amniotic fluid and she's not breathing very well.

   How do you best manage this patient?

   _____

   _____

   _____

3. You are on the scene with a 32-year-old woman who is 38 weeks pregnant and delivery is imminent. As the infant starts to crown, you notice that the amniotic sac is still intact.

   How would you best manage this patient?

   _____

   _____

   _____

Skill Drills
Skill Drill 20-1: Delivering the Baby
Test your knowledge of this skill drill by placing the photos below in the correct order. Number the first step with a "1," the second step with a "2," etc. Also, fill in the correct words in the photo captions.

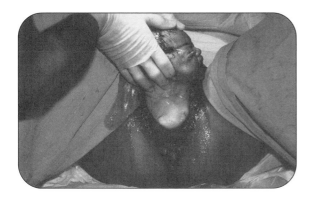

_____

Support the head and upper body as the
_____ _____ delivers, guiding the
head _____ if needed.

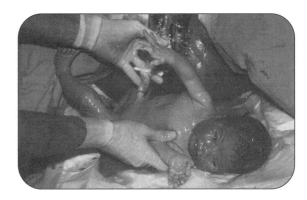

_____

Place the umbilical cord clamps _____ inch(es)
to _____ inch(es) apart, and _____
between them.

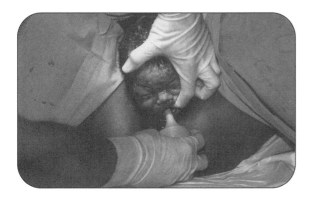

_____

Support the _____ parts of the head with
your hands as it emerges. Suction fluid from the
_____, then _____.

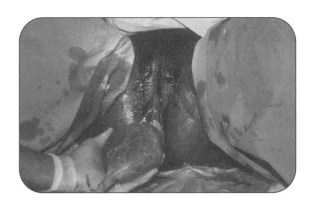

_____

Allow the _____ to deliver itself. Do not pull
on the _____ to speed delivery.

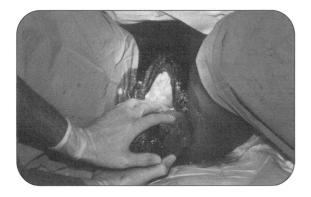

_____

As the _____ _____ appears, guide
the head _____ slightly, if needed, to
deliver the _____.

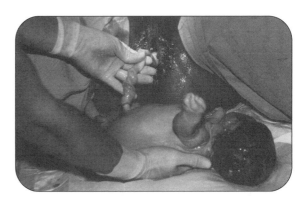

_____

Handle the slippery, delivered infant firmly but
gently, keeping the neck in _____ position
to _____ the airway.

Skill Drill 20-2: Giving Chest Compressions to an Infant
Test your knowledge of this skill drill by filling in the correct words in the photo captions.

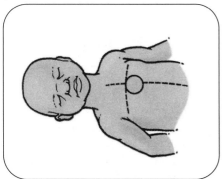
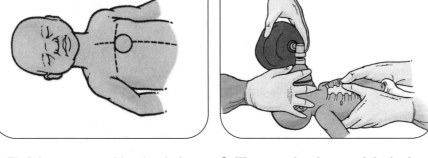
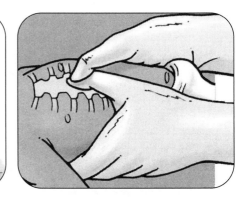

1. Find the proper position: just below the _____ _____, middle or _____ _____ of the sternum.

2. Wrap your hands around the body, with your _____ resting at that position.

3. Press your thumbs gently against the sternum, compressing _____ inch(es) to _____ inch(es) deep.

Assessment Review

Answer the following questions pertaining to the assessment of the types of emergencies discussed in this chapter.

_____ 1. After the head has been delivered and the shoulder appears:
   A. guide the head down slightly, to deliver the shoulder.
   B. apply a nasal cannula to the child.
   C. guide the head up slightly, to deliver the shoulder.
   D. pull gently.

_____ 2. When giving chest compressions to an infant:
   A. press the palm of your hand over the sternum, compressing ½ to ¾ inches deep.
   B. press the palm of your hand over the sternum, compressing 1 to 1½ inches deep.
   C. compress at a rate of 60 to 80 times a minute.
   D. compress the sternum ½ to ⅓ the depth of the chest.

_____ 3. Which exam will you perform before delivery?
   A. Focused physical
   B. Rapid physical
   C. Detailed physical
   D. All of the above

_____ 4. What should the focused physical exam include?
   A. The abdomen
   B. The length and frequency of contractions
   C. The vagal opening
   D. All of the above

_____ 5. When cutting the umbilical cord:
   A. place the clamps 7 to 10 inches apart.
   B. place the clamps 2 to 4 inches apart.
   C. tie the cord with shoelaces if you don't have any clamps.
   D. tie the cord with string if you don't have any clamps.

Emergency Care Summary

Complete the statements pertaining to emergency care for the types of emergencies discussed in this chapter by filling in the missing words.

NOTE: While the steps below are widely accepted, be sure to consult and follow your local protocol.

Delivering the Baby	Giving Chest Compressions to an Infant	Premature Infant
1. Support the _____ parts of the head with your hands as it emerges. Suction fluid from the baby's _____, then _____.	1. Find the proper position——just below the nipple line, _____ or _____ third of the sternum.	1. Keep the infant warm. Keep the infant in a place where the temperature is between ___°F and ___°F (between ___°C and ___°C).
2. As the upper shoulder appears, guide the head _____ slightly, if needed, to deliver the shoulder.	2. Wrap your hands around the body, with your thumbs resting at that position.	2. Keep the mouth and nose clear of mucus with a _____ _____.
3. Support the head and upper body as the lower shoulder delivers, guiding the head _____ if needed.	3. Press the two thumbs against the sternum, compressing ___" to ___" deep.	3. Inspect the cut end of the cord attached to the infant for _____.
4. Handle the slippery, delivered infant firmly but gently, keeping the neck in _____ position to maintain the airway.		4. Give _____ into a small tent over the infant's head.
5. Place the umbilical cord clamps ___" to ___" apart, and cut between them.		5. Do not infect the infant. Wear a mask to help prevent you from breathing on the infant.
6. Allow the placenta to deliver itself. Do not pull on the cord to speed delivery.		6. Notify the hospital of a neonatal transport.

CHAPTER

21 Kinematics of Trauma

Workbook Activities

The following activities have been designed to help you. Your instructor may require you to complete some or all of these activities as a regular part of your EMT-B training program. You are encouraged to complete any activity that your instructor does not assign as a way to enhance your learning in the classroom.

Chapter Review

The following exercises provide an opportunity to refresh your knowledge of this chapter.

Matching

Match each of the terms in the left column to the appropriate definition in the right column.

_____ **1.** Cavitation

_____ **2.** Deceleration

_____ **3.** Kinetic energy

_____ **4.** Mechanism of injury

_____ **5.** Potential energy

_____ **6.** Blunt trauma

_____ **7.** Penetrating trauma

_____ **8.** Work

A. impact on the body by objects that cause injury without penetrating soft tissue or internal organs and cavities

B. force acting over a distance

C. product of mass, gravity, and height

D. injury caused by objects that pierce the surface of the body

E. how trauma occurs

F. energy of moving object

G. slowing

H. emanation of pressure waves that can damage nearby structures

Multiple Choice

Read each item carefully, then select the best response.

_____ 1. The following are concepts of energy typically associated with injury, except:

A. potential energy.

B. thermal energy.

C. kinetic energy.

D. work.

_____ 2. The energy of a moving object is called:

A. potential energy.

B. thermal energy.

C. kinetic energy.

D. work.

_____ 3. Energy may be:

A. created.

B. destroyed.

C. converted.

D. all of the above

_____ 4. The amount of kinetic energy that is converted to do work on the body dictates the _____ of the injury.

A. location

B. severity

C. cause

D. speed

_____ 5. Potential energy is mostly associated with the energy of:

A. falling objects.

B. motor vehicle crashes.

C. pedestrian vs. bicycle crashes.

D. gun shot wounds.

_____ 6. Motor vehicle crashes are classified traditionally as:

A. frontal.

B. rollover.

C. lateral.

D. all of the above

_____ 7. The three collisions in a frontal impact include all of the following, except:

A. car vs. object.

B. passenger vs. vehicle.

C. flying objects vs. passengers.

D. internal organs vs. solid structures of the body.

_____ 8. The mechanism of injury provides information about the severity of the collision and therefore has a(n) _____ effect on patient care.

A. direct

B. positive

C. indirect

D. negative

_____ **9.** Your index of suspicion for the presence of life-threatening injuries should automatically increase if you see:

    A. seats torn from their mountings.

    B. collapsed steering wheels.

    C. intrusion into the passenger compartment.

    D. all of the above

_____ **10.** In a motor vehicle crash, as the passenger's head hits the windshield, the brain continues to move forward until it strikes the inside of the skull resulting in a _____ injury.

    A. compression

    B. laceration

    C. lateral

    D. motion

_____ **11.** Your quick initial assessment of the patient and the evaluation of the _____ can help to direct lifesaving care and provide critical information to the hospital staff.

    A. scene

    B. index of suspicion

    C. mechanism of injury

    D. abdominal area

_____ **12.** A contusion to a patient's forehead along with a spiderwebbed windshield suggests possible injury to the:

    A. nose.

    B. brain.

    C. face.

    D. heart.

_____ **13.** Significant mechanisms of injury include:

    A. moderate intrusions from a lateral impact.

    B. severe damage from the rear.

    C. collisions in which rotation is involved.

    D. all of the above

_____ **14.** Significant clues to the possibility of severe injuries include:

    A. death of a passenger.

    B. a blown out tire.

    C. broken glass.

    D. a deployed airbag.

_____ **15.** When properly applied, seat belts are successful in:

    A. restraining the passengers in a vehicle.

    B. preventing a second collision inside the motor vehicle.

    C. decreasing the severity of the third collision.

    D. all of the above

_____ **16.** Air bags decrease injury to all of the following, except:

    A. chest.

    B. heart.

    C. face.

    D. head.

_____ **17.** Signs of most injuries sustained in a motor vehicle crash can be found by simply inspecting the _____ during extrication of the patient.

    A. head and neck

    B. chest

    C. interior of the vehicle

    D. torso

_____ **18.** _____ impacts are probably the number one cause of death associated with motor vehicle crashes.

   A. Frontal

   B. Lateral

   C. Rear-end

   D. Rollover

_____ **19.** The most common life-threatening event in a rollover is _____ or partial ejection of the passenger from the vehicle.

   A. "sandwiching"

   B. centrifugal force

   C. ejection

   D. spinal cord injury

_____ **20.** A fall from more than _____ times the patient's height is considered to be significant.

   A. two

   B. three

   C. four

   D. five

Questions 21-25 are derived from the following scenario: A young boy was riding his bicycle down the street when he hit a parked car.

_____ **21.** How many collisions took place?

   A. 1

   B. 2

   C. 3

   D. 4

_____ **22.** What was the first collision?

   A. The bike hitting the car

   B. The bide rider hitting his bike/car

   C. The bike rider's internal organs against the solid structures of the body.

   D. None of the above

_____ **23.** What was the second collision?

   A. The bike hitting the car

   B. The bide rider hitting his bike/car

   C. The bike rider's internal organs against the solid structures of the body.

   D. None of the above

_____ **24.** "For every action, there is an equal and opposite reaction" is:

   A. Newton's first law

   B. Newton's second law

   C. Newton's third law

   D. A statement that is not true.

_____ **25.** What will rate your index of suspicion for this collision?

   A. The MOI

   B. The NOI

   C. How loudly he's crying

   D. A quick visual assessment

Fill-in

Read each item carefully, then complete the statement by filling in the missing word(s).

1. _____ are the leading cause of death and disability in the United States among children and young adults.

2. Certain injury _____ occur with certain types of injury _____.

3. _____ _____ occurs to the body when the body's tissues are exposed to energy levels beyond their tolerance.

4. The formula for calculating kinetic energy is _____.

5. _____ of the crash scene may provide valuable information to the staff and treating physicians of the trauma center.

6. Air bags provide the final capture point of the passengers and decrease the severity of _____ injuries.

7. Seat belts that buckle automatically at the shoulder but require the passengers to buckle the lap portion can result in the body _____ forward underneath the shoulder restraint when the lap portion is not attached.

True/False

If you believe the statement to be more true than false, write the letter "T" in the space provided. If you believe the statement to be more false than true, write the letter "F."

_____ 1. Work is defined as force acting over distance.
_____ 2. Energy can be both created and destroyed.
_____ 3. The energy of a moving object is called potential energy.
_____ 4. Rear-end collisions often cause whiplash injuries.
_____ 5. The cervical spine has little tolerance for lateral bending.
_____ 6. The injury potential of a fall is related to the height from which the patient fell.
_____ 7. Injuries are the leading cause of death and injuries among 1- to 34-year-olds in the United States.

Short Answer

Complete this section with short written answers using the space provided.

1. Describe potential energy.

_____

_____

2. List the three series of collisions typical with motor vehicles.

_____

_____

3. List the three factors to consider when evaluating a fall.

_____

_____

4. Describe the phenomenon of cavitation as it relates to an injury from a bullet.

_____

_____

5. Why is it important to try to determine the type of gun and ammunition used when you are caring for a gunshot victim?

_____

_____

Word Fun

The following crossword puzzle is an activity provided to reinforce correct spelling and understanding of medical terminology associated with emergency care and the EMT-B. Use the clues in the column to complete the puzzle.

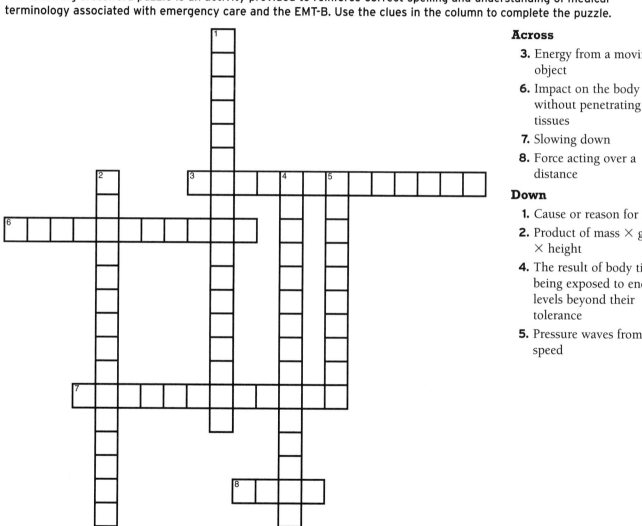

Across

3. Energy from a moving object

6. Impact on the body without penetrating soft tissues

7. Slowing down

8. Force acting over a distance

Down

1. Cause or reason for injury

2. Product of mass × gravity × height

4. The result of body tissues being exposed to energy levels beyond their tolerance

5. Pressure waves from speed

Ambulance Calls

The following case scenarios provide an opportunity to explore the concerns associated with patient management. Read each scenario, then answer each question in detail.

1. You are dispatched to a one-car MVC. As you arrive, you notice that the car hit a large deer that is lying in the road, dead. Highway speed limits on this road are 65 mph. The driver was restrained with a lap belt only, and his vehicle was not equipped with airbags. He is complaining of head and neck pain, and tells you that he doesn't remember what happened.

   How do you best manage this patient?

   _____

   _____

   _____

2. As December 25th approaches, many people in your community enjoy hanging lights and decorating their front yards. You are dispatched to assist a man who fell from a ladder as he was hanging lights from the roof of his two-story home. You arrive to find an unconscious middle-aged man lying in the snow. He is breathing and has a pulse. The call to 9-1-1 was placed after the man was found by a neighbor.

   _____

   _____

   _____

3. You are called to the residence of a 19-year-old male who was stabbed in the abdomen with an ice pick. The scene is safe and the patient is lying on the floor with the ice pick impaled in his left lower quadrant. Bystanders tell you he did not fall. He is alert and complaining of severe pain.

   How would you best manage this patient?

   _____

   _____

   _____

CHAPTER

22 Bleeding

Workbook Activities

The following activities have been designed to help you. Your instructor may require you to complete some or all of these activities as a regular part of your EMT-B training program. You are encouraged to complete any activity that your instructor does not assign as a way to enhance your learning in the classroom.

Chapter Review

The following exercises provide an opportunity to refresh your knowledge of this chapter.

Matching

Match each of the terms in the left column to the appropriate definition in the right column.

_____ **1.** Pulmonary artery

_____ **2.** Heart

_____ **3.** Ventricle

_____ **4.** Aorta

_____ **5.** Atrium

_____ **6.** Pulmonary vein

_____ **7.** Coagulation

_____ **8.** Ecchymosis

_____ **9.** Epistaxis

_____ **10.** Hematoma

_____ **11.** Hemophilia

_____ **12.** Hemorrhage

_____ **13.** Hypovolemic shock

A. mass of blood in the soft tissues beneath the skin

B. formation of clot to plug opening in injured blood vessel and stop blood flow

C. upper chamber

D. a congenital condition in which a patient lacks one or more of the blood's normal clotting factors

E. hollow muscular organ

F. largest artery in body

G. a condition in which low blood volume results in inadequate perfusion

H. oxygenated blood travels through this

I. bruising

J. deoxygenated blood travels through this

K. lower chamber

L. bleeding

M. nosebleed

Multiple Choice

Read each item carefully, then select the best response.

_____ **1.** The function of the blood is to _____ all of the body's cells and tissues.

    A. deliver oxygen to

    B. deliver nutrients to

    C. carry waste products away from

    D. all of the above

_____ **2.** The cardiovascular system consists of:

    A. a pump.

    B. a container.

    C. fluid.

    D. all of the above

_____ **3.** Blood leaves each chamber of a normal heart through a:

    A. vein.

    B. artery.

    C. one-way valve.

    D. capillary.

_____ **4.** Blood enters into the right atrium from the:

    A. coronary arteries.

    B. lungs.

    C. vena cava.

    D. coronary veins.

_____ **5.** Blood enters into the left atrium from the:

    A. coronary arteries.

    B. lungs.

    C. vena cava.

    D. coronary veins.

_____ **6.** The only arteries in the body to carry deoxygenated blood are the:

    A. pulmonary arteries.

    B. coronary arteries.

    C. femoral arteries.

    D. subclavian arteries.

_____ **7.** The _____ is the thickest chamber of the heart.

    A. right atrium

    B. right ventricle

    C. left atrium

    D. left ventricle

_____ **8.** The _____ link(s) the arterioles and the venules.

    A. aorta

    B. capillaries

    C. vena cava

    D. valves

_____ **9.** At the arterial end of the capillaries, the muscles dilate and constrict in response to conditions such as:

   A. fright.

   B. a specific need for oxygen.

   C. a need to dispose of metabolic wastes.

   D. all of the above

_____ **10.** Blood contains all of the following, except:

   A. white cells.

   B. plasma.

   C. cerebrospinal fluid.

   D. platelets.

_____ **11.** _____ is the circulation of blood within an organ or tissue in adequate amounts to meet the cells' current needs for oxygen, nutrients, and waste removal.

   A. Anatomy

   B. Perfusion

   C. Physiology

   D. Conduction

_____ **12.** The _____ only require a minimal blood supply when at rest.

   A. lungs

   B. kidneys

   C. muscles

   D. heart

_____ **13.** The term _____ means constantly adapting to changing conditions.

   A. perfusion

   B. conduction

   C. dynamic

   D. autonomic

_____ **14.** _____ is inadequate tissue perfusion.

   A. Shock

   B. Hyperperfusion

   C. Hypertension

   D. Contraction

_____ **15.** The brain and spinal cord usually cannot go for more than _____ minutes without perfusion, or the nerve cells will be permanently damaged.

   A. 30 to 45

   B. 12 to 20

   C. 8 to 10

   D. 4 to 6

_____ **16.** An organ or tissue that is considerably _____ is much better able to resist damage from hypoperfusion.

   A. warmer

   B. colder

   C. younger

   D. older

_____ **17.** The body will not tolerate an acute blood loss of greater than _____ of blood volume.

   A. 10%

   B. 20%

   C. 30%

   D. 40%

_____ **18.** If the typical adult loses more than 1 L of blood, significant changes in vital signs such as _____ will occur.

    A. increased heart rate

    B. increased respiratory rate

    C. decreased blood pressure

    D. all of the above

_____ **19.** _____ is a condition in which low blood volume results in inadequate perfusion and even death.

    A. Hypovolemic shock

    B. Metabolic shock

    C. Septic shock

    D. Psychogenic shock

_____ **20.** You should consider bleeding to be serious if all of the following conditions are present, except:

    A. blood loss is rapid.

    B. there is no mechanism of injury.

    C. the patient has a poor general appearance.

    D. assessment reveals signs and symptoms of shock.

_____ **21.** Significant blood loss demands your immediate attention as soon as the _____ has been managed.

    A. fractures

    B. extrication

    C. airway

    D. none of the above

_____ **22.** The process of blood clotting and plugging the hole is called:

    A. conglomeration.

    B. configuration.

    C. coagulation.

    D. coalition.

_____ **23.** Even though the body is very efficient at controlling bleeding on its own, it may fail in situations such as:

    A. when medications interfere with normal clotting.

    B. when damage to the vessel may be so large that a clot cannot completely block the hole.

    C. when sometimes only part of the vessel wall is cut, preventing it from constricting.

    D. all of the above

_____ **24.** A lack of one or more of the blood's clotting factors is called:

    A. a deficiency.

    B. hemophilia.

    C. platelet anomaly.

    D. anemia.

_____ **25.** You respond to a 25-year-old man who has cut his arm with a circular saw. The bleeding appears to be bright red and spurting. The patient is alert and oriented and converses with you freely. He appears to be stable at this point. What is your first step in controlling his bleeding?

    A. Direct pressure

    B. Maintaining the airway

    C. BSI precautions

    D. Elevation

_____ **26.** When applying a bandage to hold a dressing in place, stretch the bandage tight enough to control bleeding, but not so tight as to decrease _____ to the extremity.

    A. blood flow

    B. pulses

    C. oxygen

    D. CRTs

_____ **27.** If bleeding continues after applying a pressure dressing, you should do all of the following, except:

  A. remove the dressing and apply another sterile dressing.

  B. apply manual pressure through the dressing.

  C. add more gauze pads over the first dressing.

  D. secure both dressings tighter with a roller bandage.

_____ **28.** When using an air splint to control bleeding in a fractured extremity, you should reassess the _____ frequently.

  A. airway

  B. breathing

  C. circulation in the injured extremity

  D. fracture site

_____ **29.** Contraindications to the use of the PASG include:

  A. pulmonary edema.

  B. pregnancy.

  C. penetrating chest injuries.

  D. all of the above

_____ **30.** You and your partner respond to a patient who has had his hand nearly severed by a drill press. As you approach the patient he is pale and there appears to be a lot of blood on the floor. The wound continues to bleed massively. You consider applying a tourniquet but know they are rarely needed and often:

  A. resolve the problem.

  B. decrease the problem.

  C. create more problems.

  D. all of the above

_____ **31.** In the above call, all methods of trying to control the bleeding are unsuccessful and it is decided to apply the tourniquet. You know that the patient will probably lose the injured extremity but you feel it necessary to save his life. You know you must be sure to:

  A. use the narrowest bandage possible to minimize the area restricted.

  B. cover the tourniquet with a bandage.

  C. never pad underneath the tourniquet.

  D. not loosen the tourniquet after you have applied it.

_____ **32.** You are called to the local playground for an 8-year-old girl who has an uncontrolled nose bleed. The child is crying uncontrollably and will not converse with you. The baby sitter is not able to tell you if there was any trauma involved but there is a bump on the temporal portion of her head. The baby sitter does state that she has had a cold for several days, but can give you no further information on the medical history. All the children playing with her deny any trauma. What could be the possible cause(s) of the bleeding?

  A. A skull fracture

  B. Sinusitis

  C. Coagulation disorders

  D. All of the above

_____ **33.** You respond to a 33-year-old man who was hit in the ear by a line drive during a softball game. He is complaining of a severe headache, ringing in his ears, and being dizzy. He has blood draining from his ear. Why would you not apply pressure to control bleeding?

  A. It should be collected to be reinfused at the hospital.

  B. It could collect within the head and increase the pressure in the brain.

  C. It is contaminated.

  D. You could fracture the skull with the pressure needed to staunch the flow of blood.

_____ **34.** When treating a patient with signs and symptoms of hypovolemic shock and no outward signs of bleeding, always consider the possibility of bleeding into the:

  A. thoracic cavity.

  B. abdomen.

  C. skull.

  D. chest.

_____ **35.** Nontraumatic internal bleeding may be caused by:

  A. an ulcer.

  B. a ruptured ectopic pregnancy.

  C. an aneurysm.

  D. all of the above

_____ **36.** The most common symptom of internal abdominal bleeding is:

  A. bruising around the abdomen.

  B. distention of the abdomen.

  C. rigidity of the abdomen.

  D. acute abdominal pain.

_____ **37.** Signs and symptoms of internal bleeding in both trauma and medical patients include:

  A. hematochezia.

  B. melena.

  C. hemoptysis.

  D. all of the above

_____ **38.** The first sign of hypovolemic shock is a change in:

  A. respirations.

  B. heart rate.

  C. mental status.

  D. blood pressure.

Labeling

Label the following diagrams with the correct terms. Where arrows appear, indicate the substance and its origin and destination.

    1. The Left and Right Sides of the Heart

A. _____

B. _____

C. _____

D. _____

E. _____

F. _____

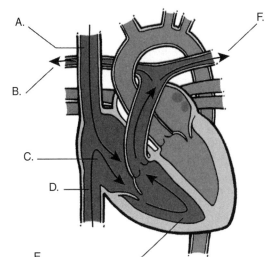

A. _____

B. _____

C. _____

D. _____

E. _____

F. _____

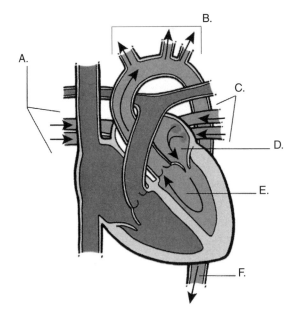

2. Perfusion

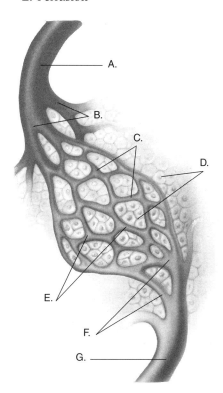

A. _____

B. _____

C. _____

D. _____

E. _____

F. _____

G. _____

3. Arterial Pressure Points

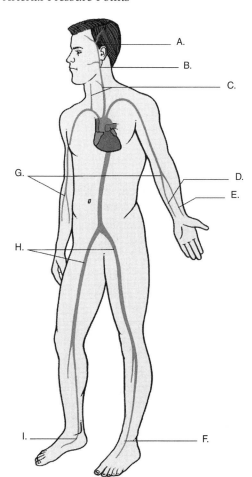

A. _____

B. _____

C. _____

D. _____

E. _____

F. _____

G. _____

H. _____

I. _____

Fill-in

Read each item carefully, then complete the statement by filling in the missing word(s).

1. The circulation of blood in adequate amounts to meet cellular needs is called _____.

2. The heart is a(n) _____ muscle, which is controlled by the autonomic nervous system.

3. Exchange of oxygen and carbon dioxide occurs in the _____.

4. Shock is a condition related to _____.

5. Blood returning to the heart from the lower body is first collected in the _____ vena cava.

6. Blood contains four major components, which are _____ _____ _____ _____.

7. The left ventricle receives _____ blood while the right ventricle receives deoxygenated blood.

8. During a state of shock, the autonomic nervous system directs blood away from some organs and distributes it to

the _____ _____ _____ _____.

9. The brain and spinal cord generally cannot go longer than _____ minutes without adequate perfusion, or

permanent nerve cell damage may occur.

10. If the patient has a significant mechanism of injury that may affect multiple systems, he or she should receive a

_____ _____ assessment.

True/False

If you believe the statement to be more true than false, write the letter "T" in the space provided. If you believe the statement to be more false than true, write the letter "F."

_____ **1.** Venous blood tends to spurt and is difficult to control.
_____ **2.** The human body is tolerant of blood losses greater than 20% of blood volume.
_____ **3.** The first step in controlling external bleeding is application of the PASG.
_____ **4.** The first step in preparing to treat a bleeding patient is BSI.
_____ **5.** A properly applied tourniquet should be loosened by the EMT-B every ten minutes.
_____ **6.** A patient who has swallowed a lot of blood may become nauseated and vomit.
_____ **7.** You should check with medical control every time a PASG may be indicated.
_____ **8.** If a wound continues to bleed after it is bandaged, you should remove the bandage and start over again.
_____ **9.** A tourniquet is always required for massive spurting blood loss.
_____ **10.** Do not use PASG if the transport time is less than 30 minutes.

Short Answer

Complete this section with short written answers in the space provided.

1. Describe how the autonomic nervous system responds to severe bleeding.

_____

_____

2. Describe the characteristics of bleeding from each type of vessel (artery, vein, capillary).

_____

_____

3. List, in the proper sequence, the methods in which an EMT-B should attempt to control external bleeding.

_____

_____

4. List ten signs and symptoms of hypovolemic shock.

_____

_____

5. List, in the proper sequence, the general EMT-B emergency care for patients with internal bleeding.

_____

_____

Word Fun

The following crossword puzzle is an activity provided to reinforce correct spelling and understanding of medical terminology associated with emergency care and the EMT-B. Use the clues in the column to complete the puzzle.

Across

3. Low blood volume, inadequate perfusion

7. Circulatory system failure

8. Circulation of blood within the body

9. Mass of blood in soft tissues

Down

1. Nosebleed

2. Lacks normal clotting factors

4. Discoloration of skin from closed wound

5. Bleeding

6. Formation of clots to stop bleeding

Ambulance Calls

The following case scenarios provide an opportunity to explore the concerns associated with patient management. Read each scenario, then answer each question in detail.

1. You are dispatched to a school playground for an 8-year-old boy complaining of a minor laceration to his left wrist with uncontrollable hemorrhaging. The teacher tells you that he has a history of hemophilia. The blood is steady, but not spurting, and dark in color.

   How would you best manage this patient?

   _____

   _____

   _____

2. You are dispatched to a local lumberyard for an "unknown accident." Upon your arrival you see that a man sitting on the ground, his coworkers surrounding him. His shirt got caught in a machine, and his arm was severed from the elbow down. He is holding a blood-soaked towel to the end of his right arm.

   How do you best manage this patient?

   _____

   _____

   _____

3. You are dispatched to the local jail for a male prisoner who has jabbed his arm with a ballpoint pen. He is bleeding continuously from the wound that is in the area of his antecubital vein on his left arm.

   How do you best manage this patient?

   _____

   _____

   _____

Skill Drills

Skill Drill 22-1: Controlling External Bleeding

Test your knowledge of this skill drill by filling in the correct words in the photo captions.

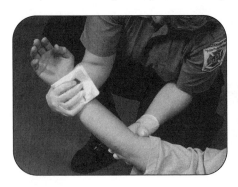 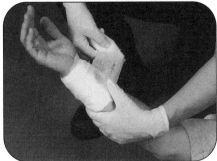 

1. Apply _____ _____ over the wound. Elevate the injury above the _____ of the _____ if no _____ is suspected.

2. Apply a _____ _____.

3. Apply pressure at the appropriate _____ _____ while continuing to hold _____.

Skill Drill 22-2: Applying a Pneumatic Antishock Garment (PASG)
Test your knowledge of this skill drill by placing the photos below in the correct order. Number the first step with a "1", the second step with a "2", etc.

_____

Inflate with the foot pump, and close the stopcocks when the patient's systolic blood pressure reaches 100 mm Hg or the Velcro crackles.

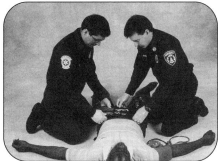

_____

Enclose both legs and the abdomen.

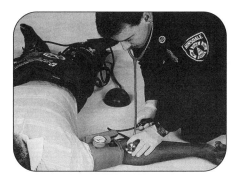

_____

Check the patient's blood pressure again. Monitor the vital signs.

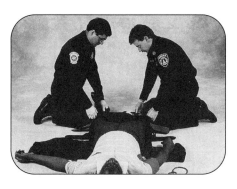

_____

Apply the garment so that the top is below the lowest rib.

_____

Open the stopcocks.

Assessment Review

Answer the following questions pertaining to the assessment of the types of emergencies discussed in this chapter.

_____ 1. When you are performing a scene size-up on a patient with external bleeding, the minimum BSI that should be taken is:

   A. gloves and gown.

   B. gown and eye protection.

   C. gloves and eye protection.

   D. gown and face mask.

_____ 2. For a patient with suspected internal bleeding, you should assess circulation by checking the pulse for:

   A. rate and quality.

   B. rate and rhythm.

   C. quality and rhythm.

   D. presence.

_____ 3. If you have completed your initial assessment and transport decision on an unresponsive patient with a significant mechanism of injury, how should the focused history and physical exam be performed?

   A. Just check the injury

   B. Rapid trauma assessment

   C. Just like any other patient

   D. Move to the ongoing assessment

_____ 4. Your external bleeding patient needs a detailed physical exam. When should it be performed?

   A. Immediately after the initial assessment

   B. During the focused physical exam

   C. When you arrive at the patient's side

   D. En route to the hospital

_____ 5. What should be included in the communications to the hospital when dealing with a patient with suspected internal bleeding?

   A. Amount of blood loss

   B. Location of the bleeding

   C. Interventions performed

   D. Patient's blood type

Emergency Care Summary

Fill in the following chart pertaining to the management of the types of emergencies discussed in this chapter.

NOTE: While the steps below are widely accepted, be sure to consult and follow your local protocol.

External Bleeding

Follow BSI precautions—minimum of gloves and eye protection.

Ensure that patient has open airway.

Maintain the _____ stabilization if MOI suggests possible spinal injury.

Provide high-flow oxygen.

Control bleeding using one of the following methods:

- Direct pressure and elevation
- Pressure dressings
- Pressure points
- Splints
- Air splints
- PASG
- Tourniquets

Controlling External Bleeding

1. Apply direct local pressure to bleeding site.
2. Elevate the bleeding extremity.
3. Create a _____ dressing.
4. Apply pressure at the appropriate pressure point while continuing to hold direct pressure.
5. If the wound continues to bleed, elevate extremity and place additional pressure over proximal pressure point.

Using PASG for Control of Massive Soft-tissue

Bleeding in the Extremities

1. Apply the garment.
2. Close and fasten both leg compartments and the abdominal compartment.
3. Open the stopcocks.
4. Inflate the compartments similar to an _____ _____.
5. Check the patient's circulation, motor function, and sensation in distal lower extremities.

Applying a Tourniquet

1. Fold a triangular bandage.
2. Wrap the bandage around the extremity _____.
3. Tie one know in the bandage. Place a stick or rod on top of the knot. Tie the ends of the bandage on the stick in a _____ knot.
4. Use the stick as a handle and twist it to tighten the tourniquet until bleeding has stopped.
5. Secure the stick in place with another triangular bandage.
6. Write "TK" and the exact time the tourniquet was applied on a piece of adhesive tape. Fasten the tape to the patient's forehead.
7. As an alternative, use a blood pressure cuff. Inflate enough to stop bleeding.

Treating Epistaxis

1. Follow BSI precautions.
2. Help the patient to sit, leaning forward.
3. Apply direct pressure for at least _____ minutes by pinching nostrils together.
4. Keep the patient calm and quiet.
5. Apply ice over the nose.
6. Maintain the pressure until bleeding is completely controlled.
7. Provide prompt transport.
8. If bleeding cannot be controlled, transport patient immediately. Treat for _____ and administer oxygen via mask if necessary.

Internal Bleeding

Steps to Caring for Patient With Internal Bleeding

1. Follow BSI precautions.
2. Maintain the airway with cervical immobilization if MOI suggests possible spinal injury.
3. Administer high-flow oxygen.
4. Control all obvious external bleeding.
5. Apply a _____ to an extremity where internal bleeding is suspected.
6. Monitor and record vital signs at least every 5 minutes.
7. Give the patient _____ by mouth.
8. Elevate the legs 6″ to 12″ in significant trauma patients.
9. Keep the patient warm.
10. Provide immediate transport for patients with signs and symptoms of shock. Report changes in condition to hospital personnel.

Using PASG for Treatment of Shock

1. Apply the garment.
2. Close and fasten both leg compartments and the _____ compartment.
3. Open the stopcocks.
4. Contact medical control for specific verbal orders to inflate or use standing order specific to inflation of PASG.
5. Inflate the compartments based on the patient's blood pressure.
6. Recheck the patient's blood pressure and inflate more based on patient _____ and blood pressure.

CHAPTER

23 Shock

Workbook Activities

The following activities have been designed to help you. Your instructor may require you to complete some or all of these activities as a regular part of your EMT-B training program. You are encouraged to complete any activity that your instructor does not assign as a way to enhance your learning in the classroom.

Chapter Review

The following exercises provide an opportunity to refresh your knowledge of this chapter.

Matching

Match each of the terms in the left column to the appropriate definition in the right column.

_____ **1.** Shock   **A.** severe allergy

_____ **2.** Perfusion   **B.** hypoperfusion

_____ **3.** Sphincters   **C.** regulates involuntary body functions

_____ **4.** Autonomic nervous system   **D.** early stage of shock

_____ **5.** Blood pressure   **E.** provides a rough measure of perfusion

_____ **6.** Anaphylaxis   **F.** severe bacterial infection

_____ **7.** Septic shock   **G.** sufficient circulation to meet cell needs

_____ **8.** Syncope   **H.** regulate blood flow in capillaries

_____ **9.** Compensated shock   **I.** fainting

Multiple Choice

Read each item carefully, then select the best response.

_____ **1.** Shock:

    A. refers to a state of collapse and failure of the cardiovascular system.

    B. results in the inadequate flow of blood to the body's cells.

    C. results in failure to rid cells of metabolic wastes.

    D. all of the above

_____ **2.** Blood flow through the capillary beds is regulated by:

    A. systolic pressure.

    B. the capillary sphincters.

    C. perfusion.

    D. diastolic pressure.

_____ **3.** The autonomic nervous system regulates involuntary functions such as:

    A. sweating.

    B. digestion.

    C. constriction and dilation of capillary sphincters.

    D. all of the above

_____ **4.** Regulation of blood flow is determined by:

    A. oxygen intake.

    B. systolic pressure.

    C. cellular need.

    D. diastolic pressure.

_____ **5.** Perfusion requires having a working cardiovascular system as well as:

    A. adequate oxygen exchange in the lungs.

    B. adequate nutrients in the form of glucose in the blood.

    C. adequate waste removal.

    D. all of the above

_____ **6.** The action of the hormones stimulates _____ to maintain pressure in the system and, as a result, perfusion of all vital organs.

    A. an increase in heart rate

    B. an increase in the strength of cardiac contractions

    C. vasoconstriction in nonessential areas

    D. all of the above

_____ **7.** Basic causes of shock include:

    A. poor pump function.

    B. blood or fluid loss.

    C. blood vessel dilation.

    D. all of the above

_____ **8.** Noncardiovascular causes of shock include respiratory insufficiency and:

    A. sepsis.

    B. metabolic.

    C. anaphylaxis.

    D. hypovolemia.

_____ **9.** You are called to the residence of a 67-year-old man who is complaining of chest pain. He is alert and oriented. During your assessment process the patient tells you he has had two previous heart attacks. He is taking medication for fluid retention. As you listen to his lungs, you notice that he has fluid in his lungs. This is known as pulmonary:

    A. edema.

    B. overload.

    C. cessation.

    D. failure.

_____ **10.** _____ develops when the heart muscle can no longer generate enough pressure to circulate the blood to all organs.

    A. Pump failure

    B. Cardiogenic shock

    C. A myocardial infarction

    D. Congestive heart failure

_____ **11.** You are called to a construction site where a 27-year-old worker has fallen from the second floor. He landed on his back and is drifting in and out of consciousness. A quick assessment reveals no bleeding or blood loss. His blood pressure is 90/60 mm Hg with a pulse rate of 110 beats/min. His airway is open and breathing is within normal limits. You realize the patient is in shock. The patient's shock is due to an injury to the:

    A. cervical vertebrae.

    B. skull.

    C. spinal cord.

    D. peripheral nerves.

_____ **12.** Neurogenic shock usually results from damage to the spinal cord at the:

    A. cervical level.

    B. thoracic level.

    C. lumbar level.

    D. sacral level.

_____ **13.** You respond to the local nursing home of an 85-year-old woman who has altered mental status. During your assessment you notice that the patient has an elevated body temperature. She is hypotensive and the pulse is tachycardic. The nursing staff tells you that she has been sick for several days and that they called because of her mental status that continued to decline. You suspect the patient is in septic shock. The shock is due to:

    A. pump failure.

    B. massive vasoconstriction.

    C. widespread dilation.

    D. increased volume.

_____ **14.** In septic shock:

    A. there is an insufficient volume of fluid in the container.

    B. the fluid that has leaked out often collects in the respiratory system.

    C. there is a larger-than-normal vascular bed to contain the smaller-than-normal volume of intravascular fluid.

    D. all of the above

_____ **15.** Neurogenic shock is caused by:

    A. a radical change in the size of the vascular system.

    B. massive vasoconstriction.

    C. low volume.

    D. fluid collecting around the spinal cord causing compression of the cord.

_____ **16.** Hypovolemic shock is a result of:

   A. widespread vasodilation.

   B. low volume.

   C. massive vasoconstriction.

   D. pump failure.

_____ **17.** An insufficient concentration of _____ in the blood can produce shock as rapidly as vascular causes.

   A. oxygen

   B. hormones

   C. epinephrine

   D. histamine

_____ **18.** In anaphylactic shock, the combination of poor oxygenation and poor perfusion is a result of:

   A. widespread vasodilation.

   B. low volume.

   C. massive vasoconstriction.

   D. pump failure.

_____ **19.** Causes of syncope include:

   A. generalized vascular dilation.

   B. the sight of blood.

   C. cardiac arrhythmias.

   D. all of the above

_____ **20.** You are called to a motor vehicle collision. Your patient is a 19-year-old female who was not wearing her seat belt. She is conscious but confused. Her airway is open and respirations are within normal limits. She pulse is slightly tachycardic but regular. The blood pressure is within normal limits. She is complaining of being thirsty and appears very anxious. You know that the last measurable factor to change to indicate shock is:

   A. mental status.

   B. blood pressure.

   C. pulse rate.

   D. respirations.

_____ **21.** You should suspect shock in all of the following except:

   A. a mild allergic reaction.

   B. multiple severe fractures.

   C. a severe infection.

   D. abdominal or chest injury.

_____ **22.** When treating a suspected shock patient, vital signs should be recorded approximately every _____ minutes.

   A. 2

   B. 5

   C. 10

   D. 15

_____ **23.** The Golden Hour refers to the first 60 minutes after:

   A. medical help arrives on scene.

   B. transport begins.

   C. the injury occurs.

   D. 9-1-1 is called.

_____ **24.** Signs of cardiogenic shock include all of the following except:

    A. cyanosis.

    B. strong, bounding pulse.

    C. nausea.

    D. anxiety.

_____ **25.** You respond to a 17-year-old football player who was hit by numerous opponents and while walking off the field became unconscious. He is currently unconscious. You take c-spine control and start your assessment. You know that in the treatment of shock you must:

    A. secure and maintain an airway.

    B. provide respiratory support.

    C. assisted ventilations.

    D. all of the above

_____ **26.** When assessing the patient who has psychogenic shock, be sure to consider possible _____ if the patient fell.

    A. extremity fractures

    B. cervical spine injury

    C. paralysis

    D. pelvic fractures

Fill-in

Read each item carefully, then complete the statement by filling in the missing word(s).

1. _____ refers to the failure of the cardiovascular system.

2. Pressure in the arteries during cardiac _____ is known as systolic pressure.

3. In shock conditions, the body redirects blood from _____ organs to _____ organs.

4. Blood pressure is a rough measurement of _____.

5. The cardiovascular system consists of the _____, _____, and _____.

6. Inadequate circulation that does not meet the body's needs is known as _____.

7. _____ are circular muscle walls in capillaries, causing the walls to _____ and _____.

8. _____ pressure occurs during cardiac relaxation, while _____ pressure occurs during cardiac contractions.

9. _____ pressure is the pressure in the blood vessels at all times.

10. The autonomic nervous system controls the _____ actions of the body.

True/False

If you believe the statement to be more true than false, write the letter "T" in the space provided. If you believe the statement to be more false than true, write the letter "F".

_____ **1.** Life-threatening allergic reactions can occur in response to almost any substance that a patient may encounter.

_____ **2.** Bleeding is the most common cause of shock following an injury.

_____ **3.** Shock occurs when oxygen and nutrients cannot get to the body's cells.

_____ **4.** A person in shock, left untreated, will most likely survive.

_____ **5.** Compensated shock is related to the last stages of shock.
_____ **6.** An injection of epinephrine is the only really effective treatment for anaphylactic shock.
_____ **7.** Septic shock is a combination of vessel and content failure.
_____ **8.** Metabolism is the cardiovascular system's circulation of blood and oxygen to all cells in different tissues and organs of the body.
_____ **9.** Shock only occurs with massive blood loss from the body.
_____ **10.** Decompensated shock occurs when the systolic blood pressure falls below 120 mm Hg.

Short Answer

Complete this section with short written answers using the space provided.

1. List the causes, signs and symptoms, and treatment of anaphylactic shock.

_____

_____

2. List the causes, signs and symptoms, and treatment of cardiogenic shock.

_____

_____

3. List the causes, signs and symptoms, and treatment of hypovolemic shock.

_____

_____

4. List the causes, signs and symptoms, and treatment of metabolic shock.

_____

_____

5. List the causes, signs and symptoms, and treatment of neurogenic shock.

_____

_____

6. List the causes, signs and symptoms, and treatment of psychogenic shock.

_____

_____

7. List the causes, signs and symptoms, and treatment of septic shock.

_____

_____

8. List the three basic physiologic causes of shock.

_____

_____

_____

9. List the signs and symptoms of decompensated shock.

_____

_____

10. List the steps for using an autoinjector.

_____

_____

Word Fun emtb.com vocab explorer

The following crossword puzzle is an activity provided to reinforce correct spelling and understanding of medical terminology associated with emergency care and the EMT-B. Use the clues in the column to complete the puzzle.

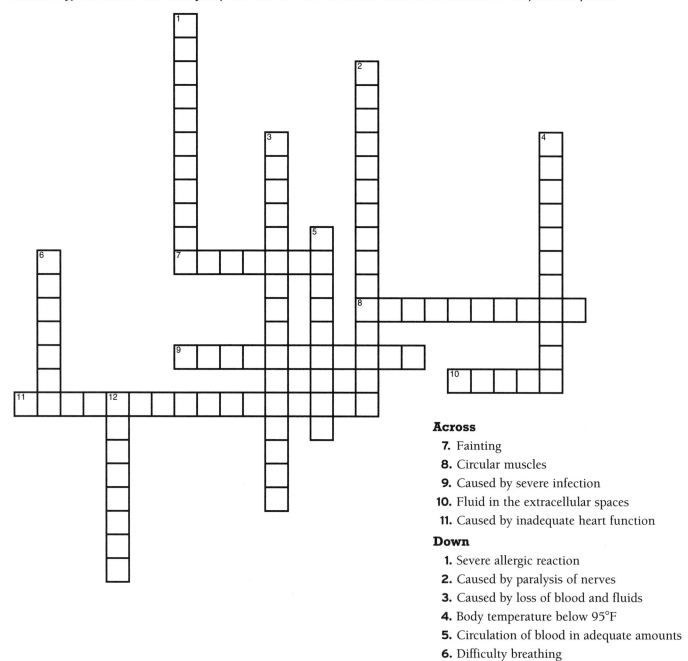

Across

7. Fainting

8. Circular muscles

9. Caused by severe infection

10. Fluid in the extracellular spaces

11. Caused by inadequate heart function

Down

1. Severe allergic reaction

2. Caused by paralysis of nerves

3. Caused by loss of blood and fluids

4. Body temperature below 95°F

5. Circulation of blood in adequate amounts

6. Difficulty breathing

12. Lack of oxygen

Ambulance Calls

The following case scenarios provide an opportunity to explore the concerns associated with patient management. Read each scenario, then answer each question in detail.

1. You are dispatched to the victim of a fall at the local community college theater. One of the students involved in rigging the theater backgrounds fell from the platform above the stage. He landed directly on his back, and is now complaining of numbness and tingling in his lower body.

How do you best manage this patient?

_____

_____

_____

2. You are dispatched to a local long-term care facility for an older man with a history of fever. You arrive to find an 80-year-old man who is responsive to painful stimuli and has the following vital signs: blood pressure 80/40 mm Hg, weak radial pulse of 140 beats/min and irregular, respirations of 60 breaths/min and shallow, and pulse oximetry of 80% on 4 L/min nasal cannula.

How do you best manage this patient?

_____

_____

_____

3. You are dispatched to a residence where a 16-year-old female was stung by a bee. Her mother tells you she is severely allergic to bees. She is voice responsive, covered in hives, and is wheezing audibly. She has a very weak radial pulse and is blue around the lips.

How would you best manage this patient?

_____

_____

_____

Skill Drills

Skill Drill 23-1: Treating Shock

Test your knowledge of this skill drill by placing the photos below in the correct order. Number the first step with a "1", the second step with a "2", etc.

_____

Splint any broken bones or joint injuries.

_____

Keep the patient supine, open the airway, and check breathing and pulse.

_____

Give high-flow oxygen if you have not already done so, and place blankets under and over the patient.

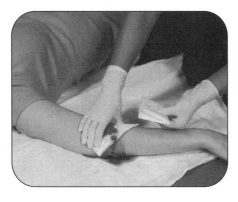

_____

Control obvious external bleeding.

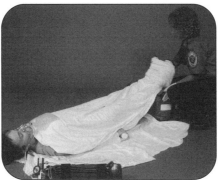

_____

If no fractures are suspected, elevate the legs 6″ to 12″.

Assessment Review

Answer the following questions pertaining to the assessment of the types of emergencies discussed in this chapter.

_____ **1.** In the scene size-up for a patient/patients who you think may be susceptible to shock, you should:

   A. ensure scene safety.

   B. determine the number of patients.

   C. consider c-spine stabilization.

   D. All of the above

_____ **2.** During the initial assessment of a patient in shock you should:

   A. treat any immediate life threats.

   B. obtain a SAMPLE history.

   C. get a complete set of vital signs.

   D. inform medical control of the situation.

_____ **3.** You have completed your initial assessment of a patient who stepped off a curb and heard her ankle pop. Your focused history should include:

    A. a rapid trauma assessment.

    B. an assessment of the affected area.

    C. a detailed physical exam.

    D. an ongoing assessment.

_____ **4.** Interventions for the treatment of shock should include:

    A. giving the patient something to drink.

    B. maintaining normal body temperature.

    C. positioning the patient in the Fowler's position for transport.

    D. delayed transport to splint fractures.

_____ **5.** You are transporting an unstable patient who you feel in going into shock. How often do you recheck his vital signs?

    A. Every 3 minutes

    B. Every 10 minutes

    C. Every 5 minutes

    D. Every 15 minutes

Emergency Care Summary

Fill in the following chart pertaining to the management of the types of emergencies discussed in this chapter.

NOTE: While the steps below are widely accepted, be sure to consult and follow your local protocol.

General Shock	Anaphylactic Shock
Minimum treatment for shock, with the exception of anaphylactic shock, is as follows:	Follow the steps for using an auto-injector (discussed in Chapter 16):
1. Provide _____ stabilization if needed.	1. Remove the auto-injector's safety cap, and quickly wipe the thigh with antiseptic.
2. Keep the patient _____, open the airway, and check breathing and pulse.	2. Place the tip of the auto-injector against the _____ part of the thigh.
3. Control obvious external bleeding.	3. Push the auto-injector firmly against the thigh, and hold it in place until all the medication is _____.
4. Splint any broken bones or joint injuries.	
5. Give high-flow oxygen if you have not already done so, and place blankets under and over the patient.	
6. If no extremity fractures are suspected, elevate the legs 6" to 12".	
7. Provide rapid transport in a position that best supports _____ and breathing.	
Treatment for cardiogenic shock also requires the following:	
For cardiogenic shock, place the patient in a position of comfort. Assist with ventilation and suction as needed. Monitor the _____ closely and provide rapid, calm transport.	

CHAPTER

24 Soft-Tissue Injuries

Workbook Activities

The following activities have been designed to help you. Your instructor may require you to complete some or all of these activities as a regular part of your EMT-B training program. You are encouraged to complete any activity that your instructor does not assign as a way to enhance your learning in the classroom.

Chapter Review

The following exercises provide an opportunity to refresh your knowledge of this chapter.

Matching

Match each of the terms in the left column to the appropriate definition in the right column.

_____ 1. Dermis

_____ 2. Sweat glands

_____ 3. Epidermis

_____ 4. Mucous membranes

_____ 5. Sebaceous glands

_____ 6. Abrasion

_____ 7. Laceration

_____ 8. Penetrating wound

_____ 9. Avulsion

_____ 10. Evisceration

A. gunshot wound

B. cool the body by discharging a substance through the pores

C. tissue hanging as a flap from wound

D. tough external layer forming a watertight covering for the body

E. razor cut

F. secrete a watery substance that lubricates the openings of the mouth and nose

G. inner layer of skin that contains the structures that give skin its characteristic appearance

H. produce oil, which waterproofs the skin and keeps it supple

I. exposed intestines

J. skinned knee

Multiple Choice

Read each item carefully, then select the best response.

_____ **1.** The _____ is (are) our first line of defense against external forces.

    A. extremities

    B. hair

    C. skin

    D. lips

_____ **2.** The skin covering the _____ is quite thick.

    A. lips

    B. scalp

    C. ears

    D. eyelids

_____ **3.** As the cells on the surface of the skin are worn away, new cells form in the _____ layer.

    A. dermal

    B. germinal

    C. epidermal

    D. subcutaneous

_____ **4.** The hair follicles, sweat glands, and sebaceous glands are found in the:

    A. dermis.

    B. germinal layer.

    C. epidermis.

    D. subcutaneous layer.

_____ **5.** The skin regulates temperature in a cold environment by:

    A. secreting sweat through sweat glands.

    B. constricting the blood vessels.

    C. dilating the blood vessels.

    D. increasing the amount of heat that is radiated from the body's surface.

_____ **6.** Closed soft-tissue injuries are characterized by all of the following except:

    A. pain at the site of injury.

    B. swelling beneath the skin.

    C. damage of the protective layer of skin.

    D. a history of blunt trauma.

_____ **7.** A(n) _____ occurs whenever a large blood vessel is damaged and bleeds.

    A. contusion

    B. hematoma

    C. crushing injury

    D. avulsion

_____ **8.** A(n) _____ is usually associated with extensive tissue damage.

    A. contusion

    B. hematoma

    C. crushing injury

    D. avulsion

_____ **9.** A hematoma can result from:

    A. a soft-tissue injury.

    B. a fracture.

    C. any injury to a large blood vessel.

    D. all of the above

_____ **10.** A(n) _____ occurs when a great amount of force is applied to the body for a long period of time.

    A. contusion

    B. hematoma

    C. crushing injury

    D. avulsion

_____ **11.** More extensive closed injuries may involve significant swelling and bleeding beneath the skin, which could lead to:

    A. compartment syndrome.

    B. contamination.

    C. hypovolemic shock.

    D. hemothorax.

_____ **12.** You respond to a 14-year-old female who was playing softball and slid into second base. She states she felt and heard a loud pop. There is no obvious bleeding but there is swelling present. Her pulse is 86 beats/min and her blood pressure is 114/74 mm Hg. You decide you can manage this situation and decide to use the RICES method of treatment. The S stands for:

    A. swelling.

    B. soft-tissue.

    C. splinting.

    D. shock.

_____ **13.** Open soft-tissue wounds include all of the following, except:

    A. abrasions.

    B. contusions.

    C. lacerations.

    D. avulsions.

_____ **14.** In doing a more detailed exam on your patient in question 12, you notice that she has an abrasion on her left knee that she sustained when she slid. The abrasion is covered with dirt and is oozing blood. You know that this injury is classified as _____:

    A. superficial.

    B. deep.

    C. full-thickness.

    D. none of the above

_____ **15.** A laceration may be:

    A. linear.

    B. deep.

    C. jagged.

    D. all of the above

_____ **16.** You decide to manage the injury for the patient in question 14. You flush the site with sterile water and it continues to bleed. What would be the best way to control the bleeding from the site?

    A. Elevation

    B. Pressure dressings

    C. Tourniquets

    D. Pressure points

_____ **17.** You respond to a 24-year-old man who has been shot. Law enforcement is on scene and the scene is safe. As you approach the victim, you notice that he is bleeding from the lower right abdominal area. He is alert and oriented but seems confused. His airway is open and he is breathing at a normal rate. His pulse is 120 beats/min, weak and regular. His blood pressure is 98/60 mm Hg. You ask the police officer about the weapon. You need the information due to the fact that the amount of damage is related to the:

    A. size of the entrance wound.

    B. size of the bullet.

    C. size of the exit wound.

    D. speed of the bullet.

_____ **18.** Because shootings usually end up in court, it is important to factually and completely document:

    A. the circumstances surrounding any gunshot injury.

    B. the patient's condition.

    C. the treatment given.

    D. all of the above

_____ **19.** All open wounds are assumed to be _____ and present a risk of infection.

    A. contaminated

    B. life-threatening

    C. minimal

    D. extensive

_____ **20.** Before you begin caring for a patient with an open wound, you should:

    A. survey the scene.

    B. follow BSI precautions.

    C. be sure the patient has an open airway.

    D. all of the above

_____ **21.** Splinting an extremity even when there is no fracture can help by:

    A. reducing pain.

    B. minimizing damage to an already-injured extremity.

    C. making it easier to move the patient.

    D. all of the above

_____ **22.** Treatment for an abdominal evisceration includes:

    A. pushing the exposed organs back into the abdominal cavity.

    B. covering the organs with dry dressings.

    C. flexing the knees and legs to relieve pressure on the abdomen.

    D. applying moist, adherent dressings.

_____ **23.** An open neck injury may result in _____ if enough air is sucked into a blood vessel.

    A. hypovolemic shock

    B. tracheal deviation

    C. air embolism

    D. subcutaneous emphysema

_____ **24.** Burns may result from:

    A. heat.

    B. toxic chemicals.

    C. electricity.

    D. all of the above

_____ **25.** Factors in helping to determine the severity of a burn include:

    A. the depth of the burn.

    B. the extent of the burn.

    C. whether or not there are critical areas involved.

    D. all of the above

_____ **26.** _____ burns involve only the epidermis.

   A. Full-thickness

   B. Second-degree

   C. Superficial

   D. Third-degree

_____ **27.** _____ burns cause intense pain.

   A. First-degree

   B. Second-degree

   C. Superficial

   D. Third-degree

_____ **28.** _____ burns may involve subcutaneous layers, muscle, bone, or internal organs.

   A. Superficial

   B. Partial-thickness

   C. Full-thickness

   D. Second-degree

_____ **29.** You respond to a house fire with the local fire department. They bring a 48-year-old woman out of the house. She is unconscious but her airway is open. Her breathing is shallow and at 30 beats/min. Her pulse is 110 beats/min strong and regular. Her blood pressure is 108/72 mm Hg. She has been burned over 40% of her body. The burned area appears to be dry and leathery. It looks charred and has pieces of fabric embedded in the flesh. You know that this type of burn is called:

   A. first-degree.

   B. second-degree.

   C. partial-thickness.

   D. third-degree.

_____ **30.** Significant airway burns may be associated with:

   A. singeing of the hair within the nostrils.

   B. hoarseness.

   C. hypoxia.

   D. all of the above

_____ **31.** The most important consideration when dealing with electrical burns is:

   A. BSI precautions.

   B. scene safety.

   C. level of responsiveness.

   D. airway.

_____ **32.** Treatment of electrical burns includes:

   A. maintaining the airway.

   B. monitoring the patient closely for respiratory or cardiac arrest.

   C. splinting any suspected injuries.

   D. all of the above

_____ **33.** All of the following, except _____, may be used as an occlusive dressing:

   A. gauze pads

   B. Vaseline gauze

   C. aluminum foil

   D. plastic

_____ **34.** Using elastic bandages to secure dressings may result in _____ if the injury swells or if improperly applied.

   A. additional tissue damage

   B. loss of a limb

   C. impaired circulation

   D. all of the above

Labeling

Label the following diagrams with the correct terms.

1. The Skin

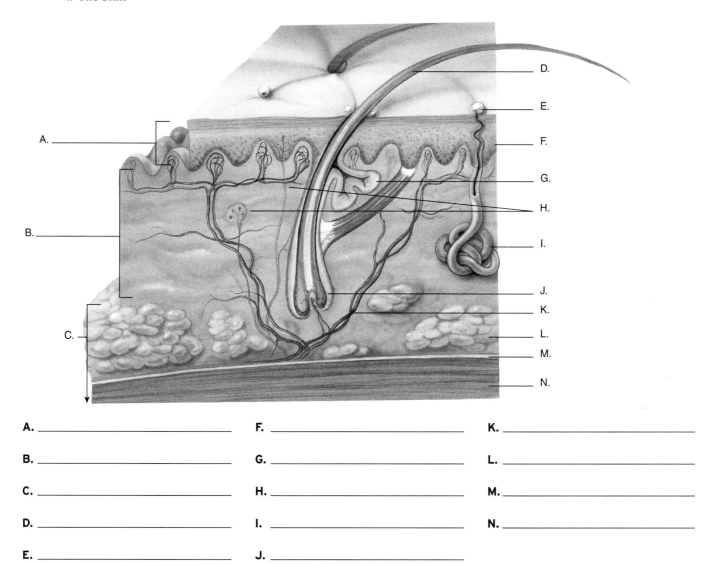

A. _____   F. _____   K. _____

B. _____   G. _____   L. _____

C. _____   H. _____   M. _____

D. _____   I. _____   N. _____

E. _____   J. _____

2. The Rule of Nines

A. _____

B. _____

C. _____

D. _____

E. _____

F. _____

G. _____

H. _____

A. _____

B. _____

C. _____

D. _____

E. _____

F. _____

G. _____

H. _____

Fill-in

Read each item carefully, then complete the statement by filling in the missing word(s).

1. Mucous membranes are _____.

2. A person will sweat in an effort to _____ the body.

3. Nerve endings are located in the _____.

4. Below the dermis lies the _____ tissue.

5. In cold weather, skin blood vessels will _____.

6. The skin protects the body by keeping _____ out and _____ in.

7. Because nerve endings are present, injury to the _____ may be painful.

8. A major function of the skin is regulating body _____.

9. The external layer of skin is the _____ and the inner layer is the _____.

10. When the vessels of the skin dilate, heat is _____ from the body.

True/False

If you believe the statement to be more true than false, write the letter "T" in the space provided. If you believe the statement to be more false than true, write the letter "F".

_____ **1.** Partial-thickness burns involve the epidermis and some portion of the dermis.

_____ **2.** Blisters are commonly seen with superficial burns.

_____ **3.** Severe burns are usually a combination of superficial, partial-thickness, and full-thickness burns.

_____ **4.** The Rule of Nines allows you to estimate the percentage of body surface area that has been burned.

_____ **5.** Two factors, depth and extent, are critical in assessing the severity of a burn.

_____ **6.** Your first responsibility with a burn patient is to stop the burning process.

_____ **7.** Burned areas should be immersed in cool water for up to 30 minutes.

_____ **8.** Electrical burns are always more severe than the external signs indicate.

_____ **9.** The universal dressing is ideal for covering large open wounds.

_____ **10.** Occlusive dressings are usually made of Vaseline gauze, aluminum foil, or plastic.

_____ **11.** Gauze pads prevent air and liquids from entering or exiting the wound.

_____ **12.** Elastic bandages can be used to secure dressings.

_____ **13.** Soft roller bandages are slightly elastic and the layers adhere somewhat to one another.

_____ **14.** Ecchymosis is associated with open wounds.

_____ **15.** A laceration is considered a closed wound.

Short Answer

Complete this section with short written answers using the space provided.

1. List the three major classifications of depth of burns.

_____

_____

2. List the three general classifications of soft-tissue injuries.

_____

_____

3. Define the acronym RICES.

R: _____

I: _____

C: _____

E: _____

S: _____

4. Describe the classifications of a critical burn for an infant or child.

_____

_____

5. What treatment should be used with a patient burned by a dry chemical?

_____

_____

6. Why are electrical burns particularly dangerous to a patient?

_____

_____

7. Describe a sucking chest wound.

_____

_____

8. List the three primary functions of dressings and bandages.

_____

_____

9. List the four types of open soft-tissue injuries.

_____

_____

10. List the five factors used to determine the severity of a burn.

_____

_____

Word Fun

The following crossword puzzle is an activity provided to reinforce correct spelling and understanding of medical terminology associated with emergency care and the EMT-B. Use the clues in the column to complete the puzzle.

Across

4. Displacement of organs outside body

6. Lining of body cavities and passages with contact to outside

7. Torn completely loose or hanging as a flap

8. Presence of infective organisms

9. Inner layer of skin

10. Bruise

11. Blood collected in soft tissues

Down

1. Discoloration associated with closed injury

2. Injury from sharp, pointed object

3. Assigns percentages to burns

5. Scraping wound

Ambulance Calls

The following case scenarios provide an opportunity to explore the concerns associated with patient management. Read each scenario, then answer each question in detail.

1. You are dispatched to a residence where a 10-year-old girl fell onto a jagged piece of metal and has a gaping laceration to the right upper arm that is spurting bright red blood. The mother tried to control bleeding with a towel, but it kept soaking through.

   How would you best manage this patient?

   _____

   _____

   _____

2. You are dispatched to the home of a 3-year-old for an unknown problem. You arrive to find a young mother screaming for your help. Apparently, she was cooking a meal for her other children when the phone rang. While she was talking, the 3-year-old child grabbed the pot handle, pulling boiling hot water onto his body.

   How do you best manage this patient?

   _____

   _____

   _____

3. You are watching TV at the station when a fire fighter comes into the room holding his left hand. His wedding ring got caught in a piece of small machinery at the stationhouse, resulting in a 'degloving' of his ring finger.

   How do you best manage this patient?

   _____

   _____

   _____

Skill Drills

Skill Drill 24-1: Controlling Bleeding from a Soft-Tissue Injury
Test your knowledge of this skill drill by filling in the correct words in the photo captions.

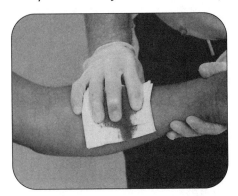

1. Apply _____ _____
   with a _____ bandage.

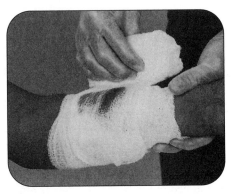

2. Maintain pressure with a
   _____ bandage.

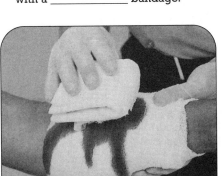

3. If bleeding continues, apply a second
   _____ and _____
   bandage over the first.

4. _____ the extremity.

Skill Drill 24-2: Stabilizing an Impaled Object
Test your knowledge of this skill drill by filling in the correct words in the photo captions.

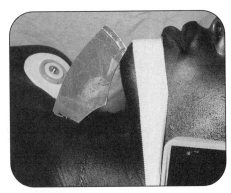

1. Do not attempt to _____ or
   _____ the object. Stabilize
   the impaled body part.

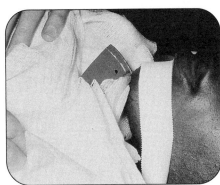

2. Control _____ and
   _____ the object in place
   using _____ _____,
   _____, and/or _____.

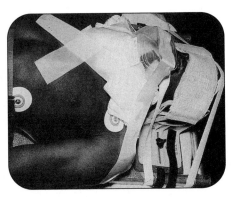

3. Tape a _____ item over the
   stabilized object to protect it from
   _____ during transport.

Skill Drill 24- 3: Caring for Burns
Test your knowledge of this skill drill by placing the photos below in the correct order. Number the first step with a "1", the second step with a "2", etc.

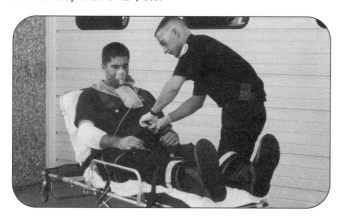

_____

Prepare for transport.

Treat for shock.

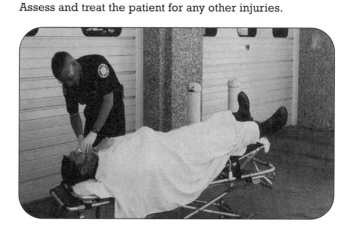

_____

Estimate the severity of the burn, then cover the area with a dry, sterile dressing or clean sheet.

Assess and treat the patient for any other injuries.

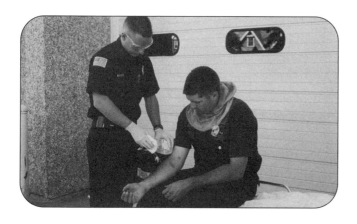

_____

Follow BSI precautions to help prevent infection.

If it is safe to do so, remove the patient from the burning area; extinguish or remove hot clothing and jewelry as needed.

If the wound(s) is still burning or hot, immerse the hot area in cool, sterile water, or cover with a wet, cool dressing.

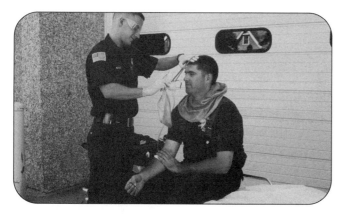

_____

Provide high-flow oxygen and continue to assess the airway.

Cover the patient with blankets to prevent loss of body heat. Transport promptly.

Assessment Review

Answer the following questions pertaining to the assessment of the types of emergencies discussed in this chapter.

_____ 1. During the initial assessment of burns, it is important to remember to:

    A. determine scene safety.

    B. obtain vital signs.

    C. prevent heat loss.

    D. estimate the amount of body surface injuries.

_____ 2. You have been dispatched to a residence for a 24-year old woman who splashed cooking grease on her arm while cooking. As you approach her she is crying and yelling that it hurts. Her pulse is 130 beats/min and regular. Her blood pressure is 126/86 mm Hg. You decide that she is not an immediate transport. The focused history and physical exam should be:

    A. rapid trauma assessment.

    B. an examination of the burned arm.

    C. a detailed physical exam.

    D. done during transport to the hospital.

_____ 3. During the rapid trauma assessment of a burn patient with a significant mechanism of injury you should:

    A. splint all fractures.

    B. determine the transport decision.

    C. open all blisters.

    D. estimate the amount of body surface injured.

_____ 4. During the intervention step of patient assessment for a burn patient, the first intervention should be:

    A. stop the burning process.

    B. provide complete spinal stabilization.

    C. treat for shock.

    D. cover burns with moist sterile dressings.

_____ 5. You respond to a patient who has been stabbed in the neck. You arrive to find the patient in police custody and bleeding moderately from the neck wound. The patient is alert, oriented, and swearing loudly. His pulse is 120 beats/min and blood pressure is 124/76 mm Hg. You start to bandage the wound. What type of bandage should you use?

    A. Triangular bandage

    B. Adhesive bandage

    C. Roller bandage

    D. Occlusive bandage

Emergency Care Summary

Fill in the following chart pertaining to the management of the types of emergencies discussed in this chapter.

NOTE: While the steps below are widely accepted, be sure to consult and follow your local protocol.

Closed Injuries	Open Injuries	Burns
1. Keep the patient as quiet comfortable as possible. 2. Apply ice (or cold packs). 3. Apply direct pressure. 4. Elevate the injured part just above the level of the patient's _____. 5. Splint the injured area.	1. Apply direct pressure with a sterile bandage. 2. Maintain pressure with a _____ bandage. 3. If bleeding continues, apply a second dressing and _____ bandage over the first. 4. Splint the extremity.	1. Follow BSI precautions. 2. Move the patient away from the burning area. 3. Innerse the burned skin in cool, _____ water. 4. Provide high-flow oxygen. 5. Cover the patient with a clean blanket. 6. Rapidly estimate the burn's severity. 7. Check for _____ injuries. 8. Treat the patient for shock. 9. Provide prompt transport.

Abdominal Injuries	Impaled Objects	Neck Wounds
1. Do not touch or move exposed organs. 2. Keep organs moist. Use moist sterile dressing, cover and secure in place. 3. If the patient's legs and knees are uninjured, flex them to relieve pressure on the _____.	1. Do not attempt to move or remove the object. 2. Control bleeding and stabilize the object in place using soft dressings _____, and/or tape. 3. Tape a rigid item over the stabilized object to protect it from movement during transport.	1. Cover wound with _____ dressing. 2. Apply manual pressure, but do not compress both _____ vessels at the same time. 3. Secure dressing over the wound.

Chemical Burns	Electrical Burns	Small Animal and Human Bites
1. Stop the burning process; safely remove any chemical from the patient, always brushing off a dry chemical. 2. Remove all of patient's clothing. 3. Flush the burn area with large amounts of water for _____ to _____ minutes after the patient says the burning has stopped.	1. Ensure that the scene is safe. 2. If indicated, begin CPR and apply the _____. 3. Treat soft-tissue areas by placing dry, sterile dressings on all burn wounds and splinting fractures.	1. Promptly stabilize the area with a splint or bandage. 2. Apply a dry, sterile dressing. 3. Provide transport to the emergency department for surgical cleansing of the wound and _____ therapy.

CHAPTER

25 Eye Injuries

Workbook Activities

The following activities have been designed to help you. Your instructor may require you to complete some or all of these activities as a regular part of your EMT-B training program. You are encouraged to complete any activity that your instructor does not assign as a way to enhance your learning in the classroom.

Chapter Review

The following exercises provide an opportunity to refresh your knowledge of this chapter.

Matching

Match each of the terms in the left column to the appropriate definition in the right column.

_____ 1. Cornea   **A.** membrane that covers the exposed surface of the eye

_____ 2. Iris   **B.** focuses light

_____ 3. Lens   **C.** muscle behind cornea

_____ 4. Pupil   **D.** transparent tissue in front of pupil and iris

_____ 5. Sclera   **E.** circular opening in iris

_____ 6. Orbit   **F.** the eyeball

_____ 7. Conjunctiva   **G.** eye socket

_____ 8. Globe   **H.** the light-sensitive area of the eye where images are projected

_____ 9. Lacrimal glands   **I.** tear glands

_____ 10. Retina   **J.** white portion of eye

Multiple Choice

Read each item carefully, then select the best response.

_____ **1.** The conjunctiva covers the:
- **A.** outer surface of the eyelids.
- **B.** exposed surface of the eye.
- **C.** lens.
- **D.** iris.

_____ **2.** The purpose of the _____, an extremely tough, fibrous tissue, is to help maintain the eye's globular shape.
- **A.** cornea
- **B.** lens
- **C.** retina
- **D.** sclera

_____ **3.** The _____ allows light to enter the eye.
- **A.** cornea
- **B.** lens
- **C.** retina
- **D.** sclera

_____ **4.** The circular muscle that adjusts the size of the opening behind the cornea to regulate the amount of light that enters the eye is called the:
- **A.** iris.
- **B.** lens.
- **C.** retina.
- **D.** sclera.

_____ **5.** You respond to a 78-year-old man who is unresponsive. His pulse is 126 beats/min and blood pressure is 184/96 mm Hg. His respirations are 22 breaths/min and regular. His wife tells you he was complaining of a severe headache. She tells you that he told her it was the worst headache he ever had and then he collapsed. During your focused exam you notice that one pupil is constricted and the other dilated. You know that a normal, uninjured eye would have pupils that:
- **A.** are round.
- **B.** are equal in size.
- **C.** react equally when exposed to light.
- **D.** all of the above

_____ **6.** Important signs and symptoms to record include all of the following, except:
- **A.** how the injury occurred.
- **B.** any changes in vision.
- **C.** any history of color blindness.
- **D.** the use of any eye medications.

_____ **7.** The delicate tissues of the eye can be burned by _____, often causing permanent damage.
- **A.** chemicals
- **B.** heat
- **C.** light rays
- **D.** all of the above

_____ **8.** You respond with the local fire department to a house fire. The fire fighters bring you a 25-year-old woman who was trying to light a propane stove when it exploded in her face. She is alert and oriented and able to converse with you. Her pulse rate is 110 beats/min and her blood pressure is 126/76 mm Hg. She is complaining that her eyes burn and it feels like she has gravel in her eyes. You decide that you need to bandage the injury. You need to remember to:

  A. cover both eyes with a sterile dressing moistened with sterile saline.

  B. apply eye shields over the dressing.

  C. transport promptly.

  D. all of the above

_____ **9.** You are called to the residence of a 48-year-old man who was watching a solar eclipse. He was looking directly at the eclipse without using protective eyewear. He states that it feels like there is something in his eye but has not been able to find anything. His vital signs are all within normal limits. Your assessment reveals no apparent injury. The patient states that he does not want to go to the hospital. You try and convince him to go because you know that retinal damage caused by exposure to extremes of light are not _____ but may result in permanent damage to vision.

  A. deep

  B. painful

  C. lacerated

  D. none of the above

_____ **10.** Superficial burns of the eyes can result from:

  A. light from prolonged exposure to a sunlamp.

  B. reflected light from a bright snow-covered area.

  C. ultraviolet rays from an arc welding unit.

  D. all of the above

_____ **11.** You respond to a 38-year-old man who is employed in a local metal shop. A piece of metal hit him in the face, lacerating the eyelid. He is alert and oriented but extremely anxious. His pulse is 140 beats/min and his blood pressure is 138/86 mm Hg. His respirations are 24 breaths/min and regular. He has heavy bleeding coming from the eye. You know that most bleeding from the eyelid can be controlled by:

  A. firm pressure.

  B. gentle pressure.

  C. pressure dressings.

  D. flushing the eyes.

_____ **12.** You respond to the same patient in question 11 but this time the patient has lacerated the globe of the eye. How much pressure would you apply to control the bleeding?

  A. Firm pressure

  B. Gentle pressure

  C. No pressure

  D. Pressure dressings

_____ **13.** Important guidelines in treating penetrating injuries of the eye include all of the following, except:

  A. Never exert pressure on or manipulate the globe in any way.

  B. If part of the eyeball is exposed, gently apply a moist, sterile dressing to prevent drying.

  C. Cover the injured eye with a protective metal eye shield or sterile dressing.

  D. Gently replace the eyeball if it is displaced out of its socket.

_____ **14.** A black eye is a result of:

  A. bleeding into tissue around the orbit.

  B. a fracture of the orbit.

  C. a torn retina.

  D. none of the above

_____ **15.** Bleeding into the anterior chamber of the eye that obscures part or all of the iris is called:

    A. hematemesis.

    B. hyphema.

    C. hyphoma.

    D. hemoptysis.

_____ **16.** Fracture of the orbit, particularly of the bones that form its floor and support of the globe is known as a:

    A. blowout fracture.

    B. retinal detachment.

    C. hyphema.

    D. black eye.

_____ **17.** Eye findings that should alert you to the possibility of a head injury include:

    A. one pupil larger than the other.

    B. bleeding under the conjunctiva.

    C. protrusion or bulging of one eye.

    D. all of the above

_____ **18.** The only time that contact lenses should be removed immediately in the field is in the case of a _____ the eye.

    A. blowout fracture of

    B. retinal detachment of

    C. chemical burn in

    D. broken contact lens in

Fill-in

Read each item carefully, then complete the statement by filling in the missing word(s).

1. The glands that produce fluids to keep the eye moist are called _____ _____.

2. A cranial nerve that transmits visual information to the brain is called an _____ _____.

3. The eye works like a _____, with the iris and pupil making adjustments to light and the retina acting like film.

4. Never remove contact lenses from an injured eye unless the injury is a _____ _____.

5. The _____ is composed of the adjacent bones of the face and skull.

6. When performing an examination, you are looking for specific _____ or conditions that may suggest the nature of the problem.

7. Large objects are prevented from penetrating the eye by the protective _____ that surrounds it.

8. _____ is inflammation and redness of the conjunctiva.

9. The delicate membrane that covers the inside of the eyelid and the surface of the eye is the _____.

10. The tough, fibrous tissue that helps maintain the eye's globular shape is called the _____.

True/False

If you believe the statement to be more true than false, write the letter "T" in the space provided. If you believe the statement to be more false than true, write the letter "F."

_____ **1.** Objects impaled in the eye should be removed before applying dressing.

_____ **2.** Vitreous humor can be replaced.

_____ **3.** Aqueous humor can be replaced.

_____ **4.** Lacrimal glands help keep the eye dry.

_____ **5.** Contact lenses should always be removed in the field.

_____ **6.** Foreign objects stuck to the cornea should be removed prior to transport.

_____ **7.** Bleeding soon after irritation or injury can result in a bright yellow conjunctiva.

_____ **8.** In a normal, uninjured eye the entire circle of the iris is visible.

_____ **9.** If a small foreign object is lying on the surface of the patient's eye, you should use a dextrose solution to gently irrigate the eye.

_____ **10.** The iris gives the eye its characteristic color.

Short Answer

Complete this section with short written answers using the space provided.

1. Describe a retinal detachment, including common signs and symptoms.

_____

_____

2. List the three guidelines for treating penetrating eye injuries.

_____

_____

3. List assessment findings in the eye that may indicate head injury.

_____

_____

Word Fun

The following crossword puzzle is an activity provided to reinforce correct spelling and understanding of medical terminology associated with emergency care and the EMT-B. Use the clues in the column to complete the puzzle.

Across

2. White portion of eye
5. Membrane that lines eyelids and surface of eye
6. Eye socket
7. Bleeding into anterior chamber of eye
8. Transmits visual sensations to the brain
9. Muscle regulating light into eye

Down

1. Break in the bones of the orbit
3. Tear producer
4. Transparent tissue in front of pupil

Ambulance Calls

The following case scenarios provide an opportunity to explore the concerns associated with patient management. Read each scenario, then answer each question in detail.

1. You are dispatched to a local school playground for an eye injury. Two of the children were "play sword fighting" with two sticks. One child lunged forward with the stick, accidentally poking the other in the eye. You arrive to find the child holding his eye and crying vigorously.

   How do you best manage this patient?

   _____

   _____

   _____

2. You are called to an industrial site where a 32-year-old man had acid splashed in his eyes. Coworkers have tried pouring water into his eyes, but he is still unable to see and is in severe pain.

   How would you best manage this patient?

   _____

   _____

   _____

3. You are dispatched to a construction site for an "injury to the face." You arrive to find a young man who was accidentally hit in the face with a ladder. He tells you that the end of the ladder hit him in the face when a coworker swung around to see someone who called his name. Your patient is complaining of pain with eye movement and double vision.

How do you best manage this patient?

_____

_____

_____

Skill Drills

Skill Drill 25-1: Removing a Foreign Object From Under the Upper Eyelid
Test your knowledge of skill drills by filling in the correct words in the photo captions.

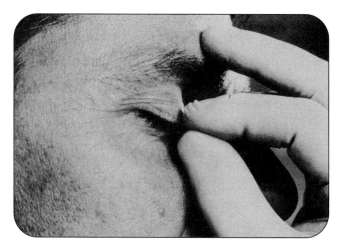

1. Have the patient look _____, grasp the _____ _____, and gently pull the lid away from the eye.

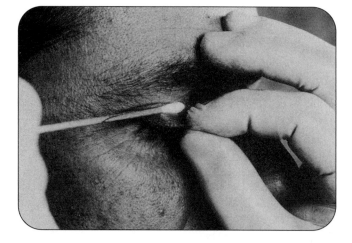

2. Place a _____ _____ on the outer surface of the upper lid.

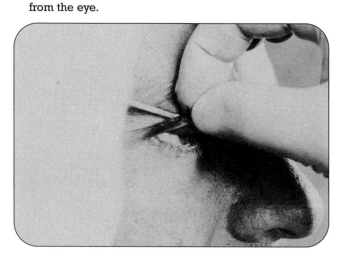

3. Pull the lid _____ and _____, folding it back over the applicator.

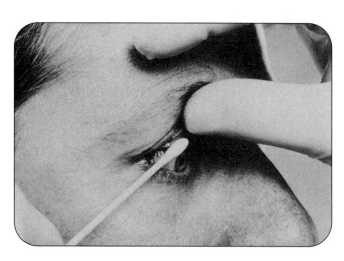

4. Gently remove the foreign object from the eyelid with a _____, _____ applicator.

Skill Drill 25-2: Stabilizing a Foreign Object Impaled in the Eye

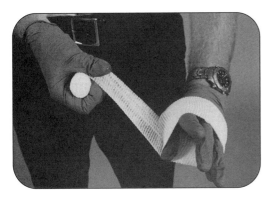

1. To prepare a doughnut ring, wrap a
_____ roll around your fingers and
thumb _____ or _____
times. Adjust the diameter by
_____ your fingers.

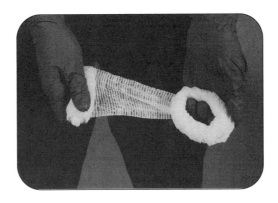

2. Wrap the remainder of the roll . . .

3. . . . working around the ring.

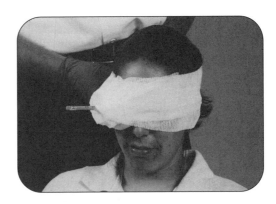

4. Place the dressing over the _____
to hold the impaled object in place, then
_____ it with a _____
dressing.

Assessment Review

Answer the following questions pertaining to the assessment of the types of emergencies discussed in this chapter.

_____ 1. You have been dispatched to an 18-year-old male patient who has a blowout fracture of the left orbit. He is alert and oriented. His respirations, pulse, and blood pressure are within normal limits. During the rapid trauma assessment, what intervention should be performed?

    A. Spinal immobilization to a backboard

    B. Airway management

    C. Hemorrhage control

    D. AVPU

_____ 2. You respond to a 23-year-old woman who was an unrestrained passenger in an automobile collision. The windshield is spider webbed and she has significant facial damage. You are in the focused history and physical exam, how do you get a SAMPLE history?

    A. Family members

    B. Friends

    C. Bystanders

    D. All of the above

_____ 3. Your patient is a factory worker who has metal shavings in his right eye. You have assessed the airway, breathing, and circulation and all are within the normal limits for a patient his age. You decide that bandaging is necessary; how would you accomplish this?

    A. Bandage the right eye only.

    B. Bandage the left eye only.

    C. Bandage both eyes.

    D. Apply a pressure bandage.

_____ 4. You respond to the local high school where a 16-year-old female working in the chemistry lab has gotten an unknown substance in her eyes. She is alert and oriented. She is able to converse with you and tells you her eyes are burning. Her pulse is 124 beats/min, respirations 18 breaths/min, and blood pressure is 118/76 mm Hg. She was cleaning the shelf in the storage room when it collapsed and several chemicals fell on top of her. How long should you irrigate the eyes?

    A. 2 to 3 minutes

    B. 5 to 20 minutes

    C. 20 to 30 minutes

    D. 30 to 45 minutes

_____ 5. Your patient is a 15-year-old who was playing baseball and was hit in the right eye. The impact resulted in a blowout fracture with the eyeball partially exposed. How should you bandage the injury?

    A. Dry sterile dressing

    B. Moist sterile dressing

    C. Occlusive dressing

    D. Multitrauma dressing

Emergency Care Summary

Fill in the following chart pertaining to the management of the types of emergencies discussed in this chapter.

NOTE: While the steps below are widely accepted, be sure to consult and follow your local protocol.

Foreign Objects	Burns	Lacerations	Blunt Trauma
1. Tell the patient to look down while you grasp the _____ of the upper eyelid with your thumb and index finger. Gently pull the eyelid away from the eyeball. 2. Gently place a cotton-tipped applicator horizontally along the center of the outer surface of the upper eyelid. 3. If you see a foreign object, gently remove it with a moistened, sterile, cotton-tipped applicator. **Stabilizing a Foreign Object Impaled in the Eye** 1. To prepare a doughnut rung, wrap a _____ around your fingers and thumb seven or eight times. 2. Wrap the remainder of the roll, working around the ring. 3. Place the dressing over the eye to hold the impaled object in place and then secure it with a gauze dressing.	**Chemical Burns** 1. Holding the eyelid open, irrigate the eye for _____ to _____ minutes. 2. Apply a clean, dry sterile dressing to cover the eye. **Thermal Burns** 1. Cover both eyes with a sterile dressing moistened with sterile _____. **Light Burns** 1. Cover each eye with a sterile, moist pad and eye shield. 2. Transport supine and prevent further exposure to bright light.	1. Never exert pressure on or manipulate the injured eye in any way. 2. If part of the eyeball is exposed, gently apply a _____ sterile dressing to prevent drying. 3. Cover the injured eye with a protective metal eye shield or sterile dressing.	1. Protect the eye from further injury with a _____ _____. 2. Cover the other eye to minimize movement on the injured side.

CHAPTER

26 Face and Throat Injuries

Workbook Activities

The following activities have been designed to help you. Your instructor may require you to complete some or all of these activities as a regular part of your EMT-B training program. You are encouraged to complete any activity that your instructor does not assign as a way to enhance your learning in the classroom.

Chapter Review

The following exercises provide an opportunity to refresh your knowledge of this chapter.

Matching

Match each of the terms in the left column to the appropriate definition in the right column.

_____ **1.** Cranium   **A.** upper jaw

_____ **2.** Occiput   **B.** posterior cranium

_____ **3.** Pinna   **C.** contains the brain

_____ **4.** Zygomas   **D.** pull/tear away

_____ **5.** Mandible   **E.** visible part of ear

_____ **6.** Avulse   **F.** cheekbones

_____ **7.** Maxilla   **G.** lower jawbone

Multiple Choice

Read each item carefully, then select the best response.

_____ 1. As an EMT-B, your objective is to:

    A. prevent further injury.

    B. manage any acute airway problems.

    C. control bleeding.

    D. all of the above

_____ 2. The head is divided into two parts: the cranium and the:

    A. brain.

    B. face.

    C. skull.

    D. medulla oblongata.

_____ 3. The brain connects to the spinal cord through a large opening at the base of the skull known as the:

    A. eustachian tube.

    B. spinous process.

    C. foramen magnum.

    D. vertebral foramina.

_____ 4. Approximately _____ of the nose is composed of bone. The remainder is composed of cartilage.

    A. 9/10

    B. 2/3

    C. 3/4

    D. 1/3

_____ 5. Motion of the mandible occurs at the:

    A. temporomandibular joint.

    B. mastoid process.

    C. chin.

    D. mandibular angle.

_____ 6. You respond to a 71-year-old woman who is unresponsive. You try to get her to respond but have no success. Her airway is open and she is breathing at a rate of 14 breaths/min. You know you can check a pulse in either side of the neck. You know that the jugular veins and several nerves run through the neck next to the trachea. What structure are you trying to locate to take a pulse?

    A. Hypothalamus

    B. Subclavian arteries

    C. Cricoid cartilage

    D. Carotid arteries

_____ 7. The _____ connects the cricoid cartilage and thyroid cartilage.

    A. larynx

    B. cricoid membrane

    C. cricothyroid membrane

    D. thyroid membrane

_____ 8. You respond to a 68-year-old man who was involved in a motor vehicle collision. He is unresponsive and as you approach you notice he is not breathing. He was unrestrained and has massive facial injuries. When you check his airway, it is obstructed. You know that upper airway obstructions can be caused by:

    A. heavy bleeding.

    B. loosened teeth or dentures.

    C. soft-tissue swelling.

    D. all of the above

_____ **9.** You are dispatched to a residential neighborhood for a 6-year-old who was bitten by the family pet. The mother meets you at the door with the child who is crying uncontrollably and has blood covering the right side of her head. You look at the child and notice that the lower right ear has been completely avulsed. You control the bleeding with direct pressure and bandage the injury. You follow the blood trail back to where the incident occurred and find the avulsed part. How do you manage the avulsed tissue?

    A. Wrap the skin in a moist, sterile dressing/place in a plastic bag in ice water/transport with patient.

    B. Place the skin in plastic "biohazard" bag and dispose of properly.

    C. Place the skin in a plastic bag filled with ice and transport to the ER.

    D. Leave it at the scene to be disposed of later.

_____ **10.** The nasal cavity is divided into two chambers by the:

    A. frontal sinus.

    B. middle turbinate.

    C. zygoma.

    D. nasal septum.

_____ **11.** You are called to the home of a 48-year-old woman who has a history of high blood pressure and now has a major nose bleed. She is alert and oriented and converses freely with you. Her respirations and pulse are within normal limits. Her blood pressure is 194/108 mm Hg. You have been able to rule out trauma. How would you manage the nose bleed?

    A. Apply a sterile dressing.

    B. Pinch the nostrils together.

    C. Put the patient in a supine position.

    D. Have the patient hold ice in her mouth.

_____ **12.** The middle ear is connected to the nasal cavity by the:

    A. frontal sinus.

    B. zygomatic process.

    C. eustachian tube.

    D. superior trachea.

_____ **13.** You are called to a motor vehicle collision that was the result of a lateral impact. The patient is conscious but extremely confused. The airway is open and the patient is breathing at 16 breaths/min and unlabored. The pulse is 110 beats/min strong and regular. The blood pressure is 136/90 mm Hg. The patient has bleeding from the left temple area. You know that serious bleeding from a facial fracture can be life-threatening and that obstruction of the upper airway may be caused by:

    A. leakage of CSF from the ears and nose.

    B. blood clots.

    C. mucus.

    D. periorbital ecchymosis.

_____ **14.** Signs of a possible facial fracture include:

    A. bleeding in the mouth.

    B. absent or loose teeth.

    C. loose and/or moveable bone fragments.

    D. all of the above

_____ **15.** The presence of air in the soft tissues of the neck that produces a crackling sensation is called:

    A. the "Rice Krispy" effect.

    B. a pneumothorax.

    C. rales.

    D. subcutaneous emphysema.

_____ **16.** Most bleeding from the neck can be controlled by:

    A. direct pressure.

    B. a pressure point.

    C. elevation.

    D. a tourniquet.

Labeling
Label the following diagrams with the correct terms.

　　1. Face/Skull

A. _____ 　**F.** _____ 　**J.** _____

B. _____ 　**G.** _____ 　**K.** _____

C. _____ 　**H.** _____ 　**L.** _____

D. _____ 　**I.** _____ 　**M.** _____

E. _____

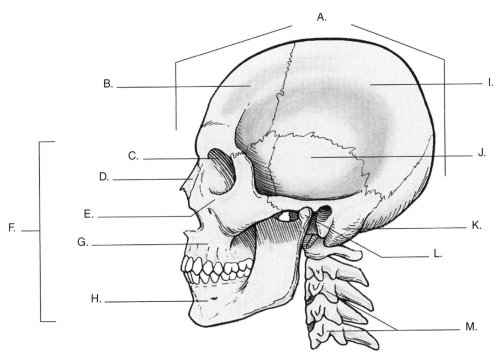

Fill-in
Read each item carefully, then complete the statement by filling in the missing word(s).

1. Pulsations in the neck are felt in the _____ vessels.

2. The _____ vertebrae are in the neck.

3. The _____ lobes of the cranium are located on the lateral portion of the head.

4. The _____ is located in the anterior portion of the neck.

5. The rings of the trachea are made of _____.

6. The Adam's Apple is more prominent in _____ than in _____.

7. The _____ _____ is a large opening at the base of the skull.

8. The _____ is the upper part of the jaw.

9. The _____ lobes lie laterally between the temporal and occipital lobes.

10. The _____ connects the larynx with the main air passage.

True/False

If you believe the statement to be more true than false, write the letter "T" in the space provided. If you believe the statement to be more false than true, write the letter "F."

_____ **1.** Injuries to the face often lead to airway problems.

_____ **2.** Care for facial injuries begins with BSI precautions and the ABCs.

_____ **3.** Exposed eye or brain injuries are covered with a dry dressing.

_____ **4.** Clear fluid in the outer ear is normal.

_____ **5.** Any crushing injury of the upper part of the neck likely involves the larynx or the trachea.

_____ **6.** Soft-tissue injuries to the face are common.

_____ **7.** The opening that the spinal core leaves the head through is called the occiput.

_____ **8.** The muscle that allows movement of the head is the temporomandibular.

_____ **9.** BSI for assessing face and throat injuries should include eye and oral protection.

_____ **10.** The airway of choice with facial injuries is the nasopharyngeal.

Short Answer

Complete this section with short written answers using the space provided.

1. Describe bleeding control methods for facial injuries.

_____

_____

2. Describe bleeding control methods for lacerations to veins or arteries in the neck.

_____

_____

Word Fun

The following crossword puzzle is a good way to review correct spelling and meaning of medical terminology associated with emergency care and the EMT-B. Use the clues in the column to complete the puzzle.

Across

1. Posterior portion of skull
4. Upper jawbone
5. Skull
8. Lower jawbone
10. Connects middle ear to nasal cavity
11. To pull or tear away
12. Blood accumulated in soft tissue

Down

2. Layers of bones in nasal cavity
3. Bony mass about 1″ posterior to ear
6. Air in the veins
7. Fleshy bulge anterior to ear canal
9. Large opening in base of skull

Ambulance Calls

The following case scenarios provide an opportunity to explore the concerns associated with patient management. Read each scenario, then answer each question in detail.

1. You are dispatched to assist a small child who was attacked by his family's dog. The dog bit his face and neck repeatedly, then grabbed him by the neck and shook him violently. The mother found the boy "making funny breathing sounds" and called for help. She has removed the dog from the area.

   How do you best manage this patient?

   _____

   _____

   _____

2. You are dispatched to a 37-year-old man with a large laceration to the right side of his neck. Bleeding is dark and heavy. He is alert, but weak.

   How would you best manage this patient?

   _____

   _____

   _____

3. You are dispatched to a Little League baseball game to assist an assault victim. Apparently, emotions were running high when two parents began to argue. You arrive to find a 40-year-old man with a bloody nose. How do you best manage this patient?

_____

_____

_____

Skill Drills
Skill Drill 26-1: Controlling Bleeding From a Neck Injury
Test your knowledge of skill drills by filling in the correct words in the photo captions.

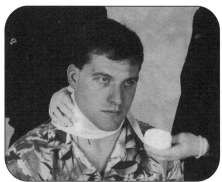

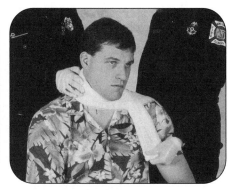

1. Apply _____ _____ to control bleeding.

2. Use _____ _____ to secure a dressing in place.

3. Wrap the bandage around and under the patient's _____.

Assessment Review

Answer the following questions pertaining to the assessment of the types of emergencies discussed in this chapter.

_____ **1.** You have responded to an automobile collision and found a 21-year-old man who has massive facial trauma. There is heavy bleeding and he is unconscious. The first thing that you do in your treatment of this patient is:
  A. take c-spine precautions.
  B. open the airway.
  C. assess his breathing.
  D. take BSI precautions.

_____ **2.** For the patient described above, how often would you reassess the vitals during your ongoing assessment?
  A. Every 3 minutes
  B. Every 5 minutes
  C. Every 10 minutes
  D. Every 15 minutes

_____ **3.** You have a patient who has severe epistaxis. You have been able to rule out trauma. How would you position this patient to help control the bleeding?
  A. Supine
  B. Prone
  C. Sitting leaning back
  D. Sitting leaning forward

_____ **4.** You have a patient who has had a tooth knocked out. You find the tooth; how would you transport it to the hospital?

    A. In the patient's saliva

    B. In dextrose

    C. In ice

    D. In a dry sterile dressing

_____ **5.** You respond to a child who has placed a pebble in his ear. He is complaining that it is hurting. You should:

    A. remove it with a Q-tip.

    B. have the child try and shake it out.

    C. leave it and transport.

    D. no load because this is not an emergency.

Emergency Care Summary

Fill in the following chart pertaining to the management of the types of emergencies discussed in this chapter.

NOTE: While the steps below are widely accepted, be sure to consult and follow your local protocol.

Nose Injuries	Ear Injuries	Facial Fractures	Neck Injuries
Controlling Epistaxis (See Skill Drill 22-4) 1. Position the patient sitting, leaning _____. 2. Apply direct pressure, pinching the fleshy part of the nostrils for 15 minutes. 3. Keep the patient calm and quiet. 4. Apply ice over the nose.	1. Place a soft, padded dressing between the ear and the scalp. 2. If the ear is avulsed, wrap it in a _____, sterile dressing and place in a plastic bag. 3. Leave any foreign object within the ear for the physician to remove. 4. Note any clear fluid coming from the ear.	1. Remove and save loose teeth and _____ fragment from the mouth and transport them with you. 2. Remove dentures and dental bridges to protect them against airway obstruction. 3. Maintain an open airway. 4. When transporting dislodged teeth, place them in a container of the patient's _____ or milk, if possible.	1. Apply direct pressure to control bleeding. 2. Use _____ _____ to secure a dressing in place. 3. Wrap the bandage around and under the patient's shoulder.

CHAPTER

27 Chest Injuries

Workbook Activities

The following activities have been designed to help you. Your instructor may require you to complete some or all of these activities as a regular part of your EMT-B training program. You are encouraged to complete any activity that your instructor does not assign as a way to enhance your learning in the classroom.

Chapter Review

The following exercises provide an opportunity to refresh your knowledge of this chapter.

Matching

Match each of the terms in the left column to the appropriate definition in the right column.

_____ **1.** Thoracic cage	**A.** chest rises
_____ **2.** Diaphragm	**B.** chest
_____ **3.** Exhalation	**C.** chest falls
_____ **4.** Inhalation	**D.** separates chest from abdomen
_____ **5.** Aorta	**E.** major artery in the chest
_____ **6.** Closed chest injury	**F.** penetrating wound
_____ **7.** Hemoptysis	**G.** rapid respirations
_____ **8.** Pericardium	**H.** unusually blunt trauma
_____ **9.** Open chest injury	**I.** coughing up blood
_____ **10.** Tachypnea	**J.** sac around the heart

Multiple Choice

Read each item carefully, then select the best response.

_____ 1. Air is supplied to the lungs via the:

A. esophagus.

B. trachea.

C. nares.

D. oropharnyx.

_____ 2. The _____ separates the thoracic cavity from the abdominal cavity.

A. diaphragm

B. mediastinum

C. xyphoid process

D. inferior border of the ribs

_____ 3. On inhalation, all of the following occur, except:

A. the intercostal muscles contract, elevating the rib cage.

B. the diaphragm contracts.

C. the pressure inside the chest increases.

D. air enters through the nose and mouth.

_____ 4. You respond to the local rodeo arena for a bull rider. The scene is safe and the patient is lying in the middle of the arena unconscious. His airway is open and he is breathing at 20 breaths/min. His pulse is 128 beats/min and blood pressure is 110/64 mm Hg. There is no obvious bleeding. Bystanders tell you he was thrown into the air and landed on the bull's head. He was not wearing a vest. You know that blunt trauma to the chest may:

A. bruise the lungs and heart.

B. fracture whole areas of the chest wall.

C. damage the aorta.

D. all of the above

_____ 5. You respond to a motor vehicle collision and have a 29-year-old woman who is complaining of chest pain. Her chest struck the steering wheel. Her airway is open, she is breathing at 24 breaths/min, and coughing up blood. Her pulse is 130 beats/min rapid and weak, and blood pressure is 90/58 mm Hg. You notice cyanosis around the lips and her fingers are also blue. When you expose the chest, she tells you it hurts and points to a bruised spot. Which of these is a symptom?

A. Cyanosis around the lips or fingertips

B. Rapid, weak pulse

C. Hemoptysis

D. Pain at the site of injury

_____ 6. Common causes of dyspnea include:

A. airway obstruction.

B. lung compression.

C. damage to the chest wall.

D. all of the above

_____ 7. You respond to an 18-year-old male who has been assaulted with a baseball bat. He was hit in the chest. He is conscious but is having difficulty breathing. His airway is open and his respirations are 30 breaths/min. His pulse is 126 beats/min and blood pressure is 114/72 mm Hg. You quickly expose the chest and there is heavy bruising. The chest wall does not expand on each side when the patient inhales. This is known as:

A. flail segment.

B. paradoxical motion.

C. pneumothorax.

D. hemoptysis.

_____ **8.** The principle reason for concern about a patient who has a chest injury is:
   A. hemoptysis.
   B. cyanosis.
   C. that the body has no means of storing oxygen.
   D. a rapid, weak pulse and low blood pressure.

_____ **9.** A _____ results when an injury allows air to enter through a hole in the chest wall or the surface of the lung as the patient attempts to breathe, causing the lung on that side to collapse.
   A. tension pneumothorax
   B. hemothorax
   C. hemopneumothorax
   D. pneumothorax

_____ **10.** A sucking chest wound should be treated:
   A. after assessing ABCs.
   B. after confirming mental status.
   C. immediately by covering with a gloved hand, then an occlusive dressing.
   D. by using a stack of gauze dressings.

_____ **11.** You respond to a 20-year-old man who was playing basketball and suddenly developed chest pain and respiratory difficulty. He is alert and oriented and complaining of chest pain and breathing at 24 breaths/min. His pulse is 140 beats/min and blood pressure is 160/90 mm Hg. Upon listening to the chest, you notice diminished breath sounds on the left. There is no obvious injury anywhere to the chest. You are thinking that the patient has a spontaneous pneumothorax. A spontaneous pneumothorax:
   A. presents with a sudden sharp chest pain.
   B. presents with increasing difficulty breathing.
   C. should be treated the same as a traumatic pneumothorax.
   D. all of the above

_____ **12.** You know that if you do not treat the patient properly and transport them rapidly to the hospital they could develop a tension pneumothorax. What effect does the development of a tension pneumothorax have on the body?
   A. Air gradually increases the pressure in the chest.
   B. It causes the complete collapse of the affected lung.
   C. It prevents blood from returning through the venae cavae to the heart.
   D. All of the above

_____ **13.** Common signs and symptoms of tension pneumothorax include:
   A. increasing respiratory distress.
   B. distended neck veins.
   C. tracheal deviation away from the injured site.
   D. all of the above

_____ **14.** A hemothorax results from blood collecting in the pleural space from:
   A. a bleeding rib cage.
   B. a bleeding lung.
   C. a bleeding great vessel.
   D. all of the above

_____ **15.** A fractured rib that penetrates into the pleural space may lacerate the surface of the lung, causing a:
   A. tension pneumothorax.
   B. hemothorax.
   C. hemopneumothorax.
   D. all of the above

_____ **16.** In what is called paradoxical movement, the detached portion of the chest wall:

    A. moves opposite of normal.

    B. moves out instead of in during inhalation.

    C. moves in instead of out during expiration.

    D. all of the above

_____ **17.** Traumatic asphyxia:

    A. is bruising of the lung.

    B. occurs when three or more adjacent ribs are fractured in two or more places.

    C. is a sudden, severe compression of the chest.

    D. all of the above

_____ **18.** Traumatic asphyxia results in a very characteristic appearance, including:

    A. distended neck veins.

    B. cyanosis.

    C. hemorrhage into the sclera of the eye.

    D. all of the above

_____ **19.** Signs and symptoms of a pericardial tamponade include:

    A. low blood pressure.

    B. a weak pulse.

    C. muffled heart tones.

    D. all of the above

_____ **20.** Large blood vessels in the chest that can result in massive hemorrhaging include all of the following, except:

    A. the pulmonary arteries.

    B. the femoral arteries.

    C. the aorta.

    D. four main pulmonary veins.

Labeling

Label the following diagrams with the correct terms.

    1. Anterior Aspect of the Chest

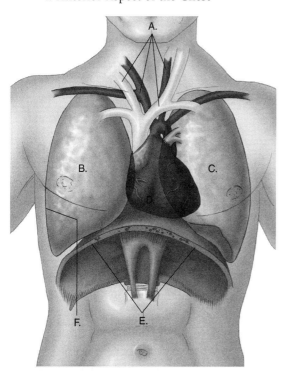

A. _____

B. _____

C. _____

D. _____

E. _____

F. _____

2. The Ribs

A. _____

B. _____

C. _____

D. _____

E. _____

F. _____

G. _____

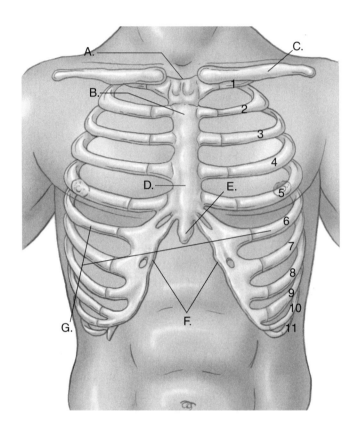

Fill-in

Read each item carefully, then complete the statement by filling in the missing word.

1. The esophagus is located in the _____ of the chest.

2. During inhalation, the pressure in the chest _____.

3. In the anterior chest, ribs connect to the _____.

4. The trachea divides into the right and left main stem _____.

5. The _____ nerves supply the diaphragm.

6. Contents of the chest are protected by the _____.

7. The chest extends from the lower end of the neck to the _____.

8. _____ line the area between the lungs and chest wall.

9. The great vessel located in the chest is the _____.

10. During inhalation, the diaphragm _____.

True/False

If you believe the statement to be more true than false, write the letter "T" in the space provided. If you believe the statement to be more false than true, write the letter "F."

_____ **1.** Dyspnea is difficulty with breathing.

_____ **2.** Tachypnea is slow respirations.

_____ **3.** Distended neck veins may be a sign of a tension pneumothorax.

_____ **4.** Rib fractures are common in children.

_____ **5.** Narrowing pulse pressure is related to spontaneous pneumothorax.

_____ **6.** Laceration of the large blood vessels in the chest can cause minimal hemorrhage.

_____ **7.** The thoracic cage extends from the lower end of the neck to the umbilicus.

_____ **8.** Patients with spinal cord injuries at C3 or above can lose their ability to breathe entirely.

_____ **9.** Almost one third of people who are killed immediately in car crashes die as a result of traumatic rupture of the myocardium.

_____ **10.** Open chest injury is caused by penetrating trauma.

Short Answer

Complete this section with short written answers using the space provided.

1. List the signs and symptoms associated with a chest injury.

_____

_____

2. Describe the two methods for sealing a sucking chest wound.

_____

_____

3. Describe the method(s) for immobilizing a flail chest wall segment.

_____

_____

4. Define traumatic asphyxia and describe its signs.

_____

_____

Word Fun

The following crossword puzzle is an activity provided to reinforce correct spelling and understanding of medical terminology associated with emergency care and the EMT-B. Use the clues in the column to complete the puzzle.

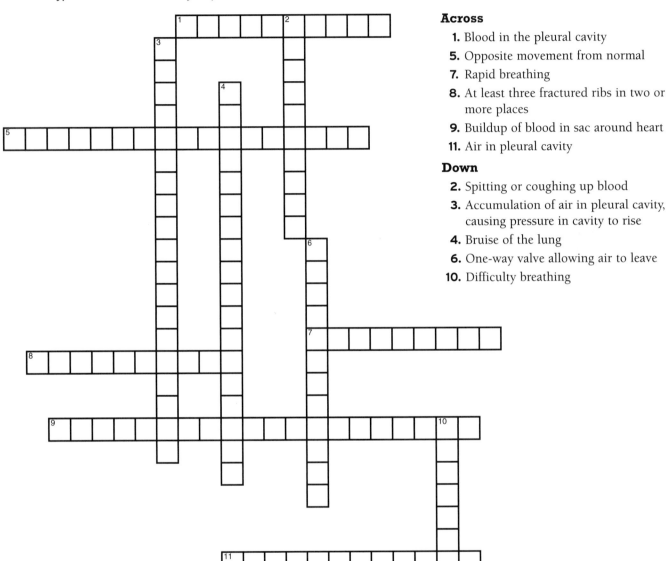

Across

1. Blood in the pleural cavity
5. Opposite movement from normal
7. Rapid breathing
8. At least three fractured ribs in two or more places
9. Buildup of blood in sac around heart
11. Air in pleural cavity

Down

2. Spitting or coughing up blood
3. Accumulation of air in pleural cavity, causing pressure in cavity to rise
4. Bruise of the lung
6. One-way valve allowing air to leave
10. Difficulty breathing

Ambulance Calls

The following case scenarios provide an opportunity to explore the concerns associated with patient management. Read each scenario, then answer each question in detail.

1. You are dispatched to an area horse ranch for an injury to a rider. You arrive to find that a horse has kicked one of the riders in the chest. He is having significant difficulty breathing and appears to be in extreme pain.

   How do you best manage this patient?

   _____

   _____

   _____

2. You are dispatched for a man trapped under a car. As you travel to the location, the dispatcher informs you that your patient is now free, but is experiencing significant chest and mid-back pain.

   How do you best manage this patient?

   _____

   _____

   _____

3. You are dispatched to a lumberyard where a 27-year-old man was crushed by a piece of heavy equipment. Coworkers pulled the equipment off the patient. He presents with distended neck veins, cyanosis, and bloodshot eyes.

   How would you best manage this patient?

   _____

   _____

   _____

Assessment Review

Answer the following questions pertaining to the assessment of the types of emergencies discussed in this chapter.

1. You respond to an accidental shooting of a 37-year-old man. During the initial assessment you find his airway to be open and breathing is labored at 24 breaths/min. His pulse is rapid and weak. Upon exposing the chest you find a sucking chest wound. You should:

   A. take a blood pressure.
   B. cover the wound.
   C. continue your assessment.
   D. transport immediately.

2. You respond to a 17-year-old female who was hit in the chest with a lawn dart. Upon arrival she is conscious and able to converse with you. Her airway is open and her breathing is becoming progressivly more difficult. Her pulse is rapid and weak. You can palpate a radial pulse. Upon examining the chest, she has a penetrating injury to the chest and there is a sucking sound as she breathes. How do you manage this wound?

   A. Apply oxygen by nasal cannula.
   B. Stabilize the c-spine.
   C. Use a gauze 4 × 4.
   D. Occlusive dressing.

3. When bandaging an open chest wound, what is the minimum number of sides that have to be taped down?

   A. 1
   B. 2
   C. 3
   D. 4

4. Dispatch sends you to a farm on the edge of town. The 57-year-old man was kicked in the chest by a horse. He walked into his house and collapsed. He is alert and oriented. His breathing is labored at 20 breaths/min, pulse is rapid and regular, and you are able to palpate a radial pulse. Upon examination of his chest you notice paradoxical movement on the right chest wall. You should:

   A. take spinal precautions.
   B. put the patient in a position of comfort.
   C. provide oxygen by nasal cannula.
   D. stabilize the flail segment.

5. A 16-year-old male patient walks into a pipe gate that hits him in the ribs on the left side. Upon your arrival he is alert and oriented. His breathing is shallow at 22 breaths/min. His pulse is regular and strong. You palpate a radial pulse. You are able to rule out spinal trauma. What position do you transport him in?

   A. Position of comfort
   B. Supine
   C. Prone
   D. Lateral recumbent

Emergency Care Summary

Fill in the following chart pertaining to the management of the types of emergencies discussed in this chapter.

NOTE: While the steps below are widely accepted, be sure to consult and follow your local protocol.

Pneumothorax	Hemothorax	Rib Fractures	Flail Chest
1. Clear and maintain an open airway.	1. Clear and maintain an open airway.	1. Clear and maintain an open airway.	1. Clear and maintain an open airway.
2. Provide high-flow oxygen.	2. Provide high-flow oxygen. Cover the patient with a blanket.	2. Provide high-flow oxygen. Cover the patient with a blanket.	2. Provide respiratory support if necessary.
3. Seal the wound with an _____ dressing, using a large enough dressing so that it is not pulled or sucked into the chest cavity.	3. Treat the patient for _____.	3. Place in a position of comfort to support breathing unless a _____ _____ is suspected.	3. Provide _____ oxygen.
4. Depending on your local protocol, you may tape the dressing down on all four sides, or create a _____ _____ by taping only three sides of the dressing.			4. Stabilize flail segment by securing (or having the patient hold) a pillow firmly against the chest wall.

CHAPTER

28 Abdomen and Genitalia Injuries

Workbook Activities

The following activities have been designed to help you. Your instructor may require you to complete some or all of these activities as a regular part of your EMT-B training program. You are encouraged to complete any activity that your instructor does not assign as a way to enhance your learning in the classroom.

Chapter Review

The following exercises provide an opportunity to refresh your knowledge of this chapter.

Matching

Match each of the terms in the left column to the appropriate definition in the right column.

_____ 1. Hollow organs   **A.** blood in urine

_____ 2. Solid organs   **B.** organs outside of the body

_____ 3. Peritonitis   **C.** abdominal lining inflammation

_____ 4. Genitourinary system   **D.** kidneys, liver, spleen

_____ 5. Filtering system   **E.** kidneys

_____ 6. Evisceration   **F.** abdomen

_____ 7. Hematuria   **G.** stomach, bladder, ureters

_____ 8. Peritoneal cavity   **H.** controls reproductive functions and the waste discharge system

Multiple Choice

Read each item carefully, then select the best response.

_____ 1. The abdomen contains several organs that make up the:
A. digestive system.
B. urinary system.
C. genitourinary system.
D. all of the above

_____ 2. Hollow organs of the abdomen include the:
A. stomach.
B. ureters.
C. bladder.
D. all of the above

_____ 3. Solid organs of the abdomen include all of the following, except the:
A. liver.
B. spleen.
C. gallbladder.
D. pancreas.

_____ 4. The first signs of peritonitis include:
A. severe abdominal pain.
B. tenderness.
C. muscular spasm.
D. all of the above

_____ 5. Late signs of peritonitis may include:
A. a soft abdomen.
B. nausea.
C. normal bowel sounds.
D. all of the above

_____ 6. _____ takes place in the solid organs.
A. Digestion
B. Excretion
C. Energy production
D. all of the above

_____ 7. Because solid organs have a rich supply of blood, any injury can result in major:
A. hemorrhaging.
B. damage.
C. pain.
D. guarding.

_____ 8. You are dispatched to an auto collision. You see a 25-year-old woman who was restrained but is complaining of abdominal pain. She is alert and oriented. Airway is open and patient is breathing normally. Her pulse is regular but weak and rapid. She has a radial pulse. You inspect the abdomen for possible bleeding. You would expect to see all of the following except:
A. pain or tenderness.
B. rigidity.
C. urticaria.
D. distention.

_____ **9.** The major soft-tissue landmark is (are) the _____, which overlie(s) the fourth lumbar vertebra.

    A. iliac crests

    B. umbilicus

    C. pubic symphysis

    D. anterior iliac spines

_____ **10.** The abdomen is divided into four:

    A. quadrants.

    B. planes.

    C. sections.

    D. angles.

_____ **11.** Injuries to the abdomen may involve:

    A. hollow organs.

    B. open injuries.

    C. solid organs.

    D. all of the above

_____ **12.** Open abdominal injuries are also known as:

    A. blunt injuries.

    B. eviscerations.

    C. penetrating injuries.

    D. peritoneal injuries.

_____ **13.** Closed abdominal injuries may result from:

    A. a stab wound.

    B. seat belts.

    C. a gunshot wound.

    D. all of the above

_____ **14.** The major complaint of patients with abdominal injury is:

    A. pain.

    B. tachycardia.

    C. rigidity.

    D. swelling.

_____ **15.** The most common sign of significant abdominal injury is:

    A. pain.

    B. tachycardia.

    C. rigidity.

    D. distention.

_____ **16.** Late signs of abdominal injury include all of the following, except:

    A. distention.

    B. increased blood pressure.

    C. rigidity.

    D. shallow respirations.

_____ **17.** Your primary concern when dealing with an unresponsive patient with an open abdominal injury is:

    A. covering the wound with a moist dressing.

    B. maintaining the airway.

    C. controlling the bleeding.

    D. monitoring vital signs.

_____ **18.** You respond to an 18-year-old high school football player who was hit in the abdominal area with a helmet. He is complaining of pain in the area. He is alert and oriented. His airway is open and his respirations are within normal limits. His pulse is rapid and regular. He has a radial pulse. You know that blunt abdominal trauma may result in:

A. severe bruises on the abdominal wall.

B. laceration of the liver or spleen.

C. rupture of the intestine.

D. all of the above

_____ **19.** For the patient in the question above, how would you provide treatment for him?

A. Log roll onto a backboard.

B. Transport rapidly.

C. Give oxygen.

D. All of the above

_____ **20.** When used alone, diagonal shoulder safety belts can cause:

A. a bruised chest.

B. a lacerated liver.

C. decapitation.

D. all of the above

_____ **21.** You are dispatched to a motor vehicle collision. Your patient is a 42-year-old restrained woman. The airbag did deploy and she has abrasions on her face. She is complaining of pain to both her chest and abdomen. Her airway is open and respirations are within normal limits. Her pulse is a little rapid but strong and regular. She has distal pulses. It is important to look at/for:

A. debris inside the vehicle.

B. the condition of the tires.

C. damage to the steering column underneath the airbag.

D. all of the above

_____ **22.** Patients with penetrating abdominal injuries often complain of:

A. pain.

B. nausea.

C. vomiting.

D. all of the above

_____ **23.** You are called to the local bar where a fight has taken place. The police department tells you that you have a 36-year-old man who has been stabbed twice in the abdomen. Upon arrival the patient is alert and oriented. His airway is open and his respirations are at 24 breaths/min, pulse is rapid, regular, and weak. He has distal pulses. With the penetrating trauma, you should assume that the object:

A. has penetrated the peritoneum.

B. entered the abdominal cavity.

C. possibly injured one or more organs.

D. all of the above

_____ **24.** When treating a patient with an evisceration, you should:

A. attempt to replace the abdominal contents.

B. cover the protruding organs with a dry, sterile dressing.

C. cover the protruding organs with moist, adherent dressings.

D. cover the protruding contents with moist, sterile gauze compresses.

_____ **25.** The solid organs of the urinary system include the:

A. kidneys.

B. ureters.

C. bladder.

D. urethra.

_____ **26.** All of the male genitalia lie outside the pelvic cavity with the exception of the:

    A. urethra.

    B. penis.

    C. seminal vesicles.

    D. testes.

_____ **27.** Suspect kidney damage if the patient has a history or physical evidence of:

    A. an abrasion, laceration, or contusion in the flank.

    B. a penetrating wound in the region of the lower rib cage or the upper abdomen.

    C. fractures on either side of the lower rib cage.

    D. all of the above

_____ **28.** Signs of injury to the kidney may include:

    A. bruises or lacerations on the overlying skin.

    B. shock.

    C. hematuria.

    D. all of the above

_____ **29.** Suspect a possible injury of the urinary bladder in all of the following findings, except:

    A. bruising to the left upper quadrant.

    B. blood at the urethral opening.

    C. blood at the tip of the penis or a stain on the patient's underwear.

    D. physical signs of trauma on the lower abdomen, pelvis, or perineum.

_____ **30.** When treating a patient with an amputation of the penile shaft, your top priority is:

    A. locating the amputated part.

    B. controlling bleeding.

    C. keeping the remaining tissue dry.

    D. delaying transport until bleeding is controlled.

_____ **31.** Treatment of injuries involving the external male genitalia includes:

    A. making the patient as comfortable as possible.

    B. using sterile, moist compresses to cover areas that have been stripped of skin.

    C. applying direct pressure with dry, sterile gauze dressings to control bleeding.

    D. all of the above

_____ **32.** In cases of sexual assault, you must treat the medical injuries but also provide:

    A. privacy.

    B. support.

    C. reassurance.

    D. all of the above

Labeling
Label the following diagrams with the correct terms.

1. Hollow Organs

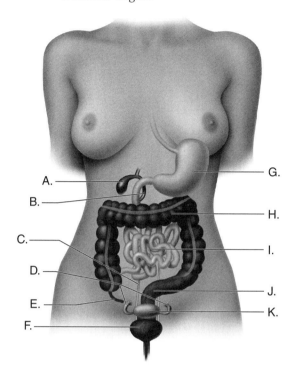

A. _____

B. _____

C. _____

D. _____

E. _____

F. _____

G. _____

H. _____

I. _____

J. _____

K. _____

2. Solid Organs

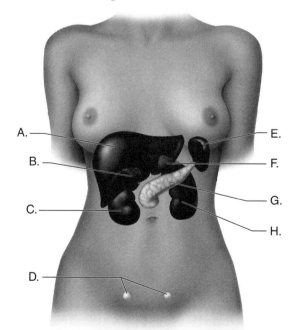

A. _____

B. _____

C. _____

D. _____

E. _____

F. _____

G. _____

H. _____

Fill-in

Read each item carefully, then complete the statement by filling in the missing word.

1. Severe bleeding may occur with injury to _____ organs.

2. The _____ system is responsible for filtering waste.

3. Kidneys are located in the _____ space.

4. Injuries to the kidneys or bladder will not have obvious _____ _____, but there are usually more subtle clues such as lower rib pain or a possible pelvic fracture.

5. When ruptured, the organs of the abdominal cavity can spill their contents into the peritoneal cavity, causing an intense inflammatory reaction called _____.

6. Blood within the peritoneal cavity does not provoke a(n) _____ _____, and may not cause pain or tenderness.

7. Closed abdominal injuries are also known as _____ _____.

8. Open abdominal injuries are also known as _____ _____.

9. Another name for the right and left upper quadrants is _____.

10. An open wound that allows internal organs or fat to protrude though the wound is called _____.

True/False

If you believe the statement to be more true than false, write the letter "T" in the space provided. If you believe the statement to be more false than true, write the letter "F."

_____ 1. Hollow organs will bleed profusely if injured.
_____ 2. The most common sign of an abdominal injury is an elevated heart rate.
_____ 3. Patients with abdominal injuries should be kept supine with head elevated.
_____ 4. Peritoneal irritation is in response to hollow organ injury.
_____ 5. Eviscerated organs should be covered with a dry dressing.
_____ 6. Injuries to the kidneys usually occur in isolation.
_____ 7. Peritonitis is an inflammation of the peritoneum.
_____ 8. The abdomen is divided into two quadrants.
_____ 9. Small children can be injured by air bags.
_____ 10. Patients with peritonitis will want to lie still with their legs drawn up.

Short Answer

Complete this section with short written answers using the space provided.

1. List the hollow organs of the abdomen and urinary system.

_____

_____

2. List the solid organs of the abdomen and urinary system.

_____

_____

3. List the signs and symptoms of an abdominal injury.

_____

_____

4. List the steps to care for a penetrating abdominal injury.

_____

_____

5. List the steps to care for an open abdominal wound with exposed organs.

_____

_____

6. List the major history or physical findings associated with possible kidney damage.

_____

_____

Word Fun

The following crossword puzzle is an activity provided to reinforce correct spelling and understanding of medical terminology associated with emergency care and the EMT-B. Use the clues in the column to complete the puzzle.

Across

4. Contracting of muscle to protect
5. Stomach, small intestines, bladder
7. Displacement of organs outside of abdomen
8. Inflammation of abdominal lining

Down

1. Liver, spleen, pancreas
2. Abdominal cavity
3. Penetrating wound of belly
6. Blood in urine

Ambulance Calls

The following case scenarios provide an opportunity to explore the concerns associated with patient management. Read each scenario, then answer each question in detail.

1. You are dispatched to a local bar where your patient, a 26-year-old man, was involved in an altercation. He has several superficial lacerations to his arms, and a knife is impaled in his right upper quadrant. He is lying supine on the floor. He is alert and bar patrons tell you that he did not fall; they helped him to the floor.

   How would you best manage this patient?

   _____

   _____

   _____

2. You are dispatched to assist police with a mentally ill patient who as threatened harm to himself and others. Police officers found the man running around his home with a knife and blood all over his lower body. The man tells you "the voices" told him to cut his penis off.

   How do best manage this patient?

   _____

   _____

   _____

3. You are dispatched to a construction site, where a man has fallen onto a piece of rebar. You arrive to find a man sitting on the ground with his legs drawn toward his chest. He tells you that he fell from a ladder onto a piece of rebar. He tells you, "Something's sticking out of me." As you visualize his abdomen, you can clearly see a portion of his bowel on the outside of his body.

   How do you best manage this patient?

   _____

   _____

   _____

Assessment Review

Answer the following questions pertaining to the assessment of the types of emergencies discussed in this chapter.

_____ 1. If you are treating a patient with an abdominal evisceration you should use a(n):
   A. moist, sterile dressing.
   B. dry, sterile dressing.
   C. adhesive dressing.
   D. occlusive dressing.

_____ 2. You have a male patient who has no immediate life threat, but does have a genitalia bleeding. You should bandage with a(n):
   A. dry dressing.
   B. moist dressing.
   C. occlusive dressing.
   D. adhesive dressing.

_____ 3. You have a patient with suspected kidney injury but no spinal injury. How should he be positioned?
   A. supine
   B. prone
   C. lateral recumbent
   D. position of comfort

_____ 4. Your patient has his penis caught in his zipper. What do you need to do to relieve pressure?
   A. Pull the pants off the patient.
   B. Force the zipper open.
   C. Remove the foreskin.
   D. Cut the zipper out of the pants.

_____ 5. Whenever possible you should always provide the sexual assault patient with:
   A. police escort.
   B. rape crisis intervention.
   C. same gender attendant.
   D. name of the assailant.

Emergency Care Summary

Fill in the following chart pertaining to the management of the types of emergencies discussed in this chapter.

NOTE: While the steps below are widely accepted, be sure to consult and follow your local protocol.

Kidney and Urinary Bladder Injuries

1. Treat for shock early and aggressively.
2. Place patient in position of comfort unless spinal injury is suspected.
3. Provide prompt transport.
4. Monitor _____ _____ en route.

External Male Genitalia Injuries

Avulsion

1. Wrap the penis in a soft, sterile dressing moistened with sterile _____ solution.
2. Use direct pressure to control any bleeding.
3. Try to save and preserve the avulsed skin, but do not delay transport for more than a few _____ to do so.

Amputation

1. Manage blood loss using local pressure with a sterile dressing on the remaining stump.
2. Wrap the amputated part in a moist, sterile dressing; place it in a plastic bag; transport in a cooled container without allowing it to come in direct contact with _____.

Fractured or Severely Angled

1. Provide prompt transport.

Lacerated Head or the Penis

1. Stop the hemorrhage with a _____ _____ and local pressure.

Penis Caught in a Zipper

1. Attempt to unzip pants.
2. If unable to unzip pants, explain to the patient that you are going to cut the zipper out of the pants to relieve pressure.

Urethral Injuries

1. Save any voided _____ for examination.

Avulsion of Scrotum Skin

1. Preserve avulsed skin in moist, sterile dressing.
2. Wrap scrotal contents or the perineal area with a sterile, moist dressing to control bleeding.

Rupture of a Testicle or Significant Accumulation of Blood Around the Testes

1. Apply ice pack to the scrotal area.

External Female Genitalia Injuries	Internal Female Genitalia Injuries, Pregnancy	Rectal Bleeding
1. Treat with moist, sterile _____. 2. Apply local pressure to control the bleeding and a diaper-type bandage to hold dressings in place.	1. Carefully place patient on her _____ side. 2. If the patient is on the backboard, tilt the board to the _____.	1. Pack the crease between the buttocks with _____ and consult with medical control to determine the need for transport.

Sexual Assault

1. Document the patient's history, assessment, treatment, and response to treatment in detail.
2. Make _____ _____ a major priority.
3. Complete the SAMPLE history in an objective, nonjudgmental way.
4. If the patient will tolerate being wrapped in a sterile burn sheet, this may help investigators to find any hair, fluid, or fiber from the alleged offender.
5. Do not examine the genitalia unless there is _____ _____. If an object has been inserted into the vagina or rectum, do not attempt to remove it.
6. Make sure the EMT-B is the same gender as the patient whenever possible.
7. Discourage the patient from bathing, voiding, or cleaning any wounds until hospital staff has completed an assessment. Handle the patient's clothes as little as possible, placing articles and any other evidence in _____ bags. Do not use _____ bags. If the female patient insists on urinating, have her do so in a sterile urine container (if available). Also, have her deposit the toilet paper in a paper bag. Seal and mark the bag for the police. This can be critical evidence.

CHAPTER

29 Musculoskeletal Care

Workbook Activities

The following activities have been designed to help you. Your instructor may require you to complete some or all of these activities as a regular part of your EMT-B training program. You are encouraged to complete any activity that your instructor does not assign as a way to enhance your learning in the classroom.

Chapter Review

The following exercises provide an opportunity to refresh your knowledge of this chapter.

Matching

Match each of the terms in the left column to the appropriate definition in the right column.

_____ **1.** Striated
_____ **2.** Tendons
_____ **3.** Smooth
_____ **4.** Joint
_____ **5.** Ligaments
_____ **6.** Closed fracture
_____ **7.** Point tenderness
_____ **8.** Displaced fracture
_____ **9.** Articular cartilage
_____ **10.** Open fracture
_____ **11.** Traction

A. any injury that makes the limb appear in an unnatural position

B. any fracture in which the skin has not been broken

C. a thin layer of cartilage, covering the articular surface of bones in synovial joints

D. involuntary muscle

E. any break in the bone in which the overlying skin has been damaged as well

F. hold joints together

G. skeletal muscle

H. the act of exerting a pulling force on a structure

I. where two bones contact

J. attach muscle to bone

K. tenderness sharply located at the site of an injury

Multiple Choice

Read each item carefully, then select the best response.

_____   **1.** Blood in the urine is known as a:

A. hematuria.

B. hemotysis.

C. hematocrit.

D. hemoglobin.

_____   **2.** Smooth muscle is found in the:

A. back.

B. blood vessels.

C. heart.

D. all of the above

_____   **3.** The bones in the skeleton produce _____ in the bone marrow.

A. red blood cells

B. minerals

C. electrolytes

D. white blood cells

_____   **4.** _____ are held together in a tough fibrous structure known as a capsule.

A. Tendons

B. Joints

C. Ligaments

D. Bones

_____   **5.** Joints are bathed and lubricated by _____ fluid.

A. cartilaginous

B. articular

C. synovial

D. cerebrospinal

_____   **6.** A _____ is a disruption of a joint in which the bone ends are no longer in contact.

A. torn ligament

B. dislocation

C. fracture dislocation

D. sprain

_____   **7.** A _____ is a joint injury in which there is both some partial or temporary dislocation of the bone ends and partial stretching or tearing of the supporting ligaments.

A. dislocation

B. strain

C. sprain

D. torn ligament

_____   **8.** A _____ is a stretching or tearing of the muscle.

A. strain

B. sprain

C. torn ligament

D. split

_____ **9.** The zone of injury includes the:

    A. adjacent nerves.

    B. adjacent blood vessels.

    C. surrounding soft tissue.

    D. all of the above

_____ **10.** A(n) _____ fractures the bone at the point of impact.

    A. direct blow

    B. indirect force

    C. twisting force

    D. high-energy injury

_____ **11.** A(n) _____ may cause a fracture or dislocation at a distant point.

    A. direct blow

    B. indirect force

    C. twisting force

    D. high-energy injury

_____ **12.** When caring for patients who have fallen, you must identify the _____ and the mechanism of injury so that you will not overlook associated injuries.

    A. site of injury

    B. height of fall

    C. point of contact

    D. twisting forces

_____ **13.** _____ produce severe damage to the skeleton, surrounding soft tissues, and vital internal organs.

    A. Direct blows

    B. Indirect forces

    C. Twisting forces

    D. High-energy injuries

_____ **14.** Regardless of the extent and severity of the damage to the skin, you should treat any injury that breaks the skin as a possible:

    A. closed fracture.

    B. open fracture.

    C. nondisplaced fracture.

    D. displaced fracture.

_____ **15.** A _____ is also known as a hairline fracture.

    A. closed fracture

    B. open fracture

    C. nondisplaced fracture

    D. displaced fracture

_____ **16.** A _____ produces actual deformity, or distortion, of the limb by shortening, rotating, or angulating it.

    A. closed fracture

    B. open fracture

    C. nondisplaced fracture

    D. displaced fracture

_____ **17.** You respond to a 19-year-old female who was kicked in the leg by a horse. She is alert and oriented. Respirations are 20 breaths/min regular and unlabored. Pulse is 110 beats/min and regular. Distal pulses are present. She has point tenderness at the site of the injury. You should compare the limb to:

  A. the opposite uninjured limb.

  B. one of your limbs or one of your partner's limbs.

  C. an injury chart.

  D. none of the above

_____ **18.** _____ is the most reliable indicator of an underlying fracture.

  A. Crepitus

  B. Deformity

  C. Point tenderness

  D. Absence of distal pulse

_____ **19.** A(n) _____ is a fracture that occurs in a growth section of a child's bone, which may prematurely stop growth if not properly treated.

  A. greenstick fracture

  B. comminuted fracture

  C. pathological fracture

  D. epiphyseal fracture

_____ **20.** A(n) _____ is an incomplete fracture that passes only partway through the shaft of a bone but may still cause severe angulation.

  A. greenstick fracture

  B. comminuted fracture

  C. pathological fracture

  D. epiphyseal fracture

_____ **21.** You are called to the local assisted living facility where a 94-year-old man has fallen. He is alert and oriented and denies passing out. His respirations are 18 breaths/min and regular. Pulse is 106 beats/min, regular, and strong. Distal pulses are present. He states he was walking and heard a pop and fell to the floor. You suspect a(n):

  A. greenstick fracture.

  B. comminuted fracture.

  C. pathological fracture.

  D. epiphyseal fracture.

_____ **22.** A(n) _____ is a fracture in which the bone is broken into two or more fragments.

  A. greenstick fracture

  B. comminuted fracture

  C. pathological fracture

  D. epiphyseal fracture

_____ **23.** Your 24-year-old patient fell off a balance beam and landed on his arm. His is complaining of pain in the upper arm and there is obvious swelling. You know that swelling is a sign of:

  A. bleeding.

  B. laceration.

  C. locked joint.

  D. compartment syndrome.

_____ **24.** Fractures are almost always associated with _____ of the surrounding soft tissue.

  A. laceration

  B. crepitus

  C. ecchymosis

  D. swelling

_____ **25.** Signs and symptoms of a dislocated joint include:

    A. marked deformity.

    B. tenderness or palpation.

    C. locked joint.

    D. all of the above

_____ **26.** Signs and symptoms of sprains include all of the following, except:

    A. point tenderness.

    B. pain prevents the patient from moving or using the limb normally.

    C. marked deformity.

    D. instability of the joint is indicated by increased motion.

_____ **27.** Assessment of patients with musculoskeletal injuries must include:

    A. initial assessment followed by a focused physical exam.

    B. evaluation of neurovascular function.

    C. applying oxygen as needed.

    D. all of the above

_____ **28.** Compartment syndrome:

    A. occurs within 6 to 12 hours after injury.

    B. usually is a result of excessive bleeding, a severely crushed extremity, or the rapid return of blood to an ischemic limb.

    C. is characterized by pain that is out of proportion to the injury.

    D. all of the above

_____ **29.** Always check neurovascular function:

    A. after any manipulation of the limb.

    B. before applying a splint.

    C. after applying a splint.

    D. all of the above

_____ **30.** You respond to a 19-year-old female who was involved in a motor vehicle collision. She is alert and oriented. Her airway is open and respirations are 18 breaths/min and unlabored. Pulse is 94 beats/min, strong, and regular. Distal pulses are present. Her upper arm has obvious deformity. You splint the upper arm. You know that splinting will help prevent:

    A. excessive bleeding of the tissues at the injury site caused by broken bone ends.

    B. laceration of the skin by broken bone ends.

    C. further damage to muscles.

    D. all of the above

_____ **31.** In-line _____ is the act of exerting a pulling force on a body structure in the direction of its normal alignment.

    A. stabilization

    B. immobilization

    C. traction

    D. direction

_____ **32.** Basic types of splints include:

    A. rigid.

    B. formable.

    C. traction.

    D. all of the above

_____ **33.** Do not use traction splints for any of the following conditions, except:

   A. injuries of the pelvis.

   B. an isolated femur fracture.

   C. partial amputation or avulsions with bone separation.

   D. lower leg or ankle injury.

_____ **34.** While transporting the patient in question 30, you continue to recheck the splint you applied. You know that improperly applying a splint can cause:

   A. compression of nerves, tissues, and blood vessels.

   B. delay in transport of a patient with a life-threatening injury.

   C. reduction of distal circulation if the splint is too tight.

   D. all of the above

_____ **35.** The _____ is one of the most commonly fractured bones in the body.

   A. scapula

   B. clavicle

   C. humerus

   D. radius

_____ **36.** Indications that blood vessels have likely been injured include:

   A. a cold, pale hand.

   B. weak or absent pulse.

   C. poor capillary refill.

   D. all of the above

_____ **37.** Signs and symptoms associated with hip dislocation include:

   A. severe pain in the hip.

   B. lateral and posterior aspects of the hip region will be tender on palpation.

   C. being able to palpate the femoral head deep within the muscles of the buttock.

   D. all of the above

_____ **38.** There is always a significant amount of blood loss, as much as _____ mL, after a fracture of the shaft of the femur.

   A. 100 to 250

   B. 250 to 500

   C. 500 to 1,000

   D. 100 to 1,500

_____ **39.** The knee is especially susceptible to _____ injuries, which occur when abnormal bending or twisting forces are applied to the joint.

   A. tendon

   B. ligament

   C. dislocation

   D. fracture-dislocation

_____ **40.** Signs and symptoms of knee ligament injury include:

   A. swelling.

   B. point tenderness.

   C. joint effusion.

   D. all of the above

_____ **41.** Although substantial ligament damage always occurs with a knee dislocation, the more urgent injury is to the _____ artery, which is often lacerated or compressed by the displaced tibia.

   A. tibial

   B. femoral

   C. popliteal

   D. dorsalis pedis

_____ **42.** Because of local tenderness and swelling, it is easy to confuse a nondisplaced or minimally displaced fracture about the knee with a:

A. tendon injury.

B. ligament injury.

C. dislocation.

D. fracture-dislocation.

_____ **43.** Fracture of the tibia and fibula are often associated with _____ as a result of the distorted positions of the limb following injury.

A. vascular injury

B. muscular injury

C. tendon injury

D. ligament injury

_____ **44.** The _____ is the most commonly injured joint.

A. knee

B. elbow

C. ankle

D. hip

Labeling

Label the following diagram with the correct terms.

1. The Human Skeleton

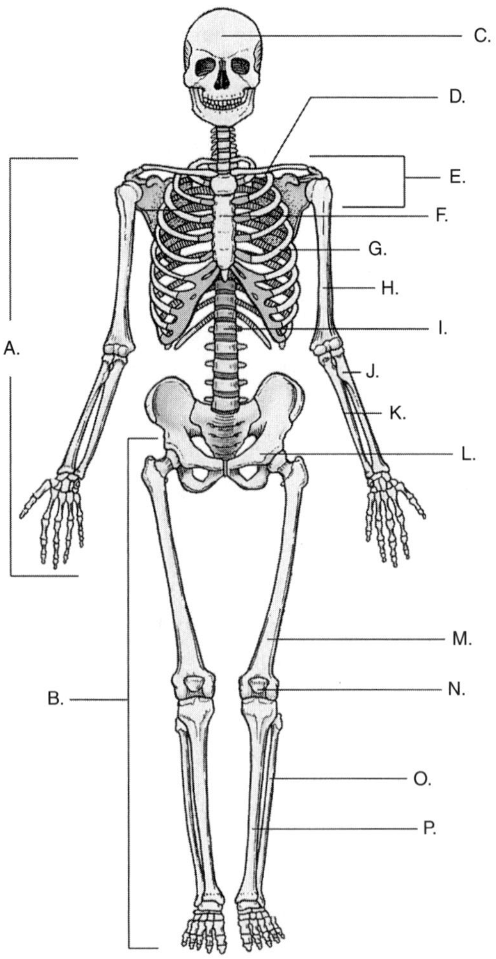

A. _____

B. _____

C. _____

D. _____

E. _____

F. _____

G. _____

H. _____

I. _____

J. _____

K. _____

L. _____

M. _____

N. _____

O. _____

P. _____

Fill-in

Read each item carefully, then complete the statement by filling in the missing word.

1. Atrophy is the _____ of muscle tissue.

2. Bone marrow produces _____ blood cells.

3. The knee and elbow are _____ and socket joints.

4. The _____ is one of the most commonly fractured bones in the body.

5. Always carefully assess the _____ _____ _____ to try to determine the amount of kinetic

   energy that an injured limb has absorbed.

6. Penetrating injury should alert you to the possibility of a(n) _____ _____.

7. The _____ _____ is the most important nerve in the lower extremity; it controls the activity of

   muscles in the thigh and below the knee.

8. The _____ is the longest and largest bone in the body.

9. A grating or grinding sensation known as _____ can be felt and sometimes even heard when fractured

   bone ends rub together.

10. A dislocated joint sometimes will spontaneously _____, or return to its normal position.

11. If you suspect that a patient has compartment syndrome, splint the affected limb, keeping it at the level of the

    heart, and provide immediate transport, checking _____ _____ frequently during transport.

True/False

If you believe the statement to be more true than false, write the letter "T" in the space provided. If you believe the statement to be more false than true, write the letter "F."

_____ 1. All extremity injuries should be splinted before moving a patient unless the patient's life is in immediate danger.

_____ 2. Splinting reduces pain and prevents the motion of bone fragments.

_____ 3. You should use traction to reduce a fracture and force all bone fragments back into alignment.

_____ 4. When applying traction, the direction of pull is always along the axis of the limb.

_____ 5. Cover wounds with a dry, sterile dressing before applying a splint.

_____ 6. When splinting a fracture, you should be careful to immobilize only the joint above the injury site.

_____ 7. One of the steps of the neurological examination is to palpate the pulse distal to the point of injury.

_____ 8. Assessment of neurovascular function should be repeated every 5 to 10 minutes until the patient arrives at the hospital.

_____ 9. A patient's ability to sense light touch in the fingers and toes distal to the injury site is a good indication that the nerve supply is intact.

_____ 10. If the hand or foot is involved in the injury, you should check motor function.

Short Answer

Complete this section with short written answers using the space provided.

1. List the four types of forces that may cause injury to a limb.

_____

_____

2. List five of the signs associated with a possible fracture.

_____

_____

3. List the four items to check when assessing neurovascular function.

_____

_____

4. List the general principles of splinting.

_____

_____

5. What are the three goals of in-line traction?

_____

_____

Word Fun

The following crossword puzzle is an activity provided to reinforce correct spelling and understanding of medical terminology associated with emergency care and the EMT-B. Use the clues in the column to complete the puzzle.

Across

3. Exerting a pulling force
5. Major lower extremity nerve
6. Hand position for splinting
7. Elevation of pressure within fibrous tissues
8. Blood in urine
11. Collarbone
12. Joint between the two pubic bones
13. Forearm bone on small finger side
14. Striated, attached to bones

Down

1. Part of the scapula that joins with the humerus
2. Broken bone with overlying skin injured
4. Bone fragments are separated
9. Discoloration from bleeding under skin
10. Grating sound of bone ends

Ambulance Calls

The following case scenarios provide an opportunity to explore the concerns associated with patient management. Read each scenario, then answer each question in detail.

1. You are dispatched to care for a 17-year-old male who jumped from the top of a three-story home into a pool. He landed directly on his feet just short of the pool. He is now complaining of low back pain and numbness and tingling of his legs.

   How do you best manage this patient?

   _____

   _____

   _____

2. You are called to a local park where an 11-year-old girl fell off the parallel bars onto her right elbow. She is cradling the arm to her chest. She has obvious swelling and deformity in the area. She has good pulse, motor, and sensation at the wrist. ABCs are normal.

How would you best manage this patient?

_____

_____

_____

3. You are dispatched to an attempted suicide at the local prison. The inmate attempted to kill himself by strangling himself with an electrical cord. When you arrive he is cyanotic and not breathing.

How do you best manage this patient?

_____

_____

_____

Skill Drills

Skill Drill 29-1: Assessing Neurovascular Status

Test your knowledge of this skill drill by filling in the correct words in the photo captions.

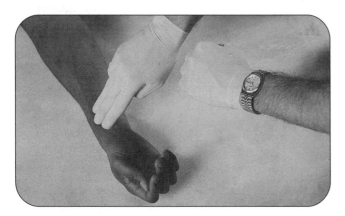

1. Palpate the _____ pulse in the upper extremity.

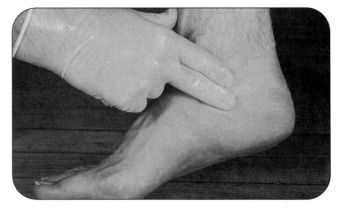

2. Palpate the _____ _____ pulse in the lower extremity.

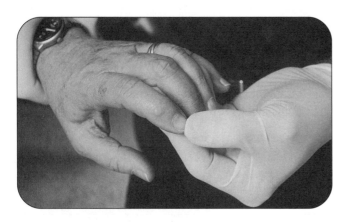

3. Assess capillary refill by blanching a fingernail or _____.

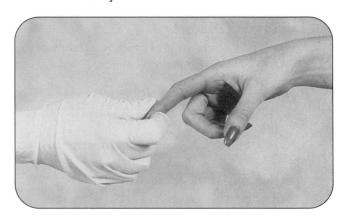

4. Assess sensation on the flesh near the _____ of the _____ finger.

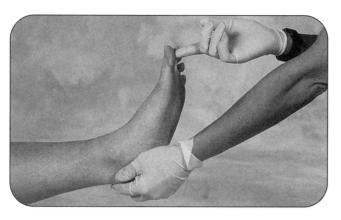

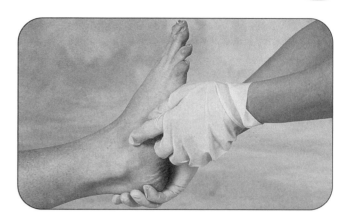

5. On the foot, first check sensation on the flesh near the
_____ of the _____ _____.

6. Also check foot sensation on the _____
_____.

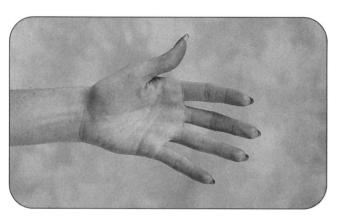

7. Evaluate motor function by asking the patient to
_____ the hand. (Perform motor tests only if the
hand or foot is not _____. _____ a test
if it causes pain.)

8. Also ask the patient to _____ _____
_____.

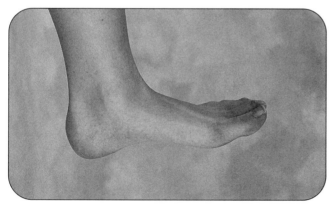

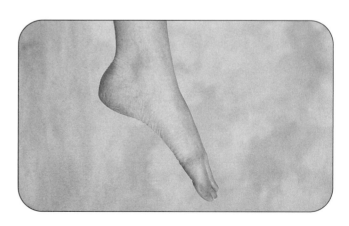

9. To evaluate motor function in the foot, ask the patient to
_____ the foot.

10. Also have the patient _____ the foot and
_____ the toes.

Skill Drill 29-2: Caring for Musculoskeletal Injuries
Test your knowledge of this skill drill by filling in the correct words in the photo captions.

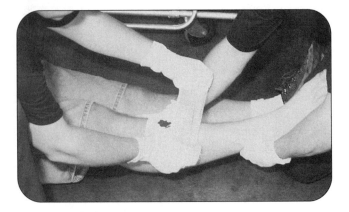

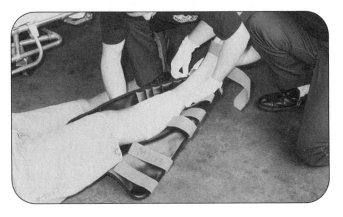

1. Cover open wounds with a _____,
   _____ dressing, and _____
   _____ to control bleeding.

2. Apply a splint and elevate the extremity about
   _____" (slightly above the level of the
   _____).

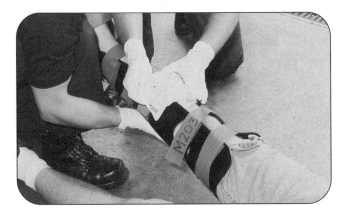

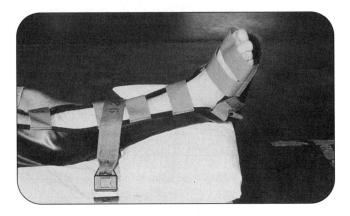

3. Apply _____ _____ if there is swelling,
   but do not place them _____ on the skin.

4. _____ the patient for transport and
   _____ the injured area.

Skill Drill 29-3: Applying a Rigid Splint
Test your knowledge of this skill drill by filling in the correct words in the photo captions.

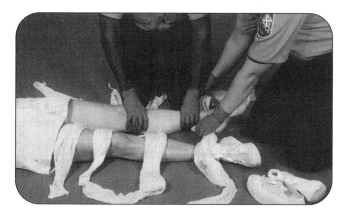

1. Provide gentle _____ and _____
   _____ for the limb.

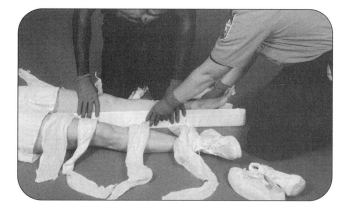

2. Second EMT-B places the splint _____ or
   _____ the limb. _____ between the
   limb and the splint as needed to ensure even pressure
   and contact.

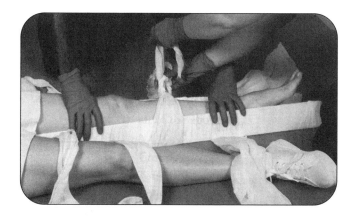

3. Secure the splint to the limb with _____.

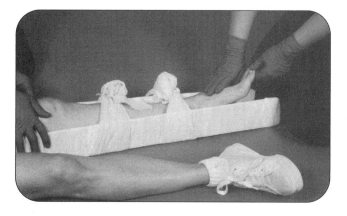

4. Assess and record _____ _____
   function.

Skill Drill 29-4: Applying a Zippered Air Splint
Test your knowledge of this skill drill by filling in the correct words in the photo captions.

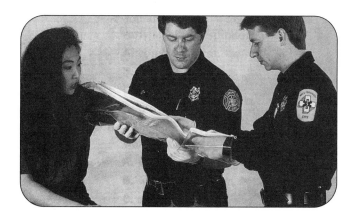

1. Support the injured limb and apply gentle
   _____ as your partner applies the open,
   deflated splint.

2. Zip up the splint, inflate it by _____ or by
   _____, and test the _____. Check and
   record _____ _____ function.

Skill Drill 29-5: Applying an Unzipped Air Splint
Test your knowledge of this skill drill by filling in the correct words in the photo captions.

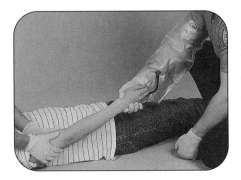

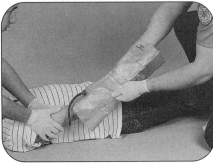

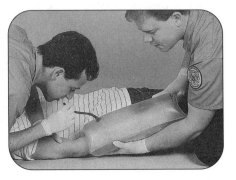

1. _____ the injured limb. Have your partner place his or her arm through the splint to grasp the patient's _____ or _____.

2. Apply gentle _____ while sliding the splint onto the injured limb.

3. _____ the splint.

Skill Drill 29-6: Applying a Vacuum Splint
Test your knowledge of this skill drill by filling in the correct words in the photo captions.

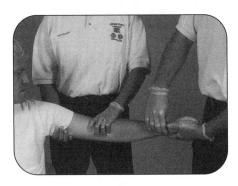

1. _____ and _____ the injury.

2. Place the splint and _____ it around the limb.

3. _____ the air _____ _____ the splint through the suction valve, and then _____ the valve.

Skill Drill 29-7: Applying a Hare Traction Splint
Test your knowledge of this skill drill by placing the photos below in the correct order. Number the first step with a "1," the second step with a "2," etc.

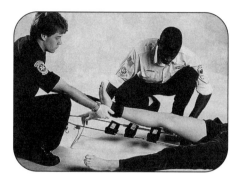

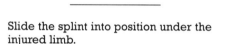

Slide the splint into position under the injured limb.

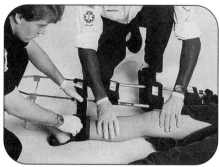

Support the injured limb as your partner fastens the ankle hitch about the foot and ankle.

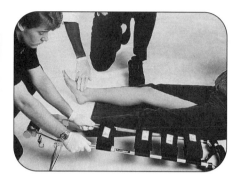

Expose the injured limb and check pulse, motor, and sensory function. Place the splint beside the uninjured limb, adjust the splint to proper length, and prepare the straps.

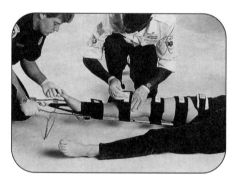

Secure and check support straps. Assess pulse, motor, and sensory functions.

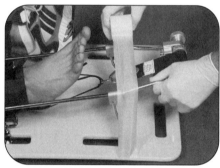

Secure the patient and splint to the backboard in a way that will prevent movement of the splint during patient movement and transport.

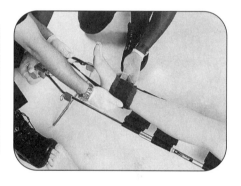

Connect the loops of the ankle hitch to the end of the splint as your partner continues to maintain traction. Carefully tighten the ratchet to the point that the splint holds adequate traction.

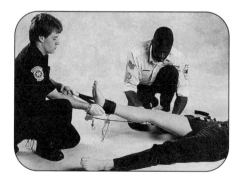

Continue to support the limb as your partner applies gentle in-line traction to the ankle hitch and foot.

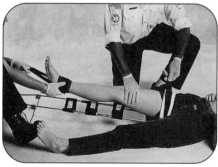

Pad the groin and fasten the ischial strap.

Skill Drill 29-8: Applying a Sager Traction Splint
Test your knowledge of this skill drill by placing the photos below in the correct order. Number the first step with a "1," the second step with a "2," etc.

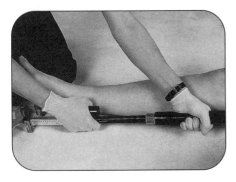

Estimate the proper length of the splint by placing it next to the injured limb. Fit the ankle pads to the ankle.

Tighten the ankle harness just above the malleoli.
Snug the cable ring against the bottom of the foot.

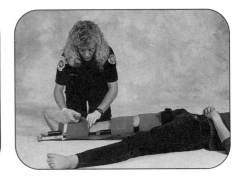

Secure the splint with elasticized cravats.

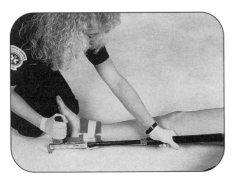

Extend the splint's inner shaft to apply traction of about 10% of body weight.

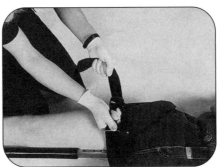

Place the splint at the inner thigh, apply the thigh strap at the upper thigh, and secure snugly.

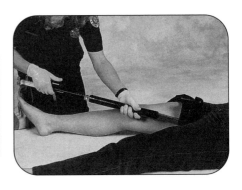

After exposing the injured area, check the patient's pulse and motor and sensory function.
Adjust the thigh strap so that it lies anteriorly when secured.

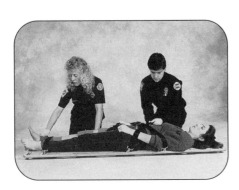

Secure the patient to a long backboard. Check pulse, motor, and sensory functions.

Skill Drill 29-9: Splinting the Hand and Wrist
Test your knowledge of this skill drill by filling in the correct words in the photo captions.

 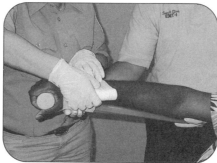 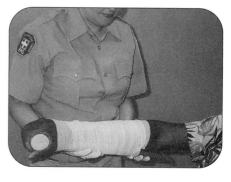

1. Move the hand into the

    _____ _____

    _____. Place a soft

    _____ _____ in the

    palm.

2. Apply a _____

    _____ splint on the

    _____ side with fingers

    _____.

3. Secure the splint with a

    _____ _____.

Assessment Review

Answer the following questions pertaining to the assessment of the types of emergencies discussed in this chapter.

_____ 1. You respond to a motorcycle accident for a 41-year-old man who is unconscious. He has obvious deformity to both lower legs and is bleeding moderately from an open fracture. His airway is open and he is making gurgling noises. Pulse is rapid and weak. Distal pulses are very weak. Your first priority with this patient is to:

    A. control bleeding.

    B. apply splints.

    C. maintain an airway.

    D. apply PASG.

_____ 2. You have loaded the patient in the question above and are on the way to the hospital. You have secured the airway and immobilized the fractures. How often would you reassess his vital signs?

    A. Every 3 minutes

    B. Every 5 minutes

    C. Every 10 minutes

    D. Every 15 minutes

_____ 3. You are called to a 16-year-old female patient who was injured in a basketball game. She is alert and oriented. Her airway is open and respirations are within normal limits. Her pulse is strong and regular. Distal pulses are present. She states that she felt her ankle pop and she immediately became nauseated. You decide to assess neurovascular status. When would you not perform the motor test?

    A. When you get a pain response

    B. When the patient walks away

    C. When you feel a distal pulse

    D. When the patient feels your touch

_____ 4. You are called to the local junior high school where a 12-year-old boy fell and hurt his wrist. There is obvious deformity. He is alert and oriented. Respirations and pulse are within normal limits. Distal pulse is present. It is important to remember to:

    A. use a zippered air splint.

    B. splint in position of function.

    C. splint the wrist only.

    D. completely cover the wrist and hand.

_____ 5. When you have applied a traction splint, the last thing that you do is:

    A. check pulse, motor, and sensation.

    B. release traction if the pulse disappears.

    C. apply elasticized straps.

    D. secure to a backboard.

Emergency Care Summary

Complete the statements pertaining to emergency care for the types of emergencies discussed in this chapter by filling in the missing words.

Please note that only two sections of the Emergency Care Summary are included here.

NOTE: While the steps below are widely accepted, be sure to consult and follow your local protocol.

Musculoskeletal Injuries

Applying a Hare Traction Splint

1. Expose the injured limb and check pulse, motor, and _____ function. Place the splint beside the uninjured limb, adjust the splint to proper length, and prepare the straps.
2. Support the injured limb as your partner fastens the ankle hitch about the foot and ankle.
3. Continue to support the limb as your partner applies gentle _____ _____ to the ankle hitch and foot.
4. Slide the splint into position under the injured limb.
5. Pad the groin and fasten the _____ strap.
6. Connect the loops of the ankle hitch to the end of the splint as your partner continues to maintain traction. Carefully tighten the ratchet to the point that the splint holds adequate traction.
7. Secure and check _____ _____. Assess pulse, motor, and sensory functions.
8. Secure the patient and splint to the backboard in a way that will prevent movement of the splint during patient movement and transport.

Applying a Sager Traction Splint

1. After exposing the injured area, check the patient's pulse and motor and sensory function. Adjust the thigh strap so that it lies _____ when secured.
2. Estimate the proper length of the splint by placing it next to the injured limb. Fit the ankle pads to the ankle.
3. Place the splint at the _____ thigh, apply the thigh strap at the _____ thigh, and secure snugly.
4. Tighten the ankle harness just above the _____. Snug the cable ring against the bottom of the foot.
5. Extend the splint's inner shaft to apply traction of about _____ of body weight.
6. Secure the splint with elasticized cravats.
7. Secure the patient to a long backboard. Check pulse, motor, and sensory functions.

CHAPTER

30 Head and Spine Injuries

Workbook Activities

The following activities have been designed to help you. Your instructor may require you to complete some or all of these activities as a regular part of your EMT-B training program. You are encouraged to complete any activity that your instructor does not assign as a way to enhance your learning in the classroom.

Chapter Review

The following exercises provide an opportunity to refresh your knowledge of this chapter.

Matching

Match each of the terms in the left column to the appropriate definition in the right column.

_____ 1. Cerebellum

_____ 2. Brain stem

_____ 3. Somatic nervous system

_____ 4. Autonomic nervous system

_____ 5. Spinal column

_____ 6. Central nervous system

_____ 7. Cerebral edema

_____ 8. Connecting nerves

_____ 9. Intervertebral disk

_____ 10. Meninges

A. consists of 33 bones

B. swelling of the brain

C. the brain and spinal cord

D. controls movement

E. the part of the central nervous system that controls virtually all the functions that are absolutely necessary for life

F. three distinct layers of tissue that surround and protect the brain and spinal cord within the skull and spinal cord

G. the part of the nervous system that regulates involuntary functions

H. the part of the nervous system that regulates voluntary activities

I. located in the brain and spinal cord, these connect the motor and sensory nerves

J. cushion that lies between the vertebrae

Multiple Choice

Read each item carefully, then select the best response.

_____ **1.** The nervous system includes:

    A. the brain.

    B. the spinal cord.

    C. billions of nerve fibers.

    D. all of the above

_____ **2.** The nervous system is divided into two parts: the central nervous system and the:

    A. autonomic nervous system.

    B. peripheral nervous system.

    C. sympathetic nervous system.

    D. somatic nervous system.

_____ **3.** The brain is divided into three major areas: the cerebrum, the cerebellum, and the:

    A. foramen magnum.

    B. meninges.

    C. brain stem.

    D. spinal column.

_____ **4.** Injury to the head and neck may indicate injury to the:

    A. thoracic spine.

    B. lumbar spine.

    C. cervical spine.

    D. sacral spine.

_____ **5.** The _____ is composed of three layers of tissue that surround the brain and spinal cord within the skull and spinal canal.

    A. meninges

    B. dura mater

    C. pia mater

    D. arachnoid space

_____ **6.** The skull is divided into two large structures: the cranium and the:

    A. occipital.

    B. face.

    C. parietal.

    D. foramen magnum.

_____ **7.** Peripheral nerves include:

    A. connecting nerves.

    B. sensory nerves.

    C. motor nerves.

    D. all of the above

_____ **8.** The brain and spinal cord float in cerebrospinal fluid (CSF), which:

    A. acts as a shock absorber.

    B. bathes the brain and spinal cord.

    C. buffers them from injury.

    D. all of the above

_____ **9.** The autonomic nervous system is composed of two parts: the sympathetic nervous system and the:

   A. peripheral nervous system.

   B. central nervous system.

   C. parasympathetic nervous system.

   D. somatic nervous system.

_____ **10.** The most prominent and the most easily palpable spinous process is at the _____ cervical vertebra at the base of the neck.

   A. 7th

   B. 6th

   C. 5th

   D. 4th

_____ **11.** You respond to a 14-year-old male who fell out of a tree at a local park. He is unresponsive. His airway is open and respirations are 16 breaths/min and regular. His pulse is strong and regular. Distal pulses are present. You manage the c-spine. Who should you ask for help in determining how the injury happened?

   A. First responders

   B. Family members

   C. Bystanders

   D. All of the above

_____ **12.** Emergency medical care of a patient with a possible spinal injury begins with:

   A. opening the airway.

   B. level of consciousness.

   C. summoning law enforcement.

   D. BSI precautions.

_____ **13.** The _____ is a tunnel running the length of the spine, which encloses and protects the spinal cord.

   A. foramen magnum

   B. spinal canal

   C. foramen foramina

   D. meninges

_____ **14.** Once the head and neck are manually stabilized, you should assess:

   A. pulse.

   B. motor function.

   C. sensation.

   D. all of the above

_____ **15.** You are called to a motor vehicle collision where you have a 27-year-old woman who has a bump on her head. You immediately take manual stabilization of the head. Her airway is open and respirations are within normal limits. Her pulse is a little fast but strong and regular. Distal pulses are present. You can release manual stabilization when:

   A. the patient's head and torso are in line.

   B. the patient is secured to a backboard with the head immobilized.

   C. the rigid cervical collar is in place.

   D. the patient arrives at the hospital.

_____ **16.** The ideal procedure for moving a patient from the ground to the backboard is the:

   A. four-person log roll.

   B. lateral slide.

   C. four-person lift.

   D. push and pull maneuver.

_____ **17.** You respond to a motor vehicle collision with a 29-year-old woman who struck the rearview mirror and has serious bleeding from the scalp. Her airway is open and respirations are normal. The pulse is a little rapid but strong and regular. Distal pulses are present and there is no deformity to the skull. Most bleeding from the scalp can be controlled by:

    A. direct pressure.

    B. elevation.

    C. pressure point.

    D. tourniquet.

_____ **18.** Exceptions to using a short spinal extrication device include all of the following, except:

    A. you or the patient is in danger.

    B. the patient is conscious and complaining of lumbar pain.

    C. you need to gain immediate access to other patients.

    D. the patient's injuries justify immediate removal.

_____ **19.** Your patient in question 17 has the same presentation except that she has an open skull fracture. You know that excessive pressure to the wound site could:

    A. increase intracranial pressure.

    B. push bone fragments into the brain.

    C. increase the size of the soft-tissue injury.

    D. all of the above

_____ **20.** A _____ is a temporary loss or alteration of a part or all of the brain's abilities to function without actual physical damage to the brain:

    A. contusion

    B. concussion

    C. hematoma

    D. subdural hematoma

_____ **21.** Symptoms of a concussion include:

    A. dizziness.

    B. weakness.

    C. visual changes.

    D. all of the above

_____ **22.** Intracranial bleeding outside of the dura and under the skull is known as a(n):

    A. concussion.

    B. intracerebral hemorrhage.

    C. subdural hematoma.

    D. epidural hematoma.

_____ **23.** The difference in signs and symptoms of traumatic vs. nontraumatic brain injuries is the:

    A. lack of altered mental status.

    B. lack of mechanism of injury.

    C. lack of swelling.

    D. increase in blood pressure.

_____ **24.** _____ is the most reliable sign of a closed head injury.

    A. Vomiting

    B. Decreased LOC

    C. Seizures

    D. Numbness and tingling in extremities

_____ **25.** _____ is one of the most common, and one of the most serious, complications of a head injury.

    A. Cyanosis

    B. Hypoxia

    C. Vomiting

    D. Cerebral edema

_____ **26.** Common causes of head injuries include all of the following, except:

    A. direct blows.

    B. motor vehicle crashes.

    C. seizure activity.

    D. sports injuries.

_____ **27.** Assessment of mental status is accomplished through the use of the mnemonic:

    A. SAMPLE.

    B. OPQRST.

    C. AVPU.

    D. AEIOU-TIPS.

_____ **28.** You respond to an 18-year-old male who fell while rock climbing. He is unconscious with an open airway. The respiration rate and pulse rates are within normal limits. His distal pulses are intact. You check his pupils and they are unequal. You know this could be a sign of:

    A. increased intracranial pressure.

    B. a congenital problem.

    C. damage to the nerves that control dilation and constriction.

    D. all of the above

_____ **29.** Patients with head injuries often have injuries to the _____ as well.

    A. face

    B. torso

    C. cervical spine

    D. extremities

_____ **30.** Proper order of treatment for traumatic head injuries includes:

    A. scene safety, airway, LOC with c-spine control, breathing, circulation.

    B. LOC with c-spine control, airway, breathing, circulation.

    C. LOC, airway, breathing, circulation, c-spine.

    D. BSI, ABCs, LOC, c-spine control.

_____ **31.** A cervical collar should be applied to a patient with a possible spinal injury based on:

    A. the mechanism of injury.

    B. the history.

    C. signs and symptoms.

    D. all of the above

_____ **32.** Helmets must be removed in all of the following cases, except:

    A. cardiac arrest.

    B. when the helmet allows for excessive movement.

    C. when there are no impending airway or breathing problems.

    D. when a shield cannot be removed.

_____ **33.** Your best choice of action for a child involved in a motor vehicle crash and found in their car seat is to:

    A. immobilize the child in the car seat.

    B. rule out spinal injury and place the child with a parent.

    C. pad sides of car seat but leave space to allow for lateral movement.

    D. move the child to a pediatric immobilization device.

Labeling

Label the following diagrams with the correct terms.

1. The Brain

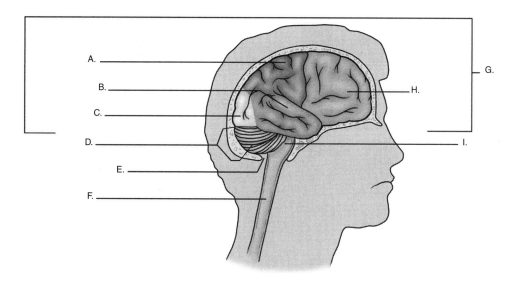

A. _____

B. _____

C. _____

D. _____

E. _____

F. _____

G. _____

H. _____

I. _____

2. The Connecting Nerves in the Spinal Cord

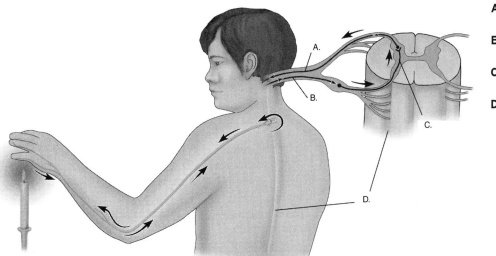

A. _____

B. _____

C. _____

D. _____

3. The Spinal Column

A. _____

B. _____

C. _____

D. _____

E. _____

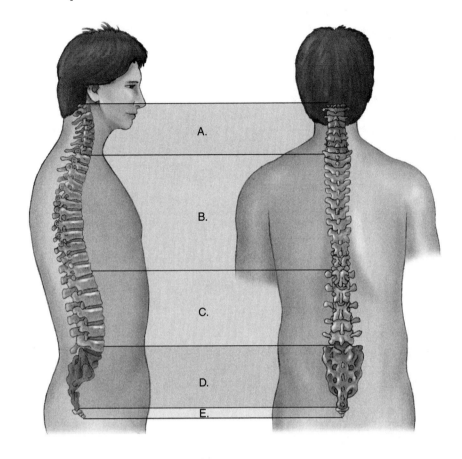

Fill-in
Read each item carefully, then complete the statement by filling in the missing word(s).

1. The _____ nerves carry information to the muscles.

2. The dura mater, arachnoid, and pia mater are layers of _____ within the skull and spinal canal.

3. The brain and spinal cord are part of the _____ nervous system.

4. Within the peripheral nervous system, there are _____ pairs of spinal nerves.

5. The _____ nerves pass through holes in the skull and transmit sensations directly to the brain.

6. Vertebrae are separated by cushions called _____ _____.

7. The skull has two large structures of bone, the _____ and the _____.

8. The _____ and _____ are the inner two layers of the meninges and are much

thinner than the dura mater.

9. The _____ nervous system reacts to stress.

10. The _____ nervous system causes the body to relax.

True/False

If you believe the statement to be more true than false, write the letter "T" in the space provided. If you believe the statement to be more false than true, write the letter "F."

_____ **1.** A distracted spine has been moved laterally.

_____ **2.** If a sensory nerve in the reflex arc detects an irritating stimulus, it will bypass the motor nerve and send a message directly to the brain.

_____ **3.** Voluntary activities are those actions we perform unconsciously.

_____ **4.** The autonomic nervous system is composed of the sympathetic nervous system and the parasympathetic nervous system.

_____ **5.** The parasympathetic nervous system reacts to stress with the fight-or-flight response whenever it is confronted with a threatening situation.

_____ **6.** All patients with suspected head and/or spine injuries should have their head realigned to an in-line neutral position.

_____ **7.** When assessing a patient for possible spinal injury, you should begin with a focused history and physical exam.

_____ **8.** Your ideal procedure for moving a patient from the ground to a backboard is the four-person log roll.

_____ **9.** You should not try to put a patient on a short board if they are in danger.

_____ **10.** To properly measure a cervical collar, use the manufacturer's specifications.

Short Answer

Complete this section with short written answers using the space provided.

1. List the five basic questions to ask a conscious patient when conducting an assessment of a head or head and spine injury.

_____

_____

2. List the reasons for not placing the head/spine injury patient's head into a neutral in-line position.

_____

_____

3. List the three major types of brain injuries.

_____

_____

4. List at least five signs and symptoms of a head injury.

_____

_____

5. List the three general principles for treating a head injury.

_____

_____

6. List the six questions to ask yourself when deciding whether or not to remove a helmet.

_____

_____

Word Fun

The following crossword puzzle is an activity provided to reinforce correct spelling and understanding of medical terminology associated with emergency care and the EMT-B. Use the clues in the column to complete the puzzle.

Across

2. Cushion between vertebrae
5. Layers of tissues surrounding the brain
6. Cerebral trauma without broken skin
8. Controls primary life functions
9. Voluntary part of CNS
10. Inability to remember after the event

Down

1. Coordinates body movement
3. Inability to remember the event
4. Join motor and sensory nerves
6. Swelling of the brain
7. Pulling the spine along its length

Ambulance Calls

The following case scenarios provide an opportunity to explore the concerns associated with patient management. Read each scenario, then answer each question in detail.

1. You are dispatched to a bicycle versus car. The driver of the car is uninjured, but the bicyclist is reported as "severely injured." You arrive to find the patient lying in the street, unconscious and with an apparent head injury. The patient was not wearing a helmet when she was struck by the car. Witnesses say she was launched into the windshield, and then landed in the road.

   How do you best manage this patient?

   _____

   _____

   _____

2. You are dispatched to assist a "young child fallen." You arrive to find a frantic parent who tells you her child was playing on the family's trampoline in the back yard. She was bouncing very high and accidentally launched herself off the trampoline, landing face down onto a concrete pad in the neighbor's yard. She responds to painful stimuli and has snoring respirations.

   How do you best manage this patient?

   _____

   _____

   _____

3. You are dispatched to a motor vehicle crash with major damage to the patient compartment. Your patient, an 18-month-old boy, is still in his car seat in the center of the back seat. He responds appropriately and there is no damage to his seat. He has no visible injuries, but a front seat passenger was killed.

   How would you best manage this patient?

   _____

   _____

   _____

Skill Drills

Skill Drill 30-1: Performing Manual In-Line Stabilization

Test your knowledge of this skill drill by filling in the correct words in the photo captions.

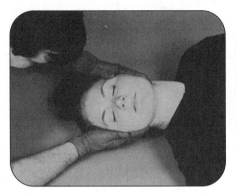

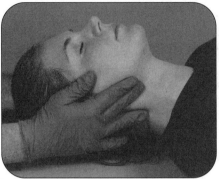

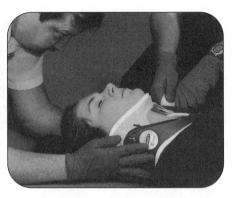

1. Kneel behind the patient and place your hands firmly around the _____ of the _____ on either _____ .

2. Support the lower jaw with your _____ and _____ fingers, and the head with your _____ . Gently _____ the head into a _____ , _____ position, aligned with the torso. Do not _____ the head or neck excessively, forcefully, or rapidly.

3. Continue to _____ the head manually while your partner places a rigid _____ _____ around the neck. Maintain _____ _____ until you have completely secured the patient to a backboard.

Skill Drill 30-2: Immobilizing a Patient to a Long Backboard

Test your knowledge of this skill drill by placing the photos below in the correct order. Number the first step with a "1," the second step with a "2," etc. Also, fill in the correct words in the photo captions.

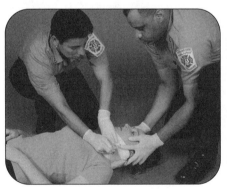

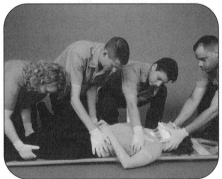

_____

Apply a _____ _____ .

_____

_____ the patient on the board.

_____

Place _____ across the patient's forehead.

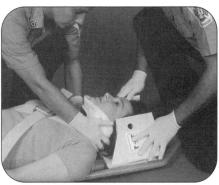

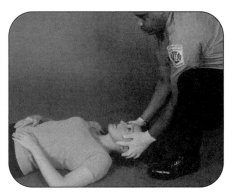

On command, rescuers _____ the patient toward themselves, quickly examine the _____, slide the backboard under the patient, and roll the patient onto the board.

Begin to secure the patient's head using a commercial immobilization device or _____ _____.

Apply and maintain _____ _____ .

Assess _____ _____ in all extremities.

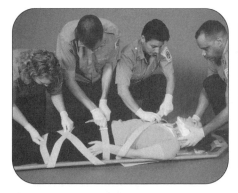

Secure the _____, _____, and _____ _____.

Rescuers _____ on one side of the patient and place _____ on the far side of the patient.

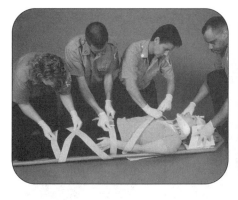

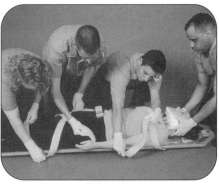

Check all _____ and readjust as needed.

Reassess _____ _____ in all extremities.

Secure the _____ _____ first.

Skill Drill 30-3: Immobilizing a Patient Found in a Sitting Position

Test your knowledge of this skill drill by placing the photos below in the correct order. Number the first step with a "1," the second step with a "2," etc.

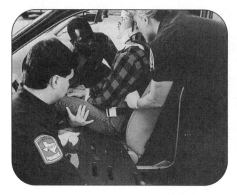

_____

Wedge a long backboard next to the patient's buttocks.

_____

Open the side flaps, and position them around the patient's torso, snug around the armpits.

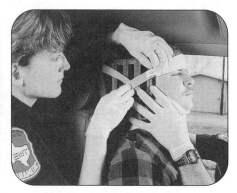

_____

Pad between the head and the device as needed.
Secure the forehead strap and fasten the lower head strap around the collar.

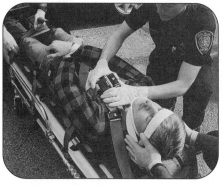

_____

Secure the immobilization devices to each other.
Reassess pulse, motor, and sensory functions in each extremity.

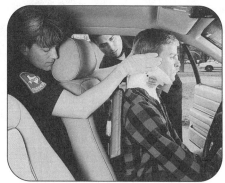

_____

Insert a short spine immobilization device between the patient's upper back and the seat.

_____

Secure the upper torso flaps, then the midtorso flaps.

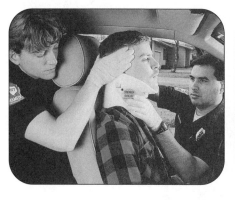

_____

Stabilize the head and neck in a neutral, in-line position.
Assess pulse, motor, and sensory function in each extremity.
Apply a cervical collar.

_____

Secure the groin (leg) straps. Check and adjust torso straps.

Turn and lower the patient onto the long board.
Lift the patient, and slip the long board under the spine device.

Skill Drill 30-4: Immobilizing a Patient Found in a Standing Position
Test your knowledge of this skill drill by filling in the correct words in the photo captions.

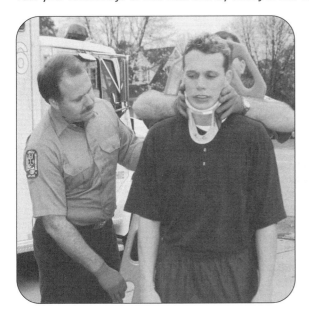

1. While _____ stabilizing the head and neck, apply a _____ _____. Position the board _____ the patient.

2. Position EMT-Bs at _____ and _____ the patient. Side EMT-Bs reach under patient's _____ and grasp _____ at or slightly above _____ level.

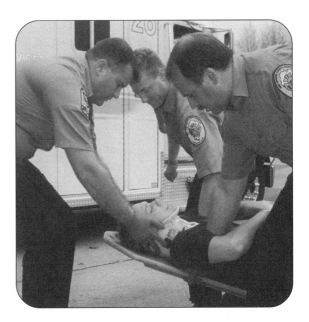

3. Prepare to lower the patient. EMT-Bs on the sides should be _____ the EMT-B at the head and _____ for his or her _____.

4. On command, _____ the backboard to the ground.

Skill Drill 30-5: Application of a Cervical Collar

Test your knowledge of this skill drill by filling in the correct words in the photo captions.

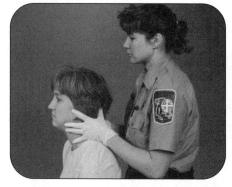

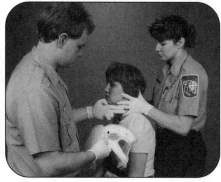

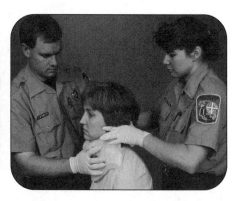

1. Apply _____ stabilization.

2. Measure the proper _____ _____.

3. Place the _____ _____ first.

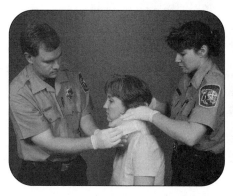

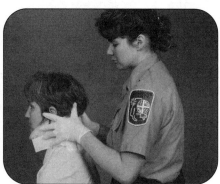

4. _____ the collar around the neck and _____ the collar.

5. Ensure proper _____ and maintain _____, _____ stabilization.

Skill Drill 30-6: Removing a Helmet

Test your knowledge of this skill drill by placing the photos below in the correct order. Number the first step with a "1," the second step with a "2," etc. Also, fill in the correct words in the photo captions.

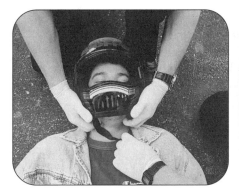

———————————

Prevent head movement by placing your _____ on either side of the helmet and fingers on the _____ _____ . Have your partner _____ the strap.

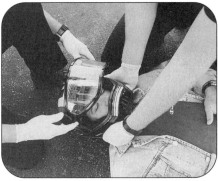

———————————

Kneel down at the patient's head with your _____ at one side. Open the face shield to assess _____ and _____ . Remove _____ if present.

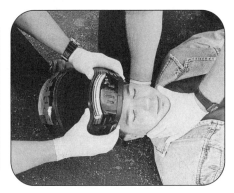

———————————

Have your partner slide the hand from the _____ to the _____ of the head to prevent it from snapping back.

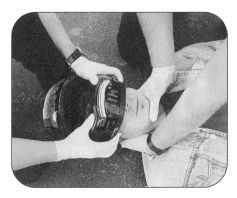

———————————

Gently slip the helmet about _____ off, then stop.

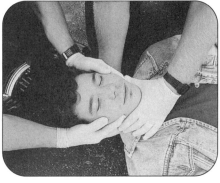

———————————

Remove the helmet and _____ the cervical spine.
Apply a _____ _____ and secure the patient to a _____ _____ .
_____ as needed to prevent neck flexion or extension.

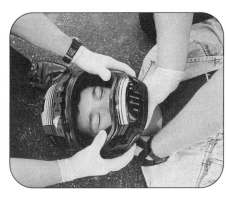

———————————

Have your partner place one hand at the _____ of the _____ and the other at the _____ .

Assessment Review

Answer the following questions pertaining to the assessment of the types of emergencies discussed in this chapter.

_____ 1. You respond to a patient who was assaulted and is unconscious. Upon reaching his side you check his airway and it is open. The breathing is at 18 breaths/min and regular. Pulse is strong and regular with distal pulses present. You want to administer oxygen. How would you give it?

    A. 2 L/min

    B. 6 L/min

    C. 10 L/min

    D. 15 L/min

_____ 2. You and your partner have determined that you need to put the patient above in full immobilization. When can your partner release manual stabilization of the head?

    A. When the C-collar is applied

    B. When the torso is secured to the board

    C. When the patient is completely secured to the board

    D. When you arrive at the hospital

_____ 3. You decide to put the patient on a long backboard and will use the logroll technique to accomplish the task. When do you check the back of the patient?

    A. After being secured to the board

    B. As the patient is rolled onto their side

    C. Before any movement is attempted

    D. When you move them to the hospital bed

_____ 4. If you respond to a patient who is in a sitting position and stable, how would you immobilize them?

    A. Lay them down and log roll them.

    B. Use a scoop stretcher.

    C. Use a short board.

    D. Have them lay down on your backboard.

_____ 5. You respond to a motorcycle accident. You decide that you need to remove the rider's helmet. What is the minimum number of people required to remove the helmet?

    A. 2

    B. 3

    C. 4

    D. 5

Emergency Care Summary

Complete the statements pertaining to emergency care for the types of emergencies discussed in this chapter by filling in the missing words.
Please note that only two sections of the Emergency Care Summary are included here.

NOTE: While the steps below are widely accepted, be sure to consult and follow your local protocol.

Head and Spine Injuries

Immobilizing a Patient Found in a Sitting Position

1. Stabilize the head and neck in a _____, _____ position. Assess pulse, motor, and sensory function in each extremity. Apply a _____ _____.

2. Insert a _____ _____ immobilization device between the patient's upper back and the seat.

3. Open the side flaps, and position them around the patient's torso, snug around the armpits.

4. Secure the upper torso flaps, then the midtorso flaps.

5. Secure the groin (leg) straps. Check and adjust torso straps.

6. Pad between the head and the device as needed. Secure the forehead strap and fasten the _____ _____ _____ around the collar.

7. Wedge a _____ _____ next to the patient's buttocks.

8. Turn and lower the patient onto the long board. Lift the patient, and slip the long board under the spine device.

9. Secure the immobilization devices to each other. Reassess pulse, motor, and sensory functions in each _____.

CHAPTER

31 Pediatric Emergencies

Workbook Activities

The following activities have been designed to help you. Your instructor may require you to complete some or all of these activities as a regular part of your EMT-B training program. You are encouraged to complete any activity that your instructor does not assign as a way to enhance your learning in the classroom.

Chapter Review

The following exercises provide an opportunity to refresh your knowledge of this chapter.

Matching

Match each of the terms in the left column to the appropriate definition in the right column.

_____ **1.** Gastrostomy tube   **A.** ages 12 to 18 years

_____ **2.** Shunt   **B.** ages 3 to 6 years

_____ **3.** Tracheostomy tube   **C.** soft openings within the skull of an infant

_____ **4.** Toddler   **D.** specialized medical practice devoted to the care of children

_____ **5.** Preschool-age children   **E.** used for breathing

_____ **6.** Adolescents   **F.** diverts excess cerebrospinal fluid

_____ **7.** Infancy   **G.** first month after birth

_____ **8.** Neonate   **H.** the first year of life

_____ **9.** Fontanels   **I.** used for feeding

_____ **10.** Pediatrics   **J.** after infancy, until about 3 years of age

Multiple Choice
Read each item carefully, then select the best response.

_____ **1.** In addition to the tongue, the _____ help(s) to produce a smaller opening to move air easily.
- **A.** tonsils
- **B.** adenoids
- **C.** soft pallet
- **D.** all of the above

_____ **2.** A respiratory rate of _____ breaths/min is normal for the newborn.
- **A.** 12 to 20
- **B.** 20 to 40
- **C.** 30 to 60
- **D.** 50 to 80

_____ **3.** Breathing requires the use of the _____ and diaphragm.
- **A.** chest muscles
- **B.** neck muscles
- **C.** subclavian muscles
- **D.** abdominal muscles

_____ **4.** You respond to an automobile accident involving a 3-year-old boy. You remove the patient from the vehicle and prepare to secure him to a long backboard. You should be careful not to cause respiratory compromise by putting a strap across the:
- **A.** thorax.
- **B.** air.
- **C.** diaphragm.
- **D.** lungs.

_____ **5.** The primary method for the body to compensate for decreased oxygenation is to:
- **A.** increase the respiratory rate.
- **B.** increase the heart rate.
- **C.** increase the blood pressure.
- **D.** increase diaphragm contractions.

_____ **6.** Signs of vasoconstriction can include:
- **A.** weak peripheral pulses.
- **B.** delayed capillary refill.
- **C.** cool hands or feet.
- **D.** all of the above

_____ **7.** Infants respond mainly to _____ stimuli.
- **A.** social
- **B.** mental
- **C.** physical
- **D.** all of the above

_____ **8.** You respond to a call for a sick 10-month-old. As you enter the residence, you would begin your initial assessment from across the room. What would you be checking?
- **A.** Work of breathing
- **B.** Skin color and alertness
- **C.** Level of activity
- **D.** All of the above

_____ **9.** Injuries in the _____ age group are more frequent.

    A. infant

    B. toddler

    C. preschool

    D. school

_____ **10.** At the _____ age, children are easily distracted with counting games and small toys.

    A. infant

    B. toddler

    C. preschool

    D. adolescent

_____ **11.** You are called to a local residence for a child on a home ventilator. The parents state he has been having trouble with his breathing. You should first determine:

    A. baseline vital signs.

    B. normal baseline status.

    C. previous injuries.

    D. over-the-counter medications.

_____ **12.** For the patient above, your first priority in treating a special-needs child includes:

    A. obtaining an extensive history.

    B. determining mode of transportation.

    C. assessing the airway.

    D. obtaining the patient's medications to take to the hospital.

_____ **13.** You are called to the home of a 13-year-old girl with altered mental status. Upon arrival, you notice the child has a tracheostomy tube. You perform your initial assessment. You know that potential problems associated with the tracheostomy tube include:

    A. obstruction of the tube by mucous plugs.

    B. bleeding.

    C. air leakage around the tube.

    D. all of the above

_____ **14.** Tubes that extend from the brain to the abdomen to drain excess cerebrospinal fluid that may accumulate near the brain are called:

    A. shunts.

    B. central lines.

    C. G-tubes.

    D. tracheostomy tubes.

Fill-in

Read each item carefully, then complete the statement by filling in the missing word.

1. The specialized medical practice devoted to the care of the young is called _____.

2. The _____ is longer and more rounded compared to the size of the mandible, or lower jaw, in younger children.

3. In a child, the _____ is softer and narrower.

4. An infant's heart rate can become as high as _____ beats or more per minute if the body needs to compensate for injury or illness.

5. _____ _____ is an early sign that the child may be compensating for decreased perfusion.

6. Infants have two soft openings within the skull called _____.

7. Most _____ are able to think abstractly and can participate in decision making.

True/False

If you believe the statement to be more true than false, write the letter "T" in the space provided. If you believe the statement to be more false than true, write the letter "F."

_____ **1.** Toddlers often resist separation and demonstrate stranger anxiety.

_____ **2.** The skeletal system contains growth plates at the ends of long bones, which enable these bones to grow during childhood.

_____ **3.** Adulthood begins at age 18.

_____ **4.** Infants are usually afraid of strangers, because they are the center of attention in most families.

_____ **5.** Preschool-age children have a rich fantasy life, which can make them particularly fearful of pain and change involving their bodies.

_____ **6.** The parent or caregiver of a special-needs child will be an important part of your assessment.

_____ **7.** Children with diabetes who receive insulin and tube feedings may become hyperglycemic quickly if tube feedings are discontinued.

_____ **8.** If a shunt becomes clogged due to infection, changes in mental status and respiratory arrest may occur.

Short Answer

Complete this section with short written answers using the space provided.

1. Discuss developmental considerations for infancy and approach for caregivers.

_____

_____

2. Give four examples of special-needs children.

_____

_____

3. Discuss developmental considerations for toddlers and approach for caregivers.

_____

_____

4. Discuss developmental considerations for the school-age child and approach for caregivers.

_____

_____

5. Discuss developmental considerations for the adolescent and approach for caregivers.

_____

_____

Word Fun

The following crossword puzzle is an activity provided to reinforce correct spelling and understanding of medical terminology associated with emergency care and the EMT-B. Use the clues in the column to complete the puzzle.

Across

2. Terrible twos stage
5. First month after birth
6. Feeding tube placed through abdominal wall
8. Tube diverting cerebrospinal fluid to abdomen

Down

1. Specialized medical practice devoted to care of children
3. First year of life
4. Two soft openings within the skull
7. Back of the head

Ambulance Calls

The following case scenarios provide an opportunity to explore the concerns associated with patient management. Read each scenario, then answer each question in detail.

1. You are dispatched to the residence of a toddler with a history of fever who is now unresponsive. You arrive to find a 13-year-old babysitter who tells you she is not sure what is wrong with the 2-year-old boy. She tells you he started "shaking all over" and she didn't know what to do. He is currently responsive to painful stimuli and warm to the touch.

How do you best manage this patient?

_____

_____

_____

2. You are dispatched to the residence of a 3-year-old child with a history of lung problems. The child, a very small boy, is cyanotic and lethargic. He is pain responsive. He has copious mucous secretions in his airway. The grandmother, who was sitting with the child, is hysterical.

How would you best manage this patient?

_____

_____

_____

3. You are dispatched to assist a young girl with shortness of breath. You arrive to find a 16-year-old female patient who is obviously having difficulty breathing. She is holding a metered-dose inhaler and is so air-starved, she cannot answer your questions.

How do you best manage this patient?

_____

_____

_____

CHAPTER

32 Pediatric Assessment and Management

Workbook Activities

The following activities have been designed to help you. Your instructor may require you to complete some or all of these activities as a regular part of your EMT-B training program. You are encouraged to complete any activity that your instructor does not assign as a way to enhance your learning in the classroom.

Chapter Review

The following exercises provide an opportunity to refresh your knowledge of this chapter.

Matching

Match each of the terms in the left column to the appropriate definition in the right column.

_____ 1. Croup
_____ 2. Wheezing
_____ 3. Epiglottitis
_____ 4. Rales
_____ 5. Stridor
_____ 6. Septum
_____ 7. Shock
_____ 8. Apnea
_____ 9. Blanching
_____ 10. Xiphoid process
_____ 11. Febrile seizure
_____ 12. Tripod position
_____ 13. Jaw-thrust maneuver

A. leaning forward on two arms stretched forward
B. seizure relating to a fever
C. absence of breathing
D. infection of the airway below the level of the vocal cords
E. turning white
F. whistling sound made from air moving through narrowed bronchioles
G. the lower tip of the sternum
H. infection of soft tissue in the area above the vocal cords
I. the central divider in the nose
J. crackling sound caused by flow of air through liquid
K. insufficient blood to body organs
L. high-pitched sound usually caused by swelling around vocal cords
M. used to open the airway with suspected spinal injuries

Multiple Choice

Read each item carefully, then select the best response.

_____ **1.** Positioning the airway in a neutral sniffing position:

   A. keeps the trachea from kinking when the neck is hyperextended.

   B. keeps the trachea from kinking when the neck is flexed.

   C. maintains the proper alignment if you have to immobilize the spine.

   D. all of the above

_____ **2.** Benefits of using a nasopharyngeal airway include all of the following, except:

   A. it is usually well tolerated.

   B. it may be used in the presence of head trauma.

   C. it is not as likely as the oropharyngeal airway to cause vomiting.

   D. it is used for conscious patients or those with altered levels of consciousness.

_____ **3.** Indications for assisting ventilations in a child include:

   A. respiratory rate of less than 12 breaths/min.

   B. respiratory rate of greater than 60 breaths/min.

   C. inadequate tidal volume.

   D. all of the above

_____ **4.** Errors in technique, when providing ventilations with a BVM device, that may result in gastric distention include:

   A. providing too much volume.

   B. squeezing the bag too forcefully.

   C. ventilating too fast.

   D. all of the above

_____ **5.** _____ is an infection of the airway below the level of the vocal cords, usually caused by a virus.

   A. Croup

   B. Tonsillitis

   C. Epiglottitis

   D. Pharyngitis

_____ **6.** Signs of severe airway obstruction include:

   A. inability to speak or cry.

   B. increasing respiratory difficulty with stridor.

   C. cyanosis.

   D. all of the above

_____ **7.** You are called to a 3-year-old in respiratory distress. As you approach the child, you would expect the early signs of respiratory distress to include all of the following, except:

   A. combativeness.

   B. anxiety.

   C. cyanosis.

   D. restlessness.

_____ **8.** If the child's condition in the question above were to get worse, you would expect the signs of increased work of breathing to include:

   A. nasal flaring.

   B. wheezing, stridor, or other abnormal airway sounds.

   C. accessory muscle use.

   D. all of the above

_____ 9. You are called to the home of a 14-year-old girl with seizures. Upon your arrival, she is actively seizing. The parents tell you she seems to have had two seizures without regaining consciousness; one stopped and then the other started. This is her third. She has been seizing for approximately 2 minutes. You know this condition as:

A. status epilepticus.

B. grand mal seizure.

C. absence seizure.

D. focal motor seizure.

_____ 10. During the postictal period, the patient may appear:

A. sleepy.

B. confused.

C. unresponsive.

D. all of the above

_____ 11. Most pediatric seizures are due to _____, which is why they are called febrile seizures.

A. infection

B. fever

C. ingestion

D. trauma

_____ 12. Signs that a patient is not breathing adequately include:

A. very slow respirations.

B. very shallow breaths.

C. cyanosis or pale lips.

D. all of the above

_____ 13. Care of the actively seizing child includes all of the following, except:

A. assessing and managing the ABCs.

B. noting the type of movement and position of the eyes.

C. cooling the patient with alcohol if there is fever.

D. making sure the patient is protected from hitting anything.

_____ 14. Nonverbal infants may demonstrate consciousness by:

A. tracking.

B. babbling and cooing.

C. crying.

D. all of the above

_____ 15. You respond to an 11-year-old boy who is unconscious. Nobody has any information on what happened. His respirations are rapid, pulse weak and rapid, and distal pulses are weak but present. His skin is cool and clammy. You know that common causes of shock in children include all of the following except:

A. heart attack.

B. head trauma.

C. dehydration.

D. pneumothorax.

_____ 16. At birth, most infants only need stimulation to:

A. cry.

B. wake up.

C. breathe.

D. move.

_____ 17. You respond to a 24-year-old woman who gave birth at home. Your partner takes responsibility for the mother. He assigns you the newborn. The baby is crying and you want to check a pulse. You know you can check the brachial; where else can you check the pulse in a neonate?

A. Carotid artery

B. Radial artery

C. Femoral artery

D. The base of the umbilical cord

_____ **18.** Respiratory problems leading to cardiopulmonary arrest in children may be caused by:

   A. a foreign body in the airway.

   B. near drowning.

   C. sudden infant death syndrome.

   D. all of the above

_____ **19.** _____ will decrease the risk of gastric distention and aspiration of vomitus by pushing the larynx back to compress and close off the esophagus.

   A. Using a BVM device

   B. The Sellick maneuver

   C. Abdominal compression

   D. none of the above

_____ **20.** While checking a pulse, you may observe other signs of circulation including:

   A. breathing.

   B. coughing.

   C. movement.

   D. all of the above

_____ **21.** When approaching a child, you should look for all of the following, except:

   A. work of breathing.

   B. pulse.

   C. level of activity.

   D. skin color.

_____ **22.** When caring for children with sports-related injuries, you should remember to:

   A. elevate the extremities.

   B. assist ventilations.

   C. immobilize the cervical spine.

   D. remove all helmets.

_____ **23.** Your single most important step in caring for a child with a head injury is to:

   A. immobilize the cervical spine.

   B. bandage all wounds.

   C. ensure an open airway.

   D. obtain a SAMPLE history.

_____ **24.** Signs of shock in children include all of the following, except:

   A. tachycardia.

   B. hypotension.

   C. poor capillary refill.

   D. mental status changes.

_____ **25.** For children who have had traumatic injuries, use a child-sized BVM device at a ventilation rate of _____ breaths/min.

   A. 10–12

   B. 12–20

   C. 15–25

   D. 25–30

_____ **26.** If _____—an abnormal airway sound made by turbulent airflow—is present, use the jaw-thrust maneuver to keep the airway open.

   A. rales

   B. rhonchi

   C. stridor

   D. wheezing

Fill-in

Read each item carefully, then complete the statement by filling in the missing word.

1. The term _____ _____ is used to describe a continuous seizure, or multiple seizures without a return to consciousness, for 30 minutes or more.

2. Because a young child might not be able to speak, your assessment of his or her condition must be based in large part on what you can _____ and _____.

3. _____ occurs when fluid losses are greater than fluid intake.

4. _____ _____ are devices that help to maintain the airway or assist in providing artificial ventilation.

5. An oropharyngeal airway should be used in neither conscious patients nor those who have a decreased level of consciousness, because both will have a _____ _____.

6. _____ _____ indicates the amount of oxygen getting to the organs of the body.

7. The _____ _____ is the amount of air that is delivered to the lungs and airways in one inhalation.

8. _____ should be considered as a possible cause of airway obstruction if a child has congestion, fever, drooling, and cold symptoms.

9. A _____ is the result of disorganized electrical activity in the brain.

10. Because a child's _____ is proportionately larger than an adult's, it exerts greater stress on the neck structures during a deceleration injury.

11. Children can lose a greater proportion of their blood volume than adults can before signs and symptoms of _____ develop.

True/False

If you believe the statement to be more true than false, write the letter "T" in the space provided. If you believe the statement to be more false than true, write the letter "F."

_____ 1. You should assist ventilations in all pediatric patients who have respiratory rates greater than 60 breaths/min.

_____ 2. Febrile seizures are self-limiting and do not need transport unless they recur.

_____ 3. Febrile seizures may not be accompanied by a postictal phase.

_____ 4. Alcohol applied to skin is a recommended method of cooling a patient.

_____ 5. Infants with severe dehydration may present with sunken fontanels.

_____ 6. Young children with severe dehydration will have delayed capillary refill.

_____ 7. Children with moderate dehydration should be rehydrated at home.

_____ 8. Children younger than 1 year should not have an AED applied.

_____ 9. Children can lose a greater proportion of blood than adults before showing signs or symptoms of shock.

Short Answer

Complete this section with short written answers using the space provided.

1. How is urine output assessed in infants?

_____

_____

2. List four signs of increased work of breathing in children.

_____

_____

Word Fun

The following crossword puzzle is an activity provided to reinforce correct spelling and understanding of medical terminology associated with emergency care and the EMT-B. Use the clues in the column to complete the puzzle.

Across

4. Absence of breathing

6. External openings of nasal passages

8. The lower tip of the sternum

Down

1. Increased respiratory rate

2. Turning white

3. Deficiency of red blood cells or hemoglobin

5. Dark green material in amniotic fluid

7. Central divider of the nose

9. Inadequate tissue perfusion

Ambulance Calls

The following case scenarios provide an opportunity to explore the concerns associated with patient management. Read each scenario, then answer each question in detail.

1. You are dispatched to assist an unconscious 11-year-old boy. You arrive to find a home filled with several teenagers who appear to have been having an unsupervised party. You see several beer cans and bottles of whiskey on the floor. They tell you that they "think" the 11-year-old was drinking beer, but they're "not sure." They noticed he was sleeping and tried to wake him up without success. You see that he has vomited and is currently lying on his back with snoring respirations. How do you best manage this patient?

_____

_____

_____

2. You are called to a residence for a 2-year-old child with difficulty breathing. The little girl has stridor and expiratory wheezes, as well as intercostal retractions. She is very upset by your arrival and clings to her mother. Her breathing worsens with agitation. Her mother tells you she is currently taking medication for an upper respiratory infection and has spent much of her life in and out of hospitals with respiratory problems.

How would you best manage this patient?

_____

_____

_____

3. It's 5:30 A.M. and you are dispatched to the home of a 6-month-old girl who is not breathing. You arrive to find a crying, young mother holding a lifeless baby. The infant is not breathing, is cold to the touch, and appears to have dependent lividity.

How do you best manage this patient?

_____

_____

_____

Skill Drills
Skill Drill 32-1: Positioning the Airway in a Child
Test your knowledge of this skill drill by filling in the correct words in the photo captions.

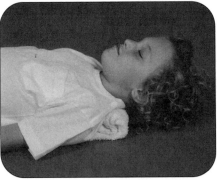

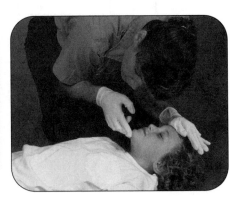

1. Position the child on a _____ surface.

2. Place a _____ towel about _____ inch(es) thick under the _____ and _____.

3. _____ the forehead to limit _____ and use the head tilt-chin lift to open the airway

Skill Drill 32-2: Inserting an Oropharyngeal Airway in a Child
Test your knowledge of this skill drill by filling in the correct words in the photo captions.

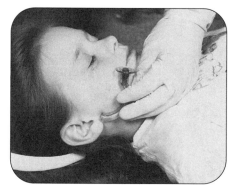

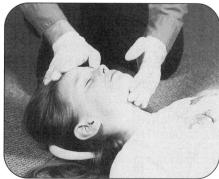

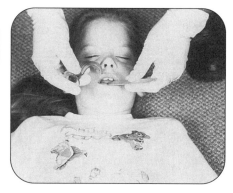

1. Determine the _____ _____ airway. Confirm the correct size _____, by placing it next to the patient's _____ .

2. Position the patient's _____ with the appropriate method.

3. Open the mouth. Insert the airway until the _____ rests against the _____ . _____ the airway.

Skill Drill 32-3: Inserting a Nasopharyngeal Airway in a Child
Test your knowledge of this skill drill by filling in the correct words in the photo captions.

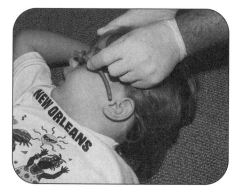

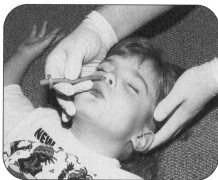

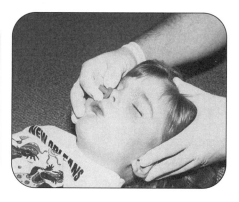

1. Determine the correct airway size by comparing its _____ to the opening of the _____ (naris). Place the airway next to the patient's _____ to confirm correct _____ . _____ the airway.

2. _____ the airway. Insert the _____ into the right naris with the bevel pointing toward the _____ .

3. Carefully move the tip forward until the _____ rests against the _____ of the nostril. Reassess the _____ .

Skill Drill 32-4: One-rescuer BVM Ventilation on a Child
Test your knowledge of this skill drill by placing the photos below in the correct order. Number the first step with a "1," the second step with a "2," etc.

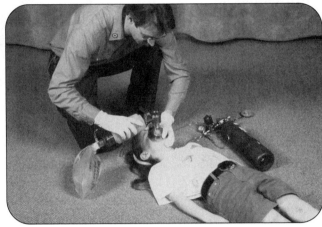

Hold the mask on the patient's face with a one-handed head tilt-chin lift technique (E-C grip).
Ensure a good mask-face seal while maintaining the airway.

Assess effectiveness of ventilation by watching bilateral rise and fall of the chest.

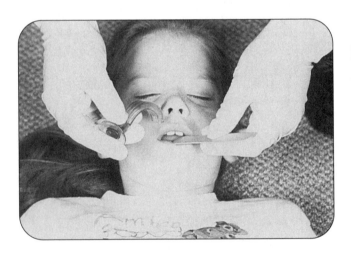

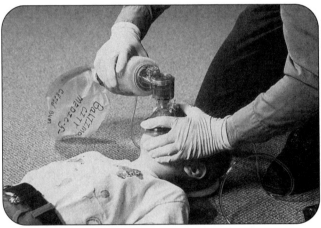

Open the airway and insert the appropriate airway adjunct.

Squeeze the bag using the correct ventilation rate of 12 to 20 breaths/min. Allow adequate time for exhalation.

Skill Drill 32-5: Removing a Foreign Body Airway Obstruction in an Unconscious Child

Test your knowledge of this skill drill by placing the photos below in the correct order. Number the first step with a "1," the second step with a "2," etc.

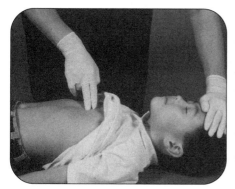

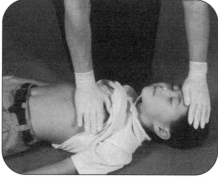

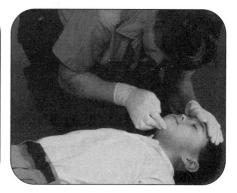

Locate the proper hand position on the chest of the child. If ventilation is still unsuccessful, begin CPR.

Administer 30 chest compressions and look inside child's mouth. If you see the object, remove it.

Inspect the airway. Remove any foreign object that you can see.

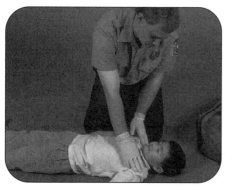

Position the child on a firm, flat surface.

Attempt rescue breathing. If unsuccessful, reposition the head and try again.

Skill Drill 32-6: Performing Infant Chest Compressions

Test your knowledge of this skill drill by filling in the correct words in the photo captions.

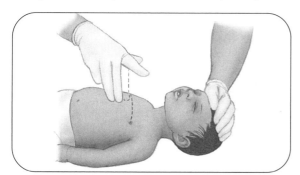

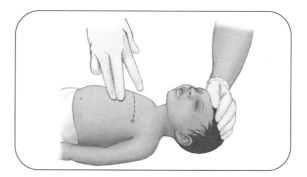

1. Position the infant on a _____ surface while _____ the airway.

Place two _____ in the _____ of the sternum just below a line between the _____ .

2. Use two fingers to _____ the chest about _____ to _____ its depth at a rate of _____ times/min.

Allow the sternum to return _____ to its _____ position between compressions.

Skill Drill 32-7: Performing CPR on a Child

Test your knowledge of this skill drill by filling in the correct words in the photo captions.

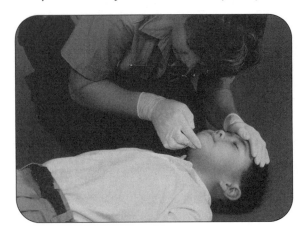

1. Place the child on a _____ surface, open the airway, and deliver two rescue breaths.

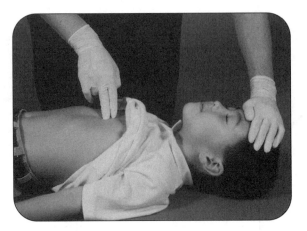

2. Place the _____ of one or both hands in the _____ of the chest, in between the nipples, avoiding the _____ process.

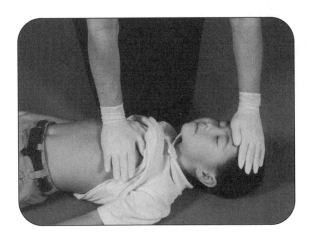

3. Compress the chest _____ to _____ the depth of the chest at a rate of _____ times/min. Coordinate compressions with ventilations in a _____ ratio (one rescuer) or 15:2 (two rescuers), pausing for _____.

4. Reassess for _____ and _____ after every _____ (about 2 minutes) of CPR. If the child resumes _____ breathing, place him or her in a position that allows for _____ reassessment of the airway and vital signs during transport.

Skill Drill 32-8: Immobilizing a Child
Test your knowledge of this skill drill by filling in the correct words in the photo captions.

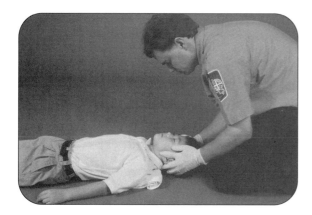

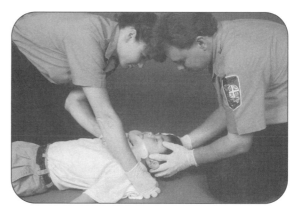

1. Use a towel under the back, from the

_____ to the _____, to maintain

the head in a _____ position.

2. Apply an appropriately sized_____

_____.

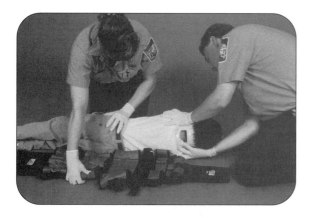

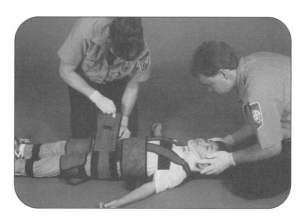

3. _____ _____ the child onto the

_____ device.

4. Secure the _____ first.

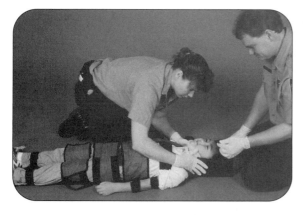

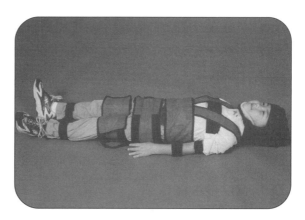

5. Secure the _____.

6. Ensure that the child is _____ in properly.

Skill Drill 32-10: Immobilizing an Infant Out of a Car Seat
Test your knowledge of this skill drill by placing the photos below in the correct order. Number the first step with a "1," the second step with a "2," etc.

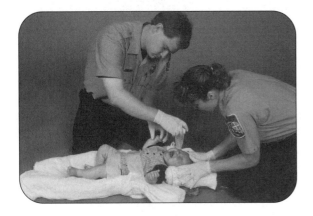

_____

Secure the head.

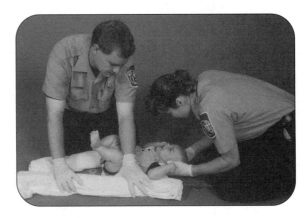

_____

Slide the infant onto the board.

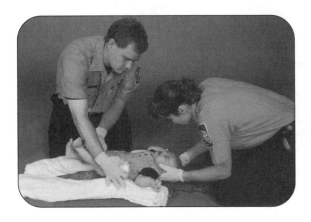

_____

Place a towel under the back, from the shoulders to the hips, to ensure neutral head position.

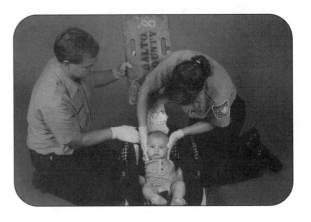

_____

Stabilize the head in neutral position.

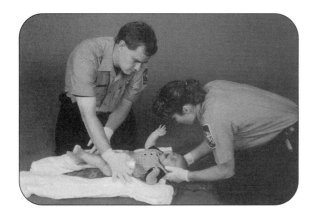

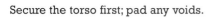

Secure the torso first; pad any voids.

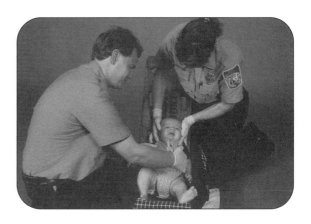

Place an immobilization device between the patient and the surface he or she is resting on.

CHAPTER

33 Geriatric Emergencies

Workbook Activities

The following activities have been designed to help you. Your instructor may require you to complete some or all of these activities as a regular part of your EMT-B training program. You are encouraged to complete any activity that your instructor does not assign as a way to enhance your learning in the classroom.

Chapter Review

The following exercises provide an opportunity to refresh your knowledge of this chapter.

Matching

Match each of the terms in the left column to the appropriate definition in the right column.

_____ 1. Aneurysm

_____ 2. Cataract

_____ 3. Delirium

_____ 4. Dementia

_____ 5. Syncope

_____ 6. Dyspnea

_____ 7. Compensated shock

_____ 8. Decompensated shock

_____ 9. Collagen

_____ 10. Vasoconstriction

A. a protein that is the chief component of connective tissue and bones

B. narrowing of a blood vessel

C. clouding of the lens of the eye

D. early shock

E. abnormal blood-filled dilation of a blood vessel

F. difficulty breathing

G. inability to focus, think logically, or maintain attention

H. late shock

I. slow onset of progressive disorientation

J. fainting

Multiple Choice

Read each item carefully, then select the best response.

_____ 1. Leading causes of death in the older population include all of the following, except:

   A. heart disease.

   B. AIDS.

   C. cancer.

   D. diabetes.

_____ 2. You are called the residence of an 86-year-old man who lives alone and seems to be confused. His airway is open and respirations are regular and unlabored. His pulse is strong and regular with good distal pulses. He is oriented and does not want to go to the hospital. You are worried about his well-being. You know that lifesaving interventions for geriatric patients may include:

   A. reviewing the home environment.

   B. providing information on preventing falls.

   C. making referrals to appropriate social services agencies.

   D. all of the above

_____ 3. Simple preventive measures can help older people to avoid:

   A. further injury.

   B. costly medical treatment.

   C. death.

   D. all of the above

_____ 4. Acute illness and trauma are more likely to involve _____ beyond those initially involved.

   A. organ systems

   B. bones

   C. fractures

   D. vessels

_____ 5. Risk factors that affect mortality in geriatric patients include all of the following, except:

   A. living alone.

   B. unsound mind.

   C. regular exercise.

   D. recent hospitalization.

_____ 6. Loss of collagen makes the skin:

   A. wrinkled.

   B. thinner.

   C. more susceptible to injury.

   D. all of the above

_____ 7. Driving and walking become more hazardous because the pupils of the eyes begin to lose the ability to:

   A. dilate.

   B. handle changes in light.

   C. constrict.

   D. detect color.

_____ 8. Problems with balance are usually related to changes in the:

   A. blood pressure.

   B. vision.

   C. inner ears.

   D. cardiovascular system.

_____ **9.** Although the alveoli became enlarged, their elasticity decreases, resulting in a decreased ability to:
   A. cough and thereby increasing the chance of infection.
   B. exchange oxygen and carbon dioxide.
   C. monitor the changes in oxygen and carbon dioxide.
   D. force carbon dioxide out of the lungs.

_____ **10.** Compensation for an increased demand on the cardiovascular system is accomplished by:
   A. increasing heart rate.
   B. increasing contraction of the heart.
   C. constricting the blood vessels to nonvital organs.
   D. all of the above

_____ **11.** Aging decreases a person's ability to _____ because of stiffer vessels.
   A. vasoconstrict
   B. vasodilate
   C. circulate blood
   D. exchange oxygen

_____ **12.** An accumulation of fatty materials in the arteries is known as:
   A. myocardial infarction.
   B. stroke.
   C. atherosclerosis.
   D. aneurysm.

_____ **13.** You respond to a 68-year-old woman who is just not feeling well. She is alert and oriented. Her airway is open and there is no indication of any respiration problem. Pulse is regular and strong. Her blood pressure is 140/86 mm Hg. When asked about medications, she gives you a box that contains multiple medications from multiple doctors. She tells you she has been taking medications for years with no ill effects. She also mentions that on her last visit the doctor seemed a little concerned about her kidney function. You know that with a decrease in renal function, levels of _____ may rise, creating the impression of an overdose.
   A. medications
   B. toxins
   C. acid
   D. alkali

_____ **14.** By age 85, a 10% reduction in brain weight can result in:
   A. increased risk of head trauma.
   B. short-term memory impairment.
   C. slower reflex times.
   D. all of the above

_____ **15.** As a person ages, fractures are more likely to occur because of a decrease in bone:
   A. cartilage.
   B. density.
   C. length.
   D. tissue.

_____ **16.** The best rule of thumb when assessing mental status is to always compare the patient's current level of consciousness or ability to function with the level or ability:
   A. of another adult in the household.
   B. before the problem began.
   C. of a person of the same age.
   D. none of the above

_____ **17.** The _____ is usually the key in helping to assess the older patient's problem.

    A. history

    B. medication

    C. environment

    D. all of the above

_____ **18.** An older patient's diminished _____ may hamper communication.

    A. sight

    B. hearing

    C. speaking ability

    D. all of the above

_____ **19.** The term applied to prescribing multiple medications is:

    A. hypermedicating.

    B. hyperpharmacy.

    C. polypharmacy.

    D. overmedicating.

_____ **20.** You are called to the residence of a 71-year-old woman. She is complaining of not feeling well. She is alert and oriented. Her airway is open and she is breathing at 22 breaths/min but seems a little pale. Her pulse is 112 beats/min, irregular and weak. Her distal pulse is diminished. You know that the sensation of pain may be _____ in a geriatric patient, leading to "silent" heart attacks.

    A. enhanced

    B. diminished

    C. overstated

    D. false

_____ **21.** You must consider the body's decreasing ability to _____ simple trauma when you are assessing and caring for an older patient.

    A. isolate

    B. separate

    C. heal

    D. recognize

_____ **22.** An isolated hip fracture in an 85-year-old patient can produce a systemic impact that results in:

    A. deterioration.

    B. shock.

    C. life-threatening conditions.

    D. all of the above

_____ **23.** When assessing an older patient who has fallen, it is important to determine why the fall occurred because it may have been the result of a medical problem such as:

    A. fainting.

    B. a cardiac rhythm disturbance.

    C. a medication interaction.

    D. all of the above

_____ **24.** Your assessment of the patient's condition and stability must include past medical conditions, even if they are not currently acute or:

    A. symptomatic.

    B. asymptomatic.

    C. complaining.

    D. on medication.

_____ **25.** A common complaint from the patient experiencing an abdominal aortic aneurysm (AAA) is pain in the:

   A. abdomen.

   B. back.

   C. leg with decreased blood flow.

   D. all of the above

_____ **26.** All of the following are true of delirium, except:

   A. it may have metabolic causes.

   B. the patient may be hypoglycemic.

   C. it develops slowly over a period of years.

   D. the memory remains mostly intact.

_____ **27.** You respond to a residence for a difficulty breathing call. Upon arrival the 86-year-old patient is pulseless and apneic. The family tells you he has a DNR. For a DNR order to be valid, it must:

   A. be signed by the patient or legal guardian.

   B. be signed by one or more physicians.

   C. be dated within the preceding 12 months.

   D. all of the above

_____ **28.** When in doubt about whether an advance directive is valid, or if one is in place, your best course of action is to:

   A. call medical control to see if an order is needed.

   B. take resuscitation action that is appropriate to the situation.

   C. wait for the family or caregivers to produce the appropriate document.

   D. none of the above

_____ **29.** Signs and symptoms of possible abuse include all of the following, except:

   A. chronic pain.

   B. no history of repeated visits to the emergency department or clinic.

   C. depression or lack of energy.

   D. self-destructive behavior.

_____ **30.** Signs of neglect include:

   A. lack of hygiene.

   B. poor dental hygiene.

   C. lack of reasonable amenities in the home.

   D. all of the above

Fill-in

Read each item carefully, then complete the statement by filling in the missing word.

1. Geriatric patients are individuals who are older than _____ years.

2. The aging body of the geriatric person may _____ serious medical conditions.

3. Common _____ about older people include the presence of mental confusion, illness, a sedentary lifestyle, and immobility.

4. Older skin feels dry due to fewer _____ _____.

5. _____ _____ is a measure of the workload of the heart.

6. An _____ is an abnormal blood-filled dilation of the wall of a blood vessel.

7. Flexion at the neck and a forward curling of the shoulders produce a condition called _____.

True/False

If you believe the statement to be more true than false, write the letter "T" in the space provided. If you believe the statement to be more false than true, write the letter "F."

_____ **1.** Vasodilation is a narrowing of a blood vessel.

_____ **2.** Cardiovascular disease is one of the leading causes of death in the older population.

_____ **3.** Mental confusion and immobility are common stereotypes about the older population.

_____ **4.** Assessment of an older patient usually takes less time than a middle-aged person.

_____ **5.** The sensation of pain in an older patient may be diminished.

_____ **6.** Older people are more prone to hypothermia than younger people.

_____ **7.** Elder abuse is on the decline in the United States.

Short Answer

Complete this section with short written answers using the space provided.

1. List the three major categories of elder abuse.

_____

_____

2. Briefly describe the three possible causes of syncope in an older patient.

_____

_____

3. List at least five informational items that may be important in assessing possible elder abuse.

_____

_____

4. Name three common signs and symptoms of a heart attack in an older patient.

_____

_____

Word Fun

The following crossword puzzle is an activity provided to reinforce correct spelling and understanding of medical terminology associated with emergency care and the EMT-B. Use the clues in the column to complete the puzzle.

Across

3. Difficulty breathing

8. Fatty materials deposited

9. Slow onset of progressive disorientation

11. Late shock

Down

1. Clouding of the lens

2. Early shock

4. Fainting

5. Generalized bone disease

6. Taking advantage (physical, emotional, etc.) of an elder

7. Chief component of connective tissue

10. Abnormal blood-filled dilation of blood vessel

Ambulance Calls

The following case scenarios provide an opportunity to explore the concerns associated with patient management. Read each scenario, then answer each question in detail.

1. You are called to the residence of an 87-year-old woman who is "not acting right." Family members tell you they think she may have taken her medication twice this morning. She is lethargic, confused, and hypotensive.

How would you best manage this patient?

_____

_____

_____

2. It's 3:00 A.M., and you are dispatched to a long-term care facility for a geriatric woman with an "unknown emergency." You arrive to find no staff around, but you hear someone crying for help. You follow the voice and find an older woman lying on the floor holding her right hip. Her bed is made and the side rails are up. When a staff member finally appears, you ask how long she's been lying there. They respond, "We don't know. How are we supposed to know that? We don't have enough help around here."

How do you best manage this patient?

_____

_____

_____

3. You are dispatched to the home of a 65-year-old woman who is complaining of severe back pain. She tells you that she tried to pick up the lawnmower when she "heard a pop and felt a crack in her back." She is experiencing intense pain in her lower back and feels some numbness in her legs.

How do you best manage this patient?

_____

_____

_____

CHAPTER
34 Geriatric Assessment and Management

Workbook Activities

The following activities have been designed to help you. Your instructor may require you to complete some or all of these activities as a regular part of your EMT-B training program. You are encouraged to complete any activity that your instructor does not assign as a way to enhance your learning in the classroom.

Chapter Review

The following exercises provide an opportunity to refresh your knowledge of this chapter.

Matching

Match each of the terms in the left column to the appropriate definition in the right column.

_____ 1. Acetabulum

_____ 2. Bacteremia

_____ 3. Central cord syndrome

_____ 4. Compression fractures

_____ 5. Polypharmacy

_____ 6. Septicemia

_____ 7. Stable spinal injury

_____ 8. Vasodilation

A. The disease state that results from the presence of microorganisms in the bloodstream

B. Stable spinal cord injuries in which often only the anterior third of the vertebra is collapsed

C. Simultaneous use of many medications

D. Spinal injury that has a low risk of leading to permanent neurologic deficit or structural deformity

E. The depression on the lateral pelvis where its three component bones join

F. Incomplete spinal cord injury in which some of the signals from the brain to the body are not received

G. The presence of bacteria in the blood

H. widening of a blood vessel

Multiple Choice

Read each item carefully, then select the best response.

_____ **1.** Scene clues that can provide important information include:

    A. the general condition of the home.

    B. the number and type of pill bottles around.

    C. any hazards that could cause a fall.

    D. all of the above

_____ **2.** You are called to the local nursing home for a 91-year-old man with altered mental status. He is very confused. His airway is open and respirations are unlabored. His pulse is strong, regular, and distal pulses are good. In order to understand a patient's baseline condition and how today's behavior differs from it, you should ask the nursing home staff questions concerning the patient's:

    A. mobility.

    B. activities of daily living.

    C. ability to speak.

    D. all of the above

_____ **3.** The S in the GEMS diamond stands for:

    A. seriousness.

    B. severity.

    C. social situation.

    D. resuscitative.

_____ **4.** You respond to a 96-year-old man who is alert and oriented. His airway is open with no respiratory distress. Pulses are present and strong and regular. You are having difficulty with communication. All of the following will help with communication except:

    A. standing directly over the patient.

    B. standing in a well lit area.

    C. turning off all radios and televisions.

    D. talking in a normal tone of voice.

_____ **5.** The average patient age 65 or older will be taking how many medications?

    A. Four

    B. Three

    C. Two

    D. One

_____ **6.** Common complaints of the geriatric patient include:

    A. diarrhea.

    B. chest pain.

    C. shortness of breath.

    D. all of the above

_____ **7.** You respond to an auto accident in which your patient is a 71-year-old woman. She is complaining of severe back pain. She is alert and oriented and respirations are normal and unlabored. Her pulses are present and strong. After applying a C-collar, you start your assessment of the injured area. Upon palpation, you notice a bulged area and the patient reacts by screaming when you touch the area. You know that you have to be extremely careful with the patient because:

    A. she has an unstable spinal injury.

    B. she has a stable spinal injury.

    C. she has no pain tolerance.

    D. you don't want to get sued.

_____ **8.** Because of the presence of arthritis, relatively small hyperextension injuries can cause the spinal cord to be squeezed, leading to a dysfunction known as:

    A. compression fractures.

    B. burst fractures.

    C. unstable spine.

    D. central cord syndrome.

_____ **9.** You are called to the residence of an 84-year-old man who fell and hit his head. He is alert and oriented with no sign of obvious injury. His respirations and pulses are within normal limits. He takes blood thinners daily. You should:

    A. assume everything is alright and release the patient.

    B. treat for a head injury and transport.

    C. check pupil response and if normal, release the patient.

    D. release him to a family member.

_____ **10.** Nearly _____ of all older patients will die within 12 months of hip fracture.

    A. 10%

    B. 15%

    C. 20%

    D. 25%

_____ **11.** You respond to a 90-year-old woman who is complaining of difficulty breathing. She is alert and oriented and respirations are at 26 breaths/min and regular. Pulse is 110 beats/min and irregular. She is not complaining of any pain; just not feeling well. You should be alert for:

    A. septicemia.

    B. aortic abdominal aneurysm.

    C. cardiovascular emergency.

    D. infectious disease.

_____ **12.** When responding to a nursing facility for a patient, what information do you need to get from the staff before you transport?

    A. patient's mobility

    B. patient's daily activities

    C. patient's ability to speak

    D. all of the above

Fill-in

Read each item carefully, then complete the statement by filling in the missing word.

1. _____ occurs when a patient takes multiple medications that can interact.

2. A spinal injury that has a low risk for leading to permanent neurologic deficit or structural deformity is called _____.

3. A spinal injury that has a high risk of permanent neurologic deficit or structural deformity is called _____.

4. The site of pelvic injuries that result from high-energy trauma in older people is the _____.

5. Because of the presence of arthritis, relatively small hyperextension injuries can cause the spinal cord to be squeezed, leading to a dysfunction known as _____ _____ _____.

6. A stable injury in which often only the anterior third of the vertebra is collapsed is the _____ _____.

7. _____ fractures typically result from a higher energy mechanism such as a motor vehicle crash or fall from substantial height.

8. The fracture that involves flexion and distraction component that cause a fracture through the entire vertebral body and bony arch is called _____ _____ _____.

9. The disease state that results from the presence of microorganisms or their toxic products in the bloodstream is _____.

10. The presence of bacteria in the blood is called _____.

True/False
If you believe the statement to be more true than false, write the letter "T" in the space provided. If you believe the statement to be more false than true, write the letter "F."

_____ 1. Approximately 20% to 30% of older people have "silent" heart attacks.

_____ 2. Most nursing facilities will send transfer records along with a patient being transported. These records contain the patient's history, medication lists, and dosages.

_____ 3. The "E" in the GEMS diamond stands for energy level.

_____ 4. You should never assume that an altered mental status is normal for a geriatric patient.

_____ 5. One of the best tools for assessing the geriatric patient is their history.

_____ 6. Geriatric patients often stop taking medications without talking to their doctor.

_____ 7. A rapid physical exam is typically performed before obtaining the vital signs.

_____ 8. Compression fractures are unstable injuries.

_____ 9. Prehospital treatment of older head-injured patients should be aimed at maintaining the heart rate.

Short Answer
Complete this section with short written answers using the space provided.

1. List the steps in immobilizing a kyphotic patient.

_____

_____

2. List the steps for immobilizing a patient with a hip fracture.

_____

_____

3. What critical information do you need to get from nursing staff at a nursing facility?

_____

_____

Word Fun emtb.com vocab explorer

The following crossword puzzle is an activity provided to reinforce correct spelling and understanding of medical terminology associated with emergency care and the EMT-B. Use the clues in the column to complete the puzzle.

Across

3. Compression fractures of the vertebrae

5. Simultaneous use of many medications

6. Depression on the lateral pelvis where its three component bones join

Down

1. Results from the presence of microorganisms in the bloodstream

2. Presence of bacteria in the blood

4. Sudden unconsciousness

Ambulance Calls

The following case scenarios provide an opportunity to explore the concerns associated with patient management. Read each scenario, and then answer each question in detail.

1. You are dispatched to an unknown medical emergency at 222 Orchid Lane. You arrive to find an older woman who greets you at the front door. She tells you her husband is laying on the bathroom floor, and he's too heavy for her to lift. As you enter the small bathroom, you notice a large laceration and hematoma on the man's forehead. He tells you that he was on the toilet and must have fallen, but doesn't remember exactly what happened.

How would you best manage this patient?

_____

_____

_____

2. You are dispatched to a private residence for a woman who has fallen. She tells you that she tripped over her granddaughter's toys and fell onto the hardwood floor. She tried to get up several times but couldn't. She denies any head, neck, back pain, or loss of consciousness. Her only complaint is pain in her right hip.

How would you best manage this patient?

_____

_____

_____

3. It is early evening, and you are dispatched to the parking lot of a popular shopping mall. Over the past several hours, local weather conditions have consisted of a combination of freezing rain and snow, making for very icy road conditions. Your patient is an older woman who has fallen and now complains of back pain and shortness of breath. She has severe kyphosis and will not tolerate traditional methods of spinal immobilization.

How would you best manage this patient?

_____

_____

_____

CHAPTER

35 Ambulance Operations

Workbook Activities

The following activities have been designed to help you. Your instructor may require you to complete some or all of these activities as a regular part of your EMT-B training program. You are encouraged to complete any activity that your instructor does not assign as a way to enhance your learning in the classroom.

Chapter Review

The following exercises provide an opportunity to refresh your knowledge of this chapter.

Matching

Match each of the terms in the left column to the appropriate definition in the right column.

_____ 1. Medivac

_____ 2. Emergency mode

_____ 3. Spotter

_____ 4. Sterilization

_____ 5. Ambulance

_____ 6. Cleaning

_____ 7. Disinfection

A. the use of lights and sirens

B. the killing of pathogenic agents by direct application of chemicals

C. the process of removing dirt, dust, blood, or other visible contaminants

D. medical evacuation of a patient by helicopter

E. a person who assists a driver in backing up an ambulance

F. specialized vehicle for treating and transporting sick and injured patients

G. removes microbial contamination

Multiple Choice

Read each item carefully, then select the best response.

_____ **1.** Ambulances today are designed according to strict government regulations based on _____ standards.

   A. local

   B. state

   C. national

   D. individual

_____ **2.** Features of the modern ambulance include all of the following, except:

   A. self-contained breathing apparatus.

   B. a patient compartment.

   C. two-way radio communication.

   D. a driver's compartment.

_____ **3.** You report to work and the first thing you do each day that you arrive is to make sure all equipment and supplies are functioning and in their assigned place. This is the _____ phase of ambulance operations.

   A. preparation

   B. dispatch

   C. arrival at scene

   D. transport

_____ **4.** The Type _____ ambulance is a standard van with a forward-control integral cab body.

   A. I

   B. II

   C. III

   D. IV

_____ **5.** Items needed to care for life-threatening conditions include:

   A. equipment for airway management.

   B. equipment for artificial ventilation.

   C. oxygen delivery devices.

   D. all of the above

_____ **6.** Oropharyngeal airways can be used for:

   A. adults.

   B. children.

   C. infants.

   D. all of the above

_____ **7.** A BVM device, when attached to oxygen supply with the oxygen reservoir in place, is able to supply almost _____ oxygen.

   A. 100%

   B. 95%

   C. 90%

   D. 85%

_____ **8.** BVM devices should be transparent so that you can:

   A. monitor the patient's respirations.

   B. notice any color changes in the patient.

   C. detect vomiting.

   D. all of the above

_____ **9.** Oxygen masks, with and without nonbreathing bags, should be transparent and disposable and in sizes for:

   A. adults.

   B. children.

   C. infants.

   D. all of the above

_____ **10.** Basic wound care supplies include all of the following, except:

   A. sterile sheets.

   B. an OB kit.

   C. an assortment of band-aids.

   D. large safety pins.

_____ **11.** Your supervisor approaches you and tells you he wants you to make up jump kits for each ambulance. He has told you that he will leave it up to you as to what goes in the kit. You would want to put everything it in that you might need within the first _____ minutes of arrival at the patient's side.

   A. 2

   B. 3

   C. 4

   D. 5

_____ **12.** Deceleration straps over the shoulders prevent the patient from continuing to move _____ in case the ambulance suddenly slows or stops.

   A. forward

   B. backwards

   C. laterally

   D. down

_____ **13.** The ambulance inspection should include checks of:

   A. fuel levels.

   B. brake fluid.

   C. wheels and tires.

   D. all of the above

_____ **14.** You are hired at the local EMS service. During your orientation, you are given a tour of the station and the ambulances you will be riding on. Your duties include station cleanup and checking the unit for mechanical problems. You should also check all medical equipment and supplies:

   A. after every call.

   B. after every emergency transport.

   C. every 12 hours.

   D. every day.

_____ **15.** For every emergency request, the dispatcher should gather and record all of the following, except:

   A. the nature of the call.

   B. the location of the patient(s).

   C. medications that the patient is currently taking.

   D. the number of patients and possible severity of their condition.

_____ **16.** During the _____ phase, the team should review dispatch information and assign specific initial duties and scene management tasks to each team member.

   A. preparation

   B. dispatch

   C. en route

   D. transport

_____ **17.** Basic requirements for the driver to safely operate an ambulance include:

    A. physical fitness.

    B. emotional fitness.

    C. proper attitude.

    D. all of the above

_____ **18.** The _____ phase may be the most dangerous part of the call.

    A. preparation

    B. en route

    C. transport

    D. on scene

_____ **19.** In order to operate an emergency vehicle safely, you must know how it responds to _____ under various conditions.

    A. steering

    B. braking

    C. acceleration

    D. all of the above

_____ **20.** You must always drive:

    A. offensively.

    B. defensively.

    C. under the speed limit.

    D. all of the above

_____ **21.** When driving with lights and siren, you are _____ drivers to yield the right-of-way.

    A. requesting

    B. demanding

    C. offering

    D. none of the above

_____ **22.** Vehicle size and _____ will greatly influence braking and stopping distances.

    A. length

    B. height

    C. weight

    D. width

_____ **23.** When on an emergency call, before proceeding past a stopped school bus with its lights flashing, you should stop before reaching the bus and wait for the driver to:

    A. make sure the children are safe.

    B. close the bus door.

    C. turn off the warning lights.

    D. all of the above

_____ **24.** The _____ is probably the most overused piece of equipment on an ambulance.

    A. stethoscope

    B. siren

    C. cardiac monitor

    D. stretcher

_____ **25.** The _____ is the most visible, effective warning device for clearing traffic in front of the vehicle.

    A. front light bar

    B. rear light bar

    C. high-beam flasher unit

    D. standard headlight

_____ **26.** If you are involved in a motor vehicle crash while operating an emergency vehicle and are found to be at fault, you may be charged:

    A. civilly.

    B. criminally.

    C. both civilly and criminally.

    D. neither civilly nor criminally.

_____ **27.** _____ crashes are the most common and usually the most serious type of collision in which ambulances are involved.

    A. T-bone

    B. Intersection

    C. Lateral

    D. Rollover

_____ **28.** You respond to a multiple vehicle collision. You and your partner are reviewing dispatch information en route to the scene. You will be in a major intersection of two state highways. As you approach the scene, you review the guidelines for sizing up the scene. The guidelines include:

    A. looking for safety hazards.

    B. evaluating the need for additional units or other assistance.

    C. evaluating the need to stabilize the spine.

    D. all of the above

_____ **29.** The main objectives in directing traffic include:

    A. warning other drivers.

    B. preventing additional crashes.

    C. keeping vehicles moving in an orderly fashion.

    D. all of the above

_____ **30.** Transferring the patient to receiving staff member occurs during the _____ phase.

    A. arrival

    B. transport

    C. delivery

    D. postrun

_____ **31.** Cleaning the vehicle inside and out, refueling the vehicle, disposing of contaminated waste, and replacing equipment and supplies all are accomplished during the _____ phase.

    A. preparation

    B. transport

    C. delivery

    D. postrun

_____ **32.** You have called for an air ambulance. While your partner is monitoring the patient, he tells you to go set up a landing zone for the helicopter. When clearing a landing site for an approaching helicopter, look for:

    A. loose debris.

    B. electric or telephone wires.

    C. poles.

    D. all of the above

Fill-in

Read each item carefully, then complete the statement by filling in the missing word.

1. A _____ _____ is a portable kit containing items that are used in the initial care of the patient.

2. The six-pointed star that identifies vehicles that meet federal specifications as licensed or certified ambulances is known as the _____ _____ _____.

3. For many decades after 1906, a _____ was the vehicle that was most often used as an ambulance.

4. _____ _____ _____ respond initially to the scene with personnel and equipment to treat the sick and injured until an ambulance can arrive.

5. An ambulance call has _____ phases.

6. Devices should be either disposable or easy to clean and _____, which means to remove radiation, chemical, or other hazardous materials.

7. Suction tubing must reach the patient's _____, regardless of the patient's position.

8. A _____ _____ provides a firm surface under the patient's torso so that you can give effective chest compressions.

True/False
If you believe the statement to be more true than false, write the letter "T" in the space provided. If you believe the answer to be more false than true, write the letter "F."

_____ 1. Equipment and supplies should be placed in the unit according to their relative importance and frequent use.

_____ 2. A CPR board is a pocket-sized reminder that the EMT-B carries to help recall CPR procedures.

_____ 3. Having the ability to exchange equipment between units or between your unit and the emergency department decreases the time that you and your unit must stay at the hospital.

_____ 4. The en route or response phase of the emergency call is the least dangerous for the EMT-B.

_____ 5. When the siren is on, you can speed up and assume that you have the right-of-way.

_____ 6. Use the "4-second rule" to help you maintain a safe following distance.

_____ 7. Always approach a helicopter from the front.

_____ 8. Fixed-wing air ambulances are generally used for short-haul patient transfers.

_____ 9. A clear landing zone of 50' by 50' is recommended for EMS helicopters.

Short Answer
Complete this section with short written answers using the space provided.

1. Describe the three basic ambulance designs.

_____

_____

2. List the phases of an ambulance call.

_____

_____

3. List the five factors that contribute to the use of excessive speed.

_____

_____

4. Describe the three basic principles that govern the use of warning lights and sirens.

_____

_____

5. List four guidelines for safe ambulance driving.

_____

_____

6. Describe the correct technique for approaching a helicopter that is "hot" (rotors turning).

_____

_____

7. List the general considerations used for selecting a helicopter landing site.

_____

_____

_____

Word Fun emtb. vocab explorer

The following crossword puzzle is an activity provided to reinforce correct spelling and understanding of medical terminology associated with emergency care and the EMT-B. Use the clues in the column to complete the puzzle.

Across

3. Specialized vehicle for transporting sick and injured patients

7. Process that removes microbial contamination

8. Remove or neutralize radiation

Down

1. Killing of pathogenic agents with chemicals

2. Portable kit used in initial care of patient

4. Medical evacuation of a patient by helicopter

5. Process of removing visible contaminants from a surface

6. Person who assists a driver in backing up an ambulance

Ambulance Calls

The following case scenarios provide an opportunity to explore the concerns associated with patient management. Read each scenario, then answer each question in detail.

1. You are cross-trained as a fire fighter and your station has been dispatched to a working structure fire. As you near the scene, another call is dispatched for "an unknown medical emergency." You are the closest unit to the medical emergency.

How would you best manage this situation?

_____

_____

_____

2. Your department's coverage area is quite large, including ALS coverage for the entire county as well as surrounding areas out of state. You are dispatched to an unfamiliar address for "CPR in progress" and are working with a newly hired partner who is not familiar with your coverage area.

How would you best manage this situation?

_____

_____

_____

3. You are called to the scene of a motor vehicle crash. The car is situated in a curve and traffic is heavy. Police are not on the scene. Your patient is alert and looking around, but stuck in the vehicle due to traffic. You see blood smeared across her face, but it appears to be minimal.

How would you best manage this situation?

_____

_____

_____

CHAPTER
36 Gaining Access

Workbook Activities

The following activities have been designed to help you. Your instructor may require you to complete some or all of these activities as a regular part of your EMT-B training program. You are encouraged to complete any activity that your instructor does not assign as a way to enhance your learning in the classroom.

Chapter Review

The following exercises provide an opportunity to refresh your knowledge of this chapter.

Matching

Match each of the terms in the left column to the appropriate definition in the right column.

_____ 1. Extrication

_____ 2. Simple access

_____ 3. Complex access

_____ 4. Access

_____ 5. Incident commander

_____ 6. Structure fire

_____ 7. Hazardous material

A. access requiring no special tools and training

B. individual who has overall command of the scene in the field

C. a substance that causes injury or death with exposure

D. access requiring special tools and training

E. removal from entrapment or a dangerous situation or position

F. gaining entry to an enclosed area to reach a patient

G. fire in a house, apartment building, or other building

Multiple Choice

Read each item carefully, then select the best response.

_____ **1.** You are dispatched to an automobile accident with reports of trapped patients. There are also reports of fluid spills. From this information you know that you will need what other rescuers at the crash scene?

  A. Firefighters

  B. Law enforcement

  C. A rescue group

  D. All of the above

_____ **2.** In the scenario above, which of the four teams is responsible for investigating the crash?

  A. Firefighter

  B. Law enforcement

  C. Rescue group

  D. EMS personnel

_____ **3.** In the scenario of question 1, which of the four teams at a crash scene is responsible for properly securing and stabilizing the vehicle?

  A. Firefighter

  B. Law enforcement

  C. Rescue group

  D. EMS personnel

_____ **4.** In the scenario of question 1, which of the four teams at a crash scene is responsible for washing down any spilled fuel?

  A. Firefighter

  B. Law enforcement

  C. Rescue group

  D. EMS personnel

_____ **5.** Before proceeding with an extrication, you should:

  A. position your unit in a safe location.

  B. make sure the scene is properly marked.

  C. determine if any additional resources will be needed.

  D. all of the above

_____ **6.** While you are gaining access to the patient and during extrication, you must make sure that the patient:

  A. remains safe.

  B. stays conscious.

  C. holds his/her head completely still.

  D. all of the above

_____ **7.** When dealing with multiple patients, you should locate and rapidly _____ each patient.

  A. treat

  B. triage

  C. transport

  D. extricate

_____ **8.** When preparing for patient removal, you should determine:

  A. how urgently the patient must be extricated.

  B. where you should be positioned during extrication.

  C. how you will best move the patient from the vehicle.

  D. all of the above

_____ **9.** Once the patient has been extricated, additional assessment should be completed:

    A. once the patient has been placed on the stretcher.

    B. inside the ambulance inclement weather.

    C. en route if the patient's condition requires rapid transport.

    D. all of the above

_____ **10.** Even when a technical rescue group includes a paramedic or physician, generally nothing but essential _____ is provided until the rescuers can bring the patient to the nearest point where a safe, stable setting exists.

    A. bandaging

    B. triage

    C. simple care

    D. splinting

_____ **11.** When called to a person lost outdoors, your role involves:

    A. standing by at the search base until the lost person is found.

    B. preparing necessary equipment.

    C. obtaining any medical history from relatives on scene.

    D. all of the above

_____ **12.** Tactical situations involve all of the following, except:

    A. an armed hostage situation.

    B. a structure fire.

    C. the presence of a sniper.

    D. an exchange of shots.

Fill-in

Read each item carefully, then complete the statement by filling in the missing word(s).

1. _____ is the final phase of extrication, and this usually results in the patient being placed on the ambulance stretcher.

2. During all phases of rescue, your primary concern is _____.

3. Good _____ among team members and clear leadership are essential to safe, efficient provision of proper emergency care.

4. You should not attempt to gain access to the patient or enter the vehicle until you are sure that the vehicle is _____.

5. When gaining access, it is up to you to identify the _____, most efficient way to gain access.

6. Moving the patient in one fast, continuous step increases the risk of _____ and confusion.

7. Search and rescue is performed by teams of firefighters wearing full turnout gear and _____ _____ _____ _____ (SCBA), and carrying tools and fully charged hose lines.

True/False

If you believe the statement to be more true than false, write the letter "T" in the space provided. If you believe the statement to be more false than true, write the letter "F."

_____ **1.** There should be no talking throughout the extrication process.

_____ **2.** You should maintain at least a 10" clearance around driver airbags that have not deployed.

_____ **3.** If you will be involved with extrication, you should wear leather gloves over your disposable gloves.

_____ **4.** Simple access is trying to get to the patient as quickly as possible using tools or other forcible entry methods.

_____ **5.** You should not try to access the patient until you are sure that the vehicle is stable and that hazards have been identified and properly controlled.

Short Answer

Complete this section with short written answers using the space provided.

1. Explain the four different basic functions that must be addressed at any crash scene.

_____

_____

2. To determine the exact location and position of the patient, you and your team should consider what questions?

_____

_____

3. List the steps for assessing and caring for a patient who is entrapped once access has been gained.

_____

_____

4. When examining the exposed area of the limb or other part of the patient that is trapped, explain what you are assessing for.

_____

_____

5. Explain the proper technique for patient removal once they are disentangled.

_____

_____

Word Fun emtb vocab explorer

The following crossword puzzle is an activity provided to reinforce correct spelling and understanding of medical terminology associated with emergency care and the EMT-B. Use the clues in the column to complete the puzzle.

Across

3. Danger zone

6. To be caught with no way out

8. Requires tools and special training

9. Cave rescue, dive rescue, etc.

Down

1. Substance that causes injury or death with exposure

2. Person responsible for overall incident management

4. Removal from trapped area

5. Fire in a house, office, or other building

7. Gaining entry to an enclosed area

Ambulance Calls

The following case scenarios provide an opportunity to explore the concerns associated with patient management. Read each scenario, then answer each question in detail.

1. You are dispatched to "chest pain" by a third party caller. No one answers the front door when you knock, and the door is locked. You hear a dog barking inside. The dispatcher informs you that the call was placed by the man's wife who is not on the premises. She told the dispatcher that her husband called her cellular phone complaining of chest pain, and he has recently been released from the hospital.

How would you best manage this situation?

_____

_____

_____

2. It is spring, and the water run off from melting snow has caused the local viaduct to swell with cold, fast moving waters. You are off duty when you hear a tone out for "small boy swept away by flood waters." You arrive to the scene before your department's on-duty responders. You see a boy of approximately 13 years old who is clinging to life in the middle of the channel, holding onto a trapped log. His mother is hysterical and screaming for you to "jump in and get him." You have no safety equipment available.

How would you best manage this patient?

_____

_____

_____

3. You are dispatched to a chemical spill where a train car derailed. The patient is the engineer who went back to look at the damage. He is lying beside the tracks and appears to be breathing from your vantage point at the staging area. HazMat team members are suiting up to go in and retrieve the patient. They will decontaminate him before bringing him to the staging area.

How will you best manage this situation and patient?

_____

_____

_____

CHAPTER

37 Special Operations

Workbook Activities

The following activities have been designed to help you. Your instructor may require you to complete some or all of these activities as a regular part of your EMT-B training program. You are encouraged to complete any activity that your instructor does not assign as a way to enhance your learning in the classroom.

Chapter Review

The following exercises provide an opportunity to refresh your knowledge of this chapter.

Matching

Match each of the terms in the left column to the appropriate definition in the right column.

_____ 1. Mass-casualty incident

_____ 2. Incident command

_____ 3. Toxicity level

_____ 4. Triage

_____ 5. Chemical Transportation

_____ 6. Protection level

_____ 7. Casualty collection area

_____ 8. Disaster

_____ 9. Hazardous materials

_____ 10. Rehabilitation area

_____ 11. Transportation area

_____ 12. Treatment officer

A. incident in which a hazardous material is no longer properly contained and isolated

B. area where patients can receive further system triage and medical care

C. individual who is in charge of and directs EMS personnel at the treatment area

D. the process of sorting patients based on the severity of injury and medical need, to establish treatment and transportation priorities

E. a measure of the amount and type of Emergency Center protective equipment that an individual (CHEMTREC) needs to avoid injury during contact with a hazardous material

F. provides protection and treatment to firefighters and other personnel working at an emergency

G. area where ambulances and crews are organized

H. an agency that assists emergency personnel in identifying and handling hazardous materials transport incidents

I. widespread event that disrupts community incident resources and functions

J. an emergency situation involving more than one patient, and which can place such demand on equipment or personnel that the system is stretched to its limit or beyond

K. an organizational system to help control, direct, and coordinate emergency responders and resources; known more generally as an incident management system (IMS)

L. a measure of the risk that a hazardous material poses to the health of an individual who comes in contact with it

Multiple Choice

Read each item carefully, then select the best response.

_____ **1.** Functions normally centered at the command post include:
   A. information.
   B. safety.
   C. liaison with other agencies and groups who are responding.
   D. all of the above

_____ **2.** In extended operations, the typical incident command structure may have multiple sectors including:
   A. operations.
   B. planning.
   C. logistics.
   D. all of the above

_____ **3.** You are dispatched to a major airline crash. Initial reports state that the plane exploded into a fireball. Multiple agencies have been alerted and are in route. With a major airplane crash, the leading agency is typically the:
   A. EMS department.
   B. law enforcement.
   C. fire department.
   D. HazMat team.

_____ **4.** As you approach the crash site there is a lot of activity with multiple agencies arriving at the same time. You report to incident commander and he tells you to report to the holding area for arriving ambulances and crews until they can be assigned a particular task. This area is know as a:
   A. staging area.
   B. treatment area.
   C. transportation area.
   D. rehabilitation area.

_____ **5.** The _____ provides protection and treatment to firefighters and other personnel working at the emergency scene.
   A. staging area
   B. treatment area
   C. transportation area
   D. rehabilitation area

_____ **6.** For the scenario in question 3, you have been asked to move forward and pickup patients for removal to a hospital. The area you report to is the:

    A. staging area.

    B. treatment area.

    C. transportation area.

    D. rehabilitation area.

_____ **7.** For the scenario in question 3, you and your partner pick up your patients and prepare to transport them to the hospital. As you load the patients into the ambulance, what should the transport officer log?

    A. Each patient's mass-casualty tag number

    B. Each patient's overall condition

    C. The hospital to which they will be taken

    D. All of the above

_____ **8.** The _____ is where a more thorough assessment is made and on-scene treatment is begun while transport is being arranged.

    A. staging area

    B. treatment area

    C. transportation area

    D. rehabilitation area

_____ **9.** Examples of mass-casualty incidents include:

    A. airplane crashes.

    B. earthquakes.

    C. railroad crashes.

    D. all of the above

_____ **10.** To make decontaminating the ambulance easier after a HazMat incident:

    A. tape the cabinet doors shut.

    B. place any equipment that will not be used en route in the front of the truck.

    C. turn on the power vent ceiling fan and patient compartment air conditioning unit fan.

    D. all of the above

_____ **11.** You respond to a HazMat call. Upon arrival, the incident commander tells you they have a critical patient whom they don't have time to decontaminate. You should do all of the following except:

    A. wear two pairs of gloves.

    B. remove goggles.

    C. wear a protective coat.

    D. wear respiratory protection.

_____ **12.** EMT-Bs providing care in the treatment area should assess and treat the patient:

    A. as contaminated.

    B. with respect.

    C. from the point where the previous caregiver left off.

    D. in the same way as a patient who has not previously been assessed or treated.

_____ **13.** To avoid entrapment and communication of contaminants, only _____ are applied, until the "clean" patient has been moved to the treatment area.

    A. pressure dressings that are needed to control bleeding

    B. bandages

    C. splints

    D. cervical collars

_____ **14.** When toxic gas, fumes, or airborne droplets or particles are involved, the safe area is upwind and at least _____ from the site of any visible cloud or other discharge.

    A. 50'

    B. 100'

    C. 150'

    D. 200'

_____ **15.** If you can see and read the placard or other warning sign, note _____, and, if included, the four-digit number that appears on it or on any orange panel near it.

    A. its color

    B. its wording

    C. any symbols that it contains

    D. all of the above

_____ **16.** Once you have reached a safe place, try to rapidly assess the situation and provide as much information as possible when calling for the HazMat team, including:

    A. your specific location.

    B. the size and shape of the containers of the hazardous material.

    C. what you have observed and been told has occurred.

    D. all of the above

_____ **17.** In the event of a leak or spill, a hazardous materials incident is often indicated by presence of:

    A. a visible cloud or strange-looking smoke resulting from the escaping substance.

    B. a leak or spill from a tank, etc., with or without HazMat placards or labels.

    C. an unusual, strong, noxious, acrid odor in the area.

    D. all of the above

_____ **18.** Safety of _____ must be your most important concern.

    A. you and your team

    B. the other responders

    C. the public

    D. all of the above

_____ **19.** In some incidents, a large number of people are _____ and may be injured or killed before the presence of a hazardous materials incident is identified.

    A. transported

    B. exposed

    C. injected

    D. decontaminated

_____ **20.** Often, the presence of hazardous materials is easily recognized from warning signs, placards, or labels found:

    A. on buildings or areas where hazardous materials are produced, used, or stored.

    B. on trucks and railroad cars that carry any amount of hazardous material.

    C. on barrels or boxes that contain hazardous material.

    D. all of the above

For the remainder of the multiple-choice section, the following answers are to be applied:

A. First priority (red)
B. Second priority (yellow)
C. Third priority (green)
D. Fourth priority (black)

Classify the following emergencies according to triage priority:

21. Shock _____

22. Major or multiple bone or joint injuries _____

23. Cardiac arrest _____

24. Minor fractures _____

25. Decreased level of consciousness _____

26. Obvious death _____

27. Airway and breathing difficulties _____

28. Burns without airway problems _____

29. Major open brain trauma _____

30. Minor soft-tissue injuries _____

Fill-in

Read each item carefully, then complete the statement by filling in the missing word(s).

1. The _____ _____ _____ is more effective when used to organize large numbers of personnel at complex incidents such as hazardous materials spills and mass-casualty incidents.

2. The incident commander usually remains at a _____ _____, the designated field command center.

3. The _____ _____ is responsible for protecting all personnel and any victims of the incident.

4. When you arrive at the scene of a possible _____ _____ _____, you must first step back and assess the situation.

5. If patients are entrapped, _____ is required.

6. A _____ is a widespread event that disrupts functions and resources of a community and threatens lives and property.

7. _____ is the sorting of patients based on the severity of their conditions to establish priorities for care based on available resources.

8. Transporting a _____ patient merely increases the size of the event.

9. Most serious injuries and deaths from hazardous materials result from _____ and _____ problems.

10. _____ is the process of removing or neutralizing and properly disposing of hazardous materials from equipment, patients, and rescue personnel.

11. When dealing with a HazMat situation, be sure to check the wind direction periodically, and _____ if a change in wind direction dictates.

12. Some substances are not hazardous; however, when mixed with another substance, they may become _____ or volatile.

13. In most cases, the package or tank must contain a certain amount of a hazardous material before a _____ is required.

True/False

If you believe the statement to be more true than false, write the letter "T" in the space provided. If you believe the statement to be more false than true, write the letter "F."

_____ **1.** When you are responding to a hazardous materials incident, you must first take time to accurately assess the scene.

_____ **2.** Moving patients from the contaminated area is your main responsibility in a hazardous materials situation.

_____ **3.** Toxicity level 1 is more dangerous than level 4.

_____ **4.** Protective clothing level A is the least level of protection.

_____ **5.** Patients with major or multiple bone or joint injuries should be assigned to the second priority triage category.

_____ **6.** Patients with severe burns should be assigned to the black triage category.

_____ **7.** A large number of hazardous gases and fluids are essentially odorless.

_____ **8.** Only the original patients who leave the hazard zone must pass through the decontamination area.

_____ **9.** Most hazardous materials have specific antidotes or treatments for exposure.

_____ **10.** The success of any incident command system depends on all personnel performing their assigned tasks and working within the system.

Short Answer

Complete this section with short written answers using the space provided.

1. Define the five levels of toxicity as classified by the NFPA.

_____

_____

2. Describe the four levels of protection and the type of protective gear required for each level.

_____

_____

3. List the eight major EMS-related positions within an incident command system.

_____

_____

4. List and define the four triage priorities.

_____

_____

5. For each of the following hazardous materials classifications, list the general category of hazard.

Class	Type
Class 1	
Class 2	
Class 3	
Class 4	
Class 5	
Class 6	
Class 7	
Class 8	
Class 9	

6. Define and describe the process of decontamination and the decontamination area.

Word Fun

The following crossword puzzle is an activity provided to reinforce correct spelling and understanding of medical terminology associated with emergency care and the EMT-B. Use the clues in the column to complete the puzzle.

Across

3. Person responsible for overall incident management

5. Designated field command center

6. Zone where contaminants are removed

8. Any harmful substance

9. Determines amount of gear to be worn around a given hazard

Down

1. Zone for loading of patients into ambulances

2. Responsible for sorting of patients

4. Measures of health risk of a substance

6. Area of least safety, exposure to harm possible

7. Sorting by priority

Ambulance Calls

The following case scenarios will give you an opportunity to explore the concerns associated with patient management. Read each scenario, then answer each question in detail.

1. You are dispatched to a multi-vehicle crash where you encounter three patients: a 4-year-old boy with bilateral femur fractures and absent radial pulse, a 27-year-old woman with a laceration to the head and a humerus fracture, and a 42-year-old man who is apneic and pulseless with an open skull fracture.

   How should you triage these patients?

   _____

   _____

   _____

2. Your response area contains a large portion of farming and other agricultural lands, including many orchards and vineyards. Right at shift change, there is a tone out for a "crop duster accident" in a remote area of your jurisdiction. It appears that 15-30 agricultural workers were accidentally sprayed with pesticides and other chemicals from a crop dusting plane. They are now experiencing a variety of signs and symptoms including: nausea, vomiting, eye and upper airway irritation.

   How would you best manage this situation?

   _____

   _____

   _____

3. You and your partner are enjoying an unusually uneventful evening at work when you receive a dispatch for "overturned semi truck." As you approach the scene, you see a semi tractor trailer that has left the roadway and rolled down a steep embankment. You see the driver attempting to climb up to the roadway where many passersby have stopped to see what happened. He is vigorously coughing, and you can see a liquid dripping from the truck's tank.

   How would you best manage this situation?

   _____

   _____

   _____

CHAPTER

38 Response to Terrorism and Weapons of Mass Destruction

Workbook Activities

The following activities have been designed to help you. Your instructor may require you to complete some or all of these activities as a regular part of your EMT-B training program. You are encouraged to complete any activity that your instructor does not assign as a way to enhance your learning in the classroom.

Chapter Review

The following exercises provide an opportunity to refresh your knowledge of this chapter.

Matching

Match each of the terms in the left column to the appropriate definition in the right column.

_____ 1. Mutagen

_____ 2. Vesicants

_____ 3. Disease vector

_____ 4. Phosgene

_____ 5. Neurotoxins

_____ 6. Bacteria

_____ 7. Volatility

_____ 8. Cyanide

_____ 9. Vapor hazard

_____ 10. Lymph nodes

A. pulmonary agent that is a product of combustion

B. describes how long a chemical agent will stay on a surface before it evaporates

C. microorganisms that reproduce by binary fission

D. agent that enters the body through the respiratory tract

E. animal that, once infected, spreads the disease to another animal

F. agent that affects the body's ability to use oxygen

G. blister agents

H. area of the lymphatic system where infection-fighting cells are housed

I. substance that mutates and damages the structures of DNA in the body's cells

J. biological agents that are the most deadly substances known to humans

Multiple Choice

Read each item carefully, then select the best response.

_____ **1.** Examples of terrorist groups include:

   A. violent religious groups.

   B. extremist political groups.

   C. technology groups.

   D. all of the above

_____ **2.** An example of a single issue group is:

   A. antiabortion groups.

   B. separatist groups.

   C. Aum Shinrikyo.

   D. KKK.

_____ **3.** When were chemical agents first introduced?

   A. Spanish-American War

   B. World War I

   C. World War II

   D. Korean War

_____ **4.** When the U. S. Department of Homeland Security Advisory System is at yellow, what does that mean?

   A. Severe risk of terrorist attack

   B. Significant risk of terrorist attack

   C. High risk of terrorist attack

   D. General risk of terrorist attack

_____ **5.** You are called to the scene of an unexplained explosion at the local shopping mall. The reports are that there are multiple injuries. You are the first unit to arrive on the scene. Your first responsibility is:

   A. scene safety.

   B. set up the incident command system.

   C. start triage.

   D. request additional resources.

_____ **6.** In the previous scenario, you have been informed that there are numerous agencies responding with many different types of apparatus. They are approaching the scene from all directions and will be arriving shortly. You need to:

   A. set up a staging area.

   B. separate the different types of apparatus.

   C. let them continue as they are.

   D. have them come in from downwind.

_____ **7.** You have been called to an explosion that has released a gas cloud into the air. What are some of the possible hazards at the location?

   A. Suspicious package

   B. Change in wind direction

   C. Contaminated patients

   D. All of the above

_____ **8.** A brownish, yellowish oily substance that is generally considered very persistent is:

   A. lewisite.

   B. phosgene oxime.

   C. sulfur mustard.

   D. vesicant.

_____ **9.** An example of a pulmonary agent is:

    A. chlorine.

    B. phosgene oxime.

    C. G agents.

    D. lewisite.

_____ **10.** The most lethal of all the nerve agents is:

    A. V agent.

    B. sarin.

    C. soman.

    D. tabun.

_____ **11.** What two medications do MARK 1 antidote kits contain?

    A. atropine and 2-PAM chloride

    B. atropine and epinephrine

    C. epinephrine and 2-PAM chloride

    D. lidocaine and atropine

_____ **12.** You are dispatched to a local farm where a 41-year-old man has been discovered unconscious. The airway is open but the patient has been vomiting. Respirations are within normal limits and distal pulses are present. The patient has muscle twitches and has urinated on himself. There is a funny odor that seems to be coming from the clothing of the patient. You would suspect:

    A. alcohol poisoning.

    B. organophosphate poisoning.

    C. nerve agent.

    D. respiratory agent.

_____ **13.** You are dispatched to a patient who is having respiratory problems. He is awake but is working so hard to breathe that he can't answer questions. His distal pulses are present and strong. There is an odor of almonds in the air. You would suspect:

    A. cyanide.

    B. sarin.

    C. soman.

    D. tabun.

_____ **14.** The period of time between the person becoming exposed to an agent and when symptoms begin is called:

    A. contagious.

    B. incubation.

    C. communicability.

    D. remission.

_____ **15.** Examples of viral hemorrhagic fevers include:

    A. Ebola.

    B. Rift Valley.

    C. Yellow Fever.

    D. all of the above.

_____ **16.** The infectious disease that has killed the most humans is:

    A. anthrax.

    B. bacteria.

    C. virus.

    D. plague.

_____ **17.** Bubonic plague infects the:

    A. respiratory system.

    B. circulatory system.

    C. lymphatic system.

    D. digestive system.

_____ **18.** The deadliest substances know to humans are:
   A. neurotoxins.
   B. hemotoxins.
   C. plagues.
   D. bacteria.

_____ **19.** The least toxic route for ricin is:
   A. oral.
   B. inhalation.
   C. injection.
   D. absorption.

_____ **20.** The EMS role in helping to determine a biological event is to:
   A. administer medications.
   B. be aware of an unusual number of calls for unexplainable flu.
   C. quarantine infected individuals.
   D. set up field hospitals.

_____ **21.** The most powerful of all radiation is:
   A. alpha.
   B. neutron.
   C. beta.
   D. gamma.

_____ **22.** To protect yourself from radiation exposure, you should:
   A. limit the time of exposure.
   B. increase distance between yourself and the source.
   C. use shielding.
   D. all of the above

Fill-in

Read each item carefully, then complete the statement by filling in the missing word.

1. The bombing of the Alfred P. Murrah Federal Building in Oklahoma City is an example of _____

_____.

2. Any agent designed to bring about mass death, casualties, and/or massive damage to property and infrastructure is

a _____ _____ _____ _____.

3. _____ _____ is when a nation has close ties with terrorist groups.

4. If there is a high risk of a terrorist attack, the Homeland Security Advisory system will be at threat level

_____.

5. _____ occurs when you come into contact with a contaminated person who has not been decontaminated.

6. _____ _____ _____ is a term used to describe how the agent most effectively enters the

body.

7. An agent that gives off very little or no vapor and enters the body through the skin is called _____

_____.

8. _____ _____ are among the most deadly chemicals developed.

9. _____ means that vapors are continuously released over a period of time.

10. _____ is the means by which a terrorist will spread the agent.

11. _____ is a germ that requires a living host to multiply and survive.

12. The group of viruses that cause the blood in the body to seep out from the tissues and blood vessels are called

_____ _____ _____.

13. _____ is a deadly bacterium that lays dormant in a spore.

14. Buboes are formed when the _____ _____ become infected and grow.

15. The most potent neurotoxin is _____.

16. _____ _____ _____ are strategically placed facilities that have been preestablished for the

mass distribution of medications and supplies.

17. Any device that is designed to disperse a radioactive device is called a _____ _____ _____.

True/False

If you believe the statement to be more true than false, write the letter "T" in the space provided. If you believe the statement to be more false than true, write the letter "F."

_____ **1.** Atlanta's Centennial Park bombing during the 1996 Summer Olympics is an example of international terrorism.

_____ **2.** WMDs are easy to obtain or create.

_____ **3.** Most acts of terrorism occur after a warning is given to the general public.

_____ **4.** Understanding and being aware of the current threat is only the beginning of responding safely.

_____ **5.** Failure to park your ambulance in a safe location can place you and your partner in danger.

_____ **6.** You should have all units responding to an explosion converge on the main entrance to the building.

_____ **7.** Vapor hazards enter the body through the pores in the skin.

_____ **8.** The primary route of exposure of vesicants is through inhalation.

_____ **9.** Phosgene and phosgene oxime are two different classes of agents.

_____ **10.** Tabun looks like baby oil.

_____ **11.** Seizures are the most common symptom of nerve agent exposure.

_____ **12.** Organophosphate is the basic ingredient in nerve agents.

_____ **13.** Cyanide binds with the body's cells, preventing oxygen from being used.

_____ **14.** When dealing with smallpox, gloves are all the BSI you need.

_____ **15.** Outbreaks of the viral hemorrhagic fevers are extremely rare worldwide.

_____ **16.** Pulmonary anthrax infections are associated with a 90% death rate if untreated.

_____ **17.** Pneumonic plague is deadlier than bubonic plague.

_____ **18.** Ricin is deadlier than botulinum.

_____ **19.** Ingestion of ricin causes necrosis of the lungs.

_____ **20.** Large containers called "life packs" are delivered during a biological event.

_____ **21.** The dirty bomb is an ineffective WMD.

_____ **22.** Being exposed to a radiation source does not make a patient contaminated or radioactive.

Short Answer

Complete this section with short written answers using the space provided.

1. What are the key questions you should ask yourself about dealing with WMDs?

_____

_____

2. List the four classes of chemical agents.

_____

_____

3. What things should you observe on every call to determine the potential for a terrorist attack?

_____

_____

4. List the signs of vesicant exposure to the skin.

_____

_____

5. What are some of the later signs and symptoms of chlorine inhalation?

_____

_____

6. What does the mnemonic SLUDGEM stand for?

_____

_____

7. What are the signs and symptoms of high doses of cyanide?

_____

_____

8. List the signs and symptoms of ricin ingestion.

_____

_____

9. List three places that radioactive waste may be found.

_____

_____

10. The best way to protect yourself from the effects of radiation is to use:

_____

_____

Word Fun

The following crossword puzzle is an activity provided to reinforce correct spelling and understanding of medical terminology associated with emergency care and the EMT-B. Use the clues in the column to complete the puzzle.

Across

4. Pulmonary agent that is a product of combustion

6. Microorganisms that reproduce by binary fission

8. Describes how long a chemical agent will stay on a surface before it evaporates

10. Agent that affects the body's ability to use oxygen

11. Blister agents

Down

1. First chemical agent ever used in warfare

2. Energy that is emitted from a strong radiological source

3. Substance that damages and changes the DNA structures in the body's cells

5. Highly contagious disease

7. Person infected with a disease that is highly communicable

9. Neurotoxin derived from mash that is left from the castor bean

Ambulance Calls

The following case scenarios provide an opportunity to explore the concerns associated with patient management. Read each scenario, then answer each question in detail.

1. You are dispatched to an explosion at a nearby shopping mall. No other information is available regarding the nature of the explosion, only that there are possibly upwards of five fatally wounded and fifty severely injured. How would you best manage this situation?

_____

_____

_____

2. Your emergency system is suddenly inundated with numerous calls for people experiencing fever, chills, headache, muscle aches, nausea/vomiting, diarrhea, severe abdominal cramping, and GI bleeding. All of the patients attended a local indoor sporting event some six hours earlier. How would you best manage this situation?

_____

_____

_____

3. You are dispatched to treat numerous patients with known exposure to cyanide. This occurred in a neighboring jurisdiction, and they have requested your assistance. The local fire department has set up a decontamination area, and you are asked to transport patients to the nearest appropriate medical facility.

What are important considerations to note about cyanide exposure?

_____

_____

_____

CHAPTER

39 Advanced Airway Management

Workbook Activities

The following activities have been designed to help you. Your instructor may require you to complete some or all of these activities as a regular part of your EMT-B training program. You are encouraged to complete any activity that your instructor does not assign as a way to enhance your learning in the classroom.

Chapter Review

The following exercises provide an opportunity to refresh your knowledge of this chapter.

Matching

Match each of the terms in the left column to the appropriate definition in the right column.

_____ 1. Pharynx

_____ 2. Larynx

_____ 3. Epiglottis

_____ 4. Inhalation

_____ 5. Alveolar/capillary

_____ 6. Capillary/cellular

_____ 7. Exhalation

_____ 8. End tidal carbon dioxide detector

_____ 9. Cricoid

_____ 10. Stylet

_____ 11. Intubation

_____ 12. Laryngoscope

_____ 13. Vallecula

A. a plastic-coated wire that adds rigidity and shape to the endotracheal tube

B. passive phase of breathing

C. gas exchanged in body

D. an instrument that provides a direct view of the vocal cords

E. active phase of breathing

F. the space between the base of the tongue and the epiglottis

G. vocal cords

H. rigid, ring-shaped cartilage

I. the placement of a tube into the trachea to maintain the airway

J. gas exchanged in the lungs

K. throat

L. leaf-shaped structure that prevents food/liquids from entering the lower airway

M. a plastic disposable indicator that signals, by color change, that an ETT is in the proper place

Multiple Choice

Read each item carefully, then select the best response.

_____ **1.** To maintain a patent airway, patients whose consciousness is altered may require:
   A. an oropharyngeal airway.
   B. a nasopharyngeal airway.
   C. suctioning.
   D. all of the above

_____ **2.** The purpose of advanced airway management is to protect and improve _____ in patients by using a tube to create a direct channel to the trachea.
   A. respiration
   B. ventilation
   C. oxygenation
   D. patency

_____ **3.** The upper airway includes all of the following, except:
   A. nose.
   B. mouth.
   C. larynx.
   D. pharynx.

_____ **4.** The _____ is located at the glottic opening and prevents food and liquid from entering the lower airway during swallowing.
   A. larynx
   B. vocal cords
   C. epiglottis
   D. carina

_____ **5.** Gas exchange occurs in the:
   A. alveoli.
   B. nares.
   C. bronchioles.
   D. trachea.

_____ **6.** The mechanical process of breathing occurs through the nose of the diaphragm and:
   A. ribs.
   B. intercostal muscles.
   C. lung parenchyma.
   D. trachea.

_____ **7.** The diaphragm and intercostal muscles _____ during inhalation, increasing the size of the chest cavity.
   A. expand
   B. contract
   C. dilate
   D. spasm

_____ **8.** During inspiration:
   A. the diaphragm contracts and the ribs move up and out.
   B. the diaphragm relaxes and the ribs move up and out.
   C. the diaphragm contracts and the ribs move down and in.
   D. the diaphragm relaxes and the ribs move down and in.

_____ **9.** During expiration:

    A. the diaphragm contracts and the ribs move up and out.

    B. the diaphragm relaxes and the ribs move up and out.

    C. the diaphragm contracts and the ribs move down and in.

    D. the diaphragm relaxes and the ribs move down and in.

_____ **10.** After _____ minutes without oxygen, cells in the brain and nervous system may die.

    A. 2 to 3

    B. 3 to 5

    C. 4 to 6

    D. 5 to 8

_____ **11.** The first step in airway management is:

    A. suctioning.

    B. c-spine control.

    C. applying oxygen.

    D. opening the airway.

_____ **12.** Patients with gastric distention are prone to:

    A. gas.

    B. vomiting.

    C. sepsis.

    D. hypoxia.

_____ **13.** A nasogastric tube can cause:

    A. nasal trauma.

    B. a basilar skull fracture.

    C. gastric distention.

    D. facial trauma.

_____ **14.** To perform the Sellick maneuver, apply pressure to the:

    A. thyroid cartilage.

    B. cricoid cartilage.

    C. cricothyroid membrane.

    D. trachea.

_____ **15.** You should not immediately intubate a patient who is unresponsive or in cardiac arrest, but you must try:

    A. to open the airway with the appropriate BLS maneuver.

    B. to clear the airway.

    C. to ventilate the patient with a BVM or oxygen-powered breathing device.

    D. all of the above

_____ **16.** _____ is the most effective way to control a patient's airway and has many advantages over other airway management techniques.

    A. An oropharyngeal adjunct

    B. Orotracheal intubation

    C. A nasopharyngeal adjunct

    D. A jaw thrust

_____ **17.** Remember that _____ is the priority for patients in cardiac arrest from ventricular fibrillation.

    A. airway

    B. breathing

    C. circulation

    D. defibrillation

_____ **18.** The purpose of a _____ is to sweep the tongue out of the way and align the airway so that you can see the vocal cords and pass the ET tube through them.
   A. laryngoscope
   B. lighted stylet
   C. Magill forceps
   D. 10-mL syringe

_____ **19.** The curved laryngoscope blade is inserted just in front of the epiglottis, into the _____, allowing you to see the glottic opening and vocal cords.
   A. uvula
   B. vallecula
   C. larynx
   D. pharynx

_____ **20.** The proper sized tube for adult male patients ranges from _____ mm.
   A. 6.5 to 8.0
   B. 7.0 to 8.0
   C. 7.5 to 8.5
   D. 8.0 to 9.0

_____ **21.** The proper sized tube for adult female patients ranges from _____ mm.
   A. 6.5 to 8.0
   B. 7.0 to 8.0
   C. 7.5 to 8.5
   D. 8.0 to 9.0

_____ **22.** A good rule of thumb is to always have a _____ mm ETT on hand; this size tube will fit most male or female adult patients.
   A. 6.5
   B. 7.0
   C. 7.5
   D. 8.0

_____ **23.** In children younger than 8 years, the circular narrowing of the trachea at the level of the _____ functions as a cuff.
   A. larynx
   B. cricoid cartilage
   C. pharynx
   D. thyroid cartilage

_____ **24.** A plastic-coated wire called a _____ may be inserted into the ETT to add rigidity and shape to the tube.
   A. Murphy eye
   B. stylet
   C. pipe cleaner
   D. vallecula

_____ **25.** You will use the _____ to test for air holes in the ETT before intubation.
   A. 10-mL syringe
   B. lighted stylet
   C. Murphy eye
   D. pilot balloon

_____ **26.** Equipment used for airway and ventilation assistance include:
   A. oxygen.
   B. a suctioning unit.
   C. a ventilation device.
   D. all of the above

_____ **27.** Do not let go of the ETT until:

    A. the distal cuff is inflated.

    B. placement is confirmed.

    C. it is secured.

    D. the stylet is removed.

_____ **28.** Confirm placement of the ETT by listening with a stethoscope over both lungs and over the _____ as you ventilate the patient through the tube.

    A. stomach

    B. heart

    C. diaphragm

    D. ribs

_____ **29.** The best way to confirm placement of an ETT is by:

    A. X-ray.

    B. auscultating over both lung fields and over the epigastrium.

    C. visualizing the cuff passing through the vocal cords.

    D. seeing the chest rise and fall.

_____ **30.** The first step in nasotracheal intubation is to:

    A. check for a gag reflex.

    B. turn the stylet on.

    C. hyperventilate the patient.

    D. use BSI precautions.

_____ **31.** Complications of endotracheal intubation include:

    A. mechanical failure.

    B. intubating the esophagus.

    C. causing soft-tissue trauma.

    D. all of the above

_____ **32.** Benefits of endotracheal intubation include all of the following, except:

    A. it provides complete protection of the airway.

    B. patient intolerance.

    C. it prevents gastric distention and aspiration.

    D. it delivers better oxygen concentration than a BVM device.

_____ **33.** Always check for a gag reflex before intubation by using:

    A. a tongue blade.

    B. the laryngoscope blade.

    C. an oral airway.

    D. all of the above

_____ **34.** Benefits of using a multi-lumen airway include all of the following, except:

    A. no mask seal necessary.

    B. requires deeply comatose patient.

    C. may be inserted blindly.

    D. ease of proper placement.

_____ **35.** Contraindications of the ETC include:

    A. patients with a gag reflex.

    B. children younger than 16 years.

    C. patients who have ingested a caustic substance.

    D. all of the above

_____ **36.** To remove the ETC, you should _____ prior to removal.
   A. turn the patient on his or her side
   B. deflate both balloon cuffs
   C. have suction available
   D. all of the above

_____ **37.** Contraindications of the PtL include all of the following, except:
   A. patients with a gag reflex.
   B. children younger than 14 years.
   C. children older than 16 years.
   D. patients who have a known esophageal disease.

_____ **38.** Potential contraindications when using an LMA include:
   A. when positive pressure ventilation with high airway pressures occur, the mask may leak.
   B. active vomiting may dislodge the device.
   C. esophageal disease.
   D. all of the above

Labeling

Label the following diagram with the correct terms.

1. The Upper and Lower Airways

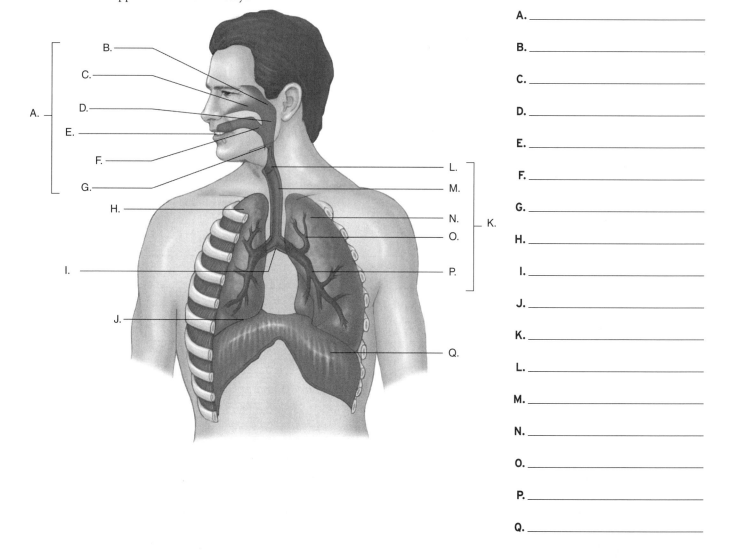

A. _____

B. _____

C. _____

D. _____

E. _____

F. _____

G. _____

H. _____

I. _____

J. _____

K. _____

L. _____

M. _____

N. _____

O. _____

P. _____

Q. _____

Fill-in

Read each item carefully, then complete the statement by filling in the missing word.

1. During _____, the diaphragm contracts.

2. The _____ branch off from the trachea.

3. During the _____ phase of breathing, the intercostal muscles relax.

4. _____ in the alveoli crosses over into the blood.

5. _____ _____ in the blood crosses over into the alveoli.

6. Cells in the brain and nervous system may die after _____ minutes without oxygen.

7. The _____ are the smallest air passages leading to the alveoli.

8. Alveoli are surrounded by _____, which bring deoxygenated blood to the lungs.

True/False

If you believe the statement to be more true than false, write the letter "T" in the space provided. If you believe the statement to be more false than true, write the letter "F."

_____ 1. The light in the laryngoscope will not work unless the blade is attached correctly.

_____ 2. Endotracheal intubation is usually most appropriate for unconscious patients.

_____ 3. There are three different sizes of ETTs.

_____ 4. The balloon cuff around the end of an ETT holds 25 mL of air.

_____ 5. Uncuffed tubes are used in children younger than 8 years.

_____ 6. When a wire stylet is used, it should stick out 1/2" beyond the tip of the ETT.

Short Answer

Complete this section with short written answers using the space provided.

1. Describe how to perform the Sellick maneuver.

_____

_____

2. List three advantages of orotracheal intubation.

_____

_____

3. List nine possible complications associated with endotracheal intubation.

_____

_____

4. List five contraindications for an Esophageal Tracheal Combitube.

_____

_____

5. List three benefits and three complications of multi-lumen airways.

_____

_____

Word Fun emtb.com vocab explorer

The following crossword puzzle is an activity provided to reinforce correct spelling and understanding of medical terminology associated with emergency care and the EMT-B. Use the clues in the column to complete the puzzle.

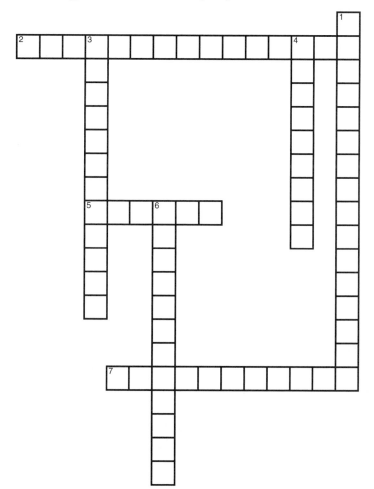

Across

2. Pressure on cricoid cartilage

5. Small metal rod inserted into an ET tube

7. Removes substances from stomach

Down

1. Rigid, ring-shaped structure at larynx

3. Vocal cord spasm

4. Between base of tongue and epiglottis

6. Used to gain direct view of patient's throat

Ambulance Calls

The following case scenarios provide an opportunity to explore the concerns associated with patient management. Read each scenario, then answer each question in detail.

1. You are dispatched to a pulseless, apneic 45-year-old man. After assessing the patient, your partner decides to perform endotracheal intubation (a newly allowed skill by your medical program director). This is the first field intubation either you or your partner has attempted without supervision of a paramedic or physician on-scene. Your partner tells you that he "thinks" he passed through the cords. As you auscultate the chest, you do not hear breath sounds, but you do hear gurgling over the epigastrium.

How would you best manage this situation?

_____

_____

_____

2. You are dispatched to "CPR in progress" at a local restaurant. You arrive to find laypersons feverishly attempting CPR. You also notice that the patient's stomach is rather distended. Witnesses tell you that the man complained of chest pain just prior to collapse when he was assisted to the floor without suffering any trauma. Your local protocols allow for placement of gastric tubes.

How would you best manage this situation?

_____

_____

_____

3. You are dispatched to the scene of a motor vehicle crash where your patient, the unrestrained 18-year-old driver, is breathing at a rate of 4 breaths/min and has weak radial pulses. He is also trapped in the vehicle and the fire department has not yet arrived.

How would you best manage this patient?

_____

_____

_____

Skill Drills
Skill Drill 39-1: Performing the Sellick Maneuver
Test your knowledge of this skill drill by filling in the correct words in the photo captions.

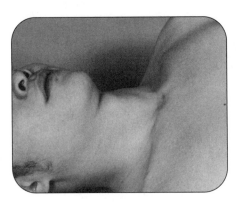
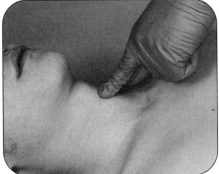

1. Visualize the _____ cartilage.

2. _____ to confirm its location.

3. Apply _____ pressure on the cricoid _____ with your thumb and index finger on either side of the _____ . Maintain pressure until the patient is

_____ .

Skill Drill 39-2: Performing Orotracheal Intubation
Test your knowledge of this skill drill by placing the photos below in the correct order. Number the first step with a "1," the second step with a "2," etc.

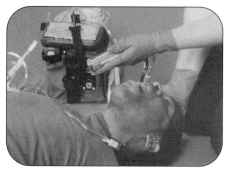

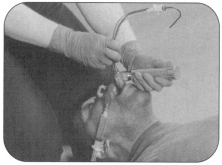

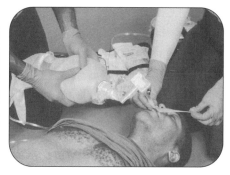

Confirm adequate preoxygenation, and remove the oral airway.

If available, have another rescuer perform the Sellick maneuver to improve visualization of the cords.

Use the head tilt-chin lift maneuver to position the nontrauma patient for insertion of the laryngoscope.

Insert the laryngoscope from the right side of mouth, and move the tongue to the left. Lift the laryngoscope away from the posterior pharynx to visualize the vocal cords. *Do not pry or use the teeth as a fulcrum.*

Insert the ET tube from the right side until the ET tube cuff passes between the vocal cords. Remove the laryngoscope and stylet. Hold the tube carefully until it is secured.

Secure the tube, and continue to ventilate.

Note and record depth of insertion (centimeter marking at the teeth), and reconfirm position after each time you move the patient.

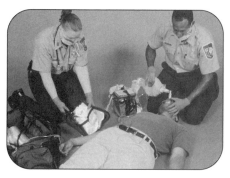

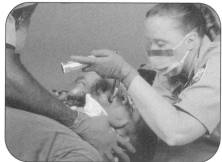

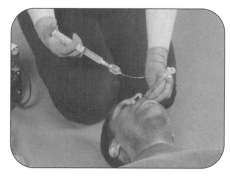

Open and clear the airway.

Place an oropharyngeal airway, and preoxygenate with a BVM device.

In a trauma patient, maintain the cervical spine in-line and neutral as your partner lies down or straddles the patient's head to visualize the vocal cords.

Inflate the balloon cuff, and remove the syringe as your partner prepares to ventilate.

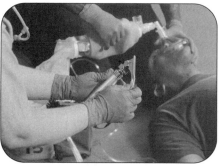

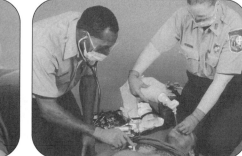

Assemble and test intubation equipment as your partner continues to ventilate.

Begin ventilating, and confirm placement of the ET tube by listening over the stomach and both lungs. Also confirm placement with an end-tidal carbon dioxide detector or EDD, if available.

CHAPTER

40 Assisting With Intravenous Therapy

Workbook Activities

The following activities have been designed to help you. Your instructor may require you to complete some or all of these activities as a regular part of your EMT-B training program. You are encouraged to complete any activity that your instructor does not assign as a way to enhance your learning in the classroom.

Chapter Review

The following exercises provide an opportunity to refresh your knowledge of this chapter.

Matching

Match each of the terms in the left column to the appropriate definition in the right column.

_____ **1.** Saline lock

_____ **2.** Proximal tibia

_____ **3.** Local reaction

_____ **4.** Access port

_____ **5.** Infiltration

_____ **6.** Vasovagal reaction

_____ **7.** Gauge

_____ **8.** Systemic complication

A. mild to moderate reaction to an irritant without systemic consequence

B. sudden hypotension and fainting associated with traumatic or medical events

C. special type of IV apparatus

D. escape of fluid into the surrounding tissue

E. moderate to severe complication affecting the systems of the body

F. measure of the interior diameter of the catheter

G. anatomic location for intraosseous catheter insertion

H. sealed hub on an administration set designed for sterile access to the IV fluid

Multiple Choice

Read each item carefully, then select the best response.

_____ **1.** IV administration requires the use of a(n):

   A. solution.

   B. administration set.

   C. catheter.

   D. all of the above

_____ **2.** When an IV solution is taken out of its protective sterile plastic bag, it must be used within:

   A. 12 hours.

   B. 24 hours.

   C. 36 hours.

   D. 48 hours.

_____ **3.** A microdrip administration set requires _____ drops to flow 1 mL.

   A. 15

   B. 30

   C. 45

   D. 60

_____ **4.** The gauge of a catheter refers to the:

   A. diameter of the needle.

   B. length of the needle.

   C. strength of the needle.

   D. use of the needle.

_____ **5.** Intraosseous IVs are started in the:

   A. external jugular.

   B. proximal tibia.

   C. hand.

   D. antecubital vein.

_____ **6.** Risk associated with starting an IV includes:

   A. infiltration.

   B. phlebitis.

   C. occlusion.

   D. all of the above

_____ **7.** An accumulation of blood in the tissues surrounding an IV site is called a:

   A. contusion.

   B. hematoma.

   C. bruise.

   D. rupture.

_____ **8.** Circulatory overload occurs most commonly in patients with:

   A. an altered mental status.

   B. diabetes.

   C. kidney dysfunction.

   D. dialysis.

_____ **9.** The best administration set to use when giving fluids to a pediatric patient is the:

    A. macrodrip.

    B. Volutrol.

    C. microdrip.

    D. whatever you have closest.

_____ **10.** If fluid replacement is not an issue, what size catheter would be best to use in a geriatric patient?

    A. 14 gauge

    B. 16 gauge

    C. 20 gauge

    D. Butterfly

Fill-in

Read each item carefully, then complete the statement by filling in the missing word.

1. A(n) _____ _____ moves fluid from the IV bag into the patient's vascular system.

2. _____ _____ are best used for rapid fluid replacement.

3. _____ _____ are a way to maintain an active IV site without having to run fluids through the vein.

4. Intraosseous IVs are started with a special needle called a _____ _____.

5. An inflammation of the vein is called _____.

6. _____ _____ occurs when part of the catheter is pinched against the needle and the needle slices through the catheter, creating a free-floating segment.

True/False

If you believe the statement to be more true than false, write the letter "T" in the space provided. If you believe the statement to be more false than true, write the letter "F."

_____ **1.** The most important thing to remember about IV techniques and fluid administration is to keep the IV equipment sterile.

_____ **2.** The more common prehospital volumes are 250 ml and 100 ml.

_____ **3.** The most common types of catheters found in the prehospital setting are central-line catheters.

_____ **4.** A 14-gauge catheter has a greater diameter than a 22-gauge catheter.

_____ **5.** Starting an external jugular IV requires a different technique than starting other IVs.

_____ **6.** An occlusion is the physical blockage of a vein.

_____ **7.** If a hematoma develops when an IV catheter insertion is attempted, the procedure should stop.

_____ **8.** Healthy adults can handle as much as 4 to 5 extra liters of fluid without compromise.

_____ **9.** For pediatric patients, the best catheters for insertions are 20, 22, 24, or 26 gauge.

_____ **10.** Fluid overload is not a problem with geriatric patients.

Short Answer

Complete this section with short written answers using the space provided.

1. List the steps for assembling IV equipment.

_____

_____

2. What are some examples of common local reactions when starting an IV?

_____

_____

3. List the common systemic complications associated with IV insertion.

_____

_____

4. What treatment should a patient suffering a vasovagal reaction be given?

_____

_____

5. What items do you check if your IV is not flowing properly?

_____

_____

Word Fun

The following crossword puzzle is an activity provided to reinforce correct spelling and understanding of medical terminology associated with emergency care and the EMT-B. Use the clues in the column to complete the puzzle.

Across

2. Blockage; usually of a tubular structure such as a blood vessel
4. Area of administration set where fluid accumulates
6. Inflammation of a vein

Down

1. Escape of fluid into the surrounding tissue
3. Measure of the interior diameter of the catheter
4. Another name for administration sets
5. Flexible, hollow structure that drains or delivers fluids

Ambulance Calls

1. You are in the patient compartment with your EMT-B partner who is taking care of a patient with congestive heart failure. Your partner has initiated IV therapy and is now giving a radio report to the receiving hospital. You notice the IV tubing is still running wide open and that nearly the entire liter of fluid has been administered over a few minutes. The patient states his shortness of breath is worsening.

How would you best manage this situation?

_____

_____

_____

2. Your fire department provides "ride alongs" for paramedic students attending school at the local community college. You watch the paramedic student as she starts the IV; she appears to follow all of the appropriate steps, but the IV fails to flow properly.

What could be the problem with this IV?

_____

_____

_____

3. You are caring for a patient who begins to complain of pain at the IV insertion site. The paramedic asks you to inspect her IV for problems. You see swelling around the catheter.

What is this condition and what does it signify?

_____

_____

_____

CHAPTER

41 Assisting With Cardiac Monitoring

Workbook Activities

The following activities have been designed to help you. Your instructor may require you to complete some or all of these activities as a regular part of your EMT-B training program. You are encouraged to complete any activity that your instructor does not assign as a way to enhance your learning in the classroom.

Chapter Review

The following exercises provide an opportunity to refresh your knowledge of this chapter.

Matching

Match each of the terms in the left column to the appropriate definition in the right column.

_____ 1. Sinus tachycardia

_____ 2. Ventricular tachycardia

_____ 3. Electrical conduction system

_____ 4. Arrhythmia

_____ 5. Sinus bradycardia

_____ 6. Ventricular fibrillation

_____ 7. Sinus rhythm

A. network of special cells in the heart through which an electrical current flows

B. consistent P waves, consistent P-R intervals, and a regular heart rate that is less than 60 beats/min

C. consistent P waves, consistent P-R intervals, and a regular heart rate that is more than 100 beats/min

D. rapid, disorganized ventricular rhythm with chaotic characteristics

E. abnormal heart rhythm

F. presence of three or more abnormal ventricular complexes in a row with a rate of more than 100 beats/min

G. rhythm in which the SA node acts as the pacemaker

Multiple Choice

Read each item carefully, then select the best response.

_____ 1. Studies have indicated a 95% or better accuracy rate in the diagnosis of myocardial infarction with the use of a:
   A. 3-lead ECG.
   B. 4-lead ECG.
   C. 6-lead ECG.
   D. 12-lead ECG.

_____ 2. The part of the electrical pathway that leads to the right side of the interventricular septum is the:
   A. left bundle.
   B. Purkinje system.
   C. AV node.
   D. right bundle.

_____ 3. In the normally functioning heart, the electrical impulse originates at the:
   A. AV node.
   B. SA node.
   C. internodal pathway.
   D. Bundle of His.

_____ 4. On the ECG paper, how many boxes equal one second?
   A. 5 big boxes
   B. 5 little boxes
   C. 2 big boxes
   D. 2 little boxes

_____ 5. Tachycardia refers to a heart rate:
   A. below 60 beats/min.
   B. above 100 beats/min.
   C. above 60 beats/min.
   D. below 100 beats/min.

_____ 6. Maximum heart rate is figured as:
   A. 220 beats/min – age (in years).
   B. 210 beats/min – age (in years).
   C. 200 beats/min – age (in years).
   D. 220 beats/min + age (in years).

_____ 7. It is not uncommon for ventricular tachycardia to deteriorate into:
   A. sinus tachycardia.
   B. ventricular fibrillation.
   C. asystole.
   D. ventricular bradycardia.

_____ 8. The reason for using a 12-lead ECG is for early identification of:
   A. vessel blockage.
   B. prolapsed valves.
   C. myocardial ischemia.
   D. embolism.

_____ 9. Which leads have to be placed exactly?
   A. Positive leads
   B. Negative leads
   C. Chest leads
   D. Limb leads

Fill-in

Read each item carefully, then complete the statement by filling in the missing word(s).

1. The heart contains a network of specialized tissue that is capable of conduction of electrical current through the heart, it is known as _____ _____ _____.

2. _____ _____ is a rhythm is which the SA node acts as the pacemaker.

3. _____ is an abnormal rhythm of the heart and is sometimes called a dysrhythmia.

4. _____ _____ is a rapid, completely disorganized ventricular rhythm with chaotic characteristics.

5. _____ refers to the complete absence of any electrical cardiac activity.

6. The leads that are placed on the arms and legs are referred to as _____ _____.

7. _____ _____ are only used in 12-lead ECGs.

True/False

If you believe the statement to be more true than false, write the letter "T" in the space provided. If you believe the statement to be more false than true, write the letter "F."

_____ 1. When the heart is deprived of oxygen, blood pressure increases.
_____ 2. The fastest pacer of the heart is the AV node.
_____ 3. The normal heart rate is between 60 and 100 beats/min.
_____ 4. Bradycardia is a heart rate above 60 beats/min.
_____ 5. Ventricular tachycardia and ventricular fibrillation account for more than 300,000 sudden cardiac deaths in the United States each year.
_____ 6. Immediate defibrillation is the most effective treatment for ventricular fibrillation.
_____ 7. One of the immediate advantages of 12-lead ECG monitoring is early identification of acute ischemia.
_____ 8. You should use alcohol preps to keep leads attached.

Short Answer

Complete this section with short written answers using the space provided.

1. Trace the electrical pathway of the heart.

_____

_____

2. List the ECG tracing through the cardiac cycle.

_____

_____

Word Fun emtb. vocab explorer

The following crossword puzzle is an activity provided to reinforce correct spelling and understanding of medical terminology associated with emergency care and the EMT-B. Use the clues in the column to complete the puzzle.

Across

2. Electrocardiogram

4. Complete absence of any electrical cardiac activity

5. An abnormal rhythm of the heart; sometimes called a dysrhythmia

6. Four leads used with a 4-lead ECG

Down

1. Refers to a fast heart rate

3. Leads that are used only with a 12-lead ECG

Ambulance Calls

The following case scenarios provide an opportunity to explore the concerns associated with patient management. Read each scenario, then answer each question in detail.

1. You are dispatched to assist with a patient who is experiencing chest pain and dizziness. You notice another EMT on the scene assist with the application of the ECG. As the paramedic's attention is focused on starting an IV, you see that the EMT has forgotten to attach one of the leads. The paramedic now begins assessing the patient's cardiac rhythm, but he appears confused by what he sees.

How would you best manage this situation?

_____

_____

_____

2. You and your paramedic partner are dispatched to a "syncopal episode." You arrive to find an older woman who is now awake, but is complaining of lightheadedness and nausea. You notice that her pulse rate is 40 beats/min. As the paramedic applies the ECG machine, you see that her rhythm is regular, with consistent, upright 'P' waves and consistent P-R intervals.

Is this considered normal, and if not, what is this arrhythmia?

_____

_____

_____

3. You and your paramedic partner are assessing a patient experiencing chest pain and shortness of breath. Your partner asks you to apply a 12-Lead ECG. As you bare the man's chest, you notice that he is very sweaty and has chest hair.

How will his skin condition and chest hair impact the effectiveness of the ECG, and how can you correct these problems?

_____

_____

_____

A BLS Review

Workbook Activities

The following activities have been designed to help you. Your instructor may require you to complete some or all of these activities as a regular part of your EMT-B training program. You are encouraged to complete any activity that your instructor does not assign as a way to enhance your learning in the classroom.

Chapter Review

The following exercises provide an opportunity to refresh your knowledge of this chapter.

Matching

Match each of the terms in the left column to the appropriate definition in the right column.

_____ **1.** Artificial ventilation

_____ **2.** Abdominal-thrust

_____ **3.** Basic life support (BLS)

_____ **4.** Advanced life support

_____ **5.** Cardiopulmonary resuscitation

A. steps used to establish artificial ventilation and circulation in a patient who is not breathing and has no pulse

B. noninvasive emergency lifesaving care used maneuver for patients in respiratory or cardiac arrest

C. procedures, such as cardiac monitoring, starting IV fluids, and using advanced airway adjuncts

D. a method of dislodging food or other (ALS) material from the throat of a choking victim

E. opening the airway and restoring breathing resuscitation (CPR) by mouth-to-mask ventilation and by the use of mechanical devices

Multiple Choice
Read each item carefully, then select the best response.

_____ **1.** Basic life support is noninvasive emergency lifesaving care that is used to treat:
   A. airway obstruction.
   B. respiratory arrest.
   C. cardiac arrest.
   D. all of the above

_____ **2.** Exhaled gas from you to the patient contains _____ oxygen.
   A. 8%
   B. 12%
   C. 16%
   D. 21%

_____ **3.** BLS differs from advanced life support by involving advanced lifesaving procedures including all of the following, except:
   A. cardiac monitoring.
   B. mouth-to-mouth.
   C. administration of IV fluids and medications.
   D. use of advanced airway adjuncts.

_____ **4.** In some instances such as _____, early BLS measures may be all that a patient needs to be resuscitated.
   A. choking
   B. near drowning
   C. lightning injuries
   D. all of the above

_____ **5.** In addition to checking level of consciousness, it is also important to protect the _____ from further injury while assessing the patient and performing CPR.
   A. spinal cord
   B. ribs
   C. internal organs
   D. facial structures

_____ **6.** In most cases, cardiac arrest in children younger than 12 to 14 years results from:
   A. choking.
   B. aspiration.
   C. congenital heart disease.
   D. respiratory arrest.

_____ **7.** Causes of respiratory arrest in infants and children include:
   A. aspiration of foreign bodies.
   B. airway infections.
   C. sudden infant death syndrome (SIDS).
   D. all of the above

_____ **8.** Signs of irreversible or biological death include clinical death along with:
   A. rigor mortis.
   B. dependent lividity.
   C. decapitation.
   D. all of the above

_____ **9.** Once you begin CPR in the field, you must continue until:

    A. the fire department arrives.

    B. the funeral home arrives.

    C. a physician arrives who assumes responsibility.

    D. law enforcement arrives and assumes responsibility.

_____ **10.** Once the patient is properly positioned, you can easily assess:

    A. airway.

    B. consciousness.

    C. disability.

    D. all of the above

_____ **11.** The chin lift has the added advantage of holding _____ in place, making obstruction by the lips less likely.

    A. loose dentures

    B. the tongue

    C. the mandible

    D. the maxilla

_____ **12.** To perform a _____, place your fingers behind the angles of the patient's lower jaw and then move the jaw forward.

    A. head tilt-chin lift maneuver

    B. jaw-thrust maneuver

    C. tongue-jaw lift maneuver

    D. all of the above

_____ **13.** Providing slow, deliberate inhalations over 2 seconds prevents:

    A. overexpansion of the lungs.

    B. rupture of the bronchial tree.

    C. gastric distention.

    D. rupture of the alveoli.

_____ **14.** A _____ is an opening that connects the trachea directly to the skin.

    A. tracheostomy

    B. stoma

    C. laryngectomy

    D. none of the above

_____ **15.** The _____ position helps to maintain a clear airway in a patient with a decreased level of consciousness who has not had traumatic injuries and is breathing on his or her own.

    A. recovery

    B. lithotomy

    C. Trendelenburg's

    D. Fowler's

_____ **16.** Excessive pressure applied to the carotid artery can:

    A. obstruct the carotid circulation.

    B. dislodge blood clots.

    C. produce marked reflex slowing of heart rate.

    D. all of the above

_____ **17.** The lower tip of the breastbone is the:

    A. xiphoid process.

    B. sternum.

    C. manubrium.

    D. intercostal space.

_____ **18.** Complications from chest compressions can include:

    A. fractured ribs.

    B. a lacerated liver.

    C. a fractured sternum.

    D. all of the above

_____ **19.** When checking for a pulse in an infant, you should palpate the _____ artery.

    A. radial

    B. brachial

    C. carotid

    D. femoral

_____ **20.** The technique for chest compressions in infants and children differs because of a number of anatomic differences, including:

    A. the position of the heart.

    B. the size of the chest.

    C. the fragile organs.

    D. all of the above

_____ **21.** The rate of compressions for an infant is at least _____ compressions per minute.

    A. 70

    B. 80

    C. 90

    D. 100

_____ **22.** The rate of compression to ventilation for infants and children is _____ for one-rescuer CPR.

    A. 1:5

    B. 30:2

    C. 15:2

    D. 2:15

_____ **23.** Sudden airway obstruction is usually easy to recognize in someone who is eating or has just finished eating because they suddenly:

    A. are unable to speak or cough.

    B. turn cyanotic.

    C. make exaggerated efforts to breathe.

    D. all of the above

_____ **24.** You should suspect an airway obstruction in the unresponsive patient if:

    A. the standard maneuvers to open the airway and ventilate the lungs are not effective.

    B. you feel resistance to blowing into the patient's lungs.

    C. pressure builds up in your mouth.

    D. all of the above

_____ **25.** You should use _____ for women in advanced stages of pregnancy, patients who are very obese, and children younger than 1 year.

    A. the Heimlich maneuver

    B. chest thrusts

    C. the abdominal-thrust maneuver

    D. any of the above

_____ **26.** For a patient with a mild airway obstruction, you should:

    A. perform the Heimlich maneuver.

    B. attempt a finger sweep to remove the foreign body.

    C. not interfere with the patient's attempt to expel the foreign body.

    D. all of the above

Fill-in

Read each item carefully, then complete the statement by filling in the missing word.

1. Permanent brain damage may occur if the brain is without oxygen for _____ to _____ minutes.

2. CPR does not require any equipment; however, you should use a _____ device to perform rescue breathing.

3. Because of the urgent need to start CPR in a pulseless, nonbreathing patient, you must complete an initial assessment as soon as possible, evaluating the patient's _____.

4. The most important element for successful CPR is immediate _____ of the airway.

5. _____ _____, such as living wills, may express the patient's wishes, but these documents are not binding for all health care providers.

6. For CPR to be effective, the patient must be lying supine on a _____ surface.

7. Without an open _____, rescue breathing will not be effective.

True/False

If you believe the statement to be more true than false, write the letter "T" in the space provided. If you believe the statement to be more false than true, write the letter "F."

_____ 1. During the initial assessment, you need to quickly evaluate the patient's airway, breathing, circulation, and level of consciousness.

_____ 2. All unconscious patients need all elements of BLS.

_____ 3. A patient who is not fully conscious often needs some degree of BLS.

_____ 4. The recovery position should be used to maintain an open airway in a patient with a head or spinal injury.

_____ 5. You should always remove a patient's dentures before initiating artificial ventilation.

_____ 6. You should not start CPR if the patient has obvious signs of irreversible death.

_____ 7. After you apply pressure to depress the sternum, you must follow with an equal period of relaxation so that the chest returns to normal position.

_____ 8. The ratio of compressions to ventilations for one-person CPR on an adult is 2:1.

_____ 9. When performed correctly, external chest compressions provide 50% of the blood normally pumped by the heart.

_____ 10. For infants, the preferred technique of artificial ventilation is mouth-to-nose-and-mouth ventilation with a mask or other barrier device.

_____ 11. You need to use less ventilatory pressure to inflate a child's lungs because the airway is smaller than that of an adult.

Short Answer

Complete this section with short written answers using the space provided.

1. List the four obvious signs of death, in addition to absence of pulse and breathing, that are used as a general rule to not start CPR.

_____

_____

2. Complete the following table regarding pediatric BLS by listing the procedure parameters or guidelines for each age group as they relate to the action noted on the left.

Procedure	Infants (younger than age 1 y)	Children (1 y to onset of puberty)
Airway		
Breathing Initial breaths Subsequent breaths		
Circulation Pulse check Compression area Compression width Compression depth Compression rate **Ratio of compressions to ventilations** **Foreign body obstruction**		

3. List the four acceptable reasons for stopping CPR.

_____

_____

4. Describe how to perform a head tilt-chin lift maneuver.

_____

_____

5. Describe how to perform a jaw-thrust maneuver.

_____

_____

6. Describe the process of chest compressions during one-rescuer adult CPR.

_____

_____

7. List and describe the method for "switching positions" during two-rescuer adult CPR.

_____

_____

8. Describe the process of abdominal thrusts for a standing patient and a supine patient.

_____

_____

9. Describe the process for chest thrusts on a standing and a supine patient.

_____

_____

10. Describe the process for removing a foreign body airway obstruction in an infant.

_____

_____

Word Fun

The following crossword puzzle is an activity provided to reinforce correct spelling and understanding of medical terminology associated with emergency care and the EMT-B. Use the clues in the column to complete the puzzle.

Across

3. Position after patient regains consciousness
4. Opening airway by moving lower bone forcibly forward
6. First level of care

Down

1. Preferred method to remove airway obstruction
2. Chest compression and artificial ventilation
5. High level of care, beyond basic

Ambulance Calls

The following case scenarios provide an opportunity to explore the concerns associated with patient management. Read each scenario, then answer each question in detail.

1. You are dispatched to a "person down". The dispatcher informs you that the caller said the patient is not breathing. Upon arrival, you find a 78-year-old woman in bed, apneic and pulseless. In the process of moving the patient to place a CPR board underneath her, you note the discoloration of her back and hips known as dependent lividity.

   How would you best manage this patient?

   _____

   _____

   _____

2. You are off-duty when you hear a dispatch for "chest pain" at a private residence near you. You arrive to find the patient's family members attempting to apply an AED they bought over the Internet.

   How would you best manage this situation?

   _____

   _____

   _____

3. You are dispatched to "unconscious male" at a private residence. You arrive to find the man lying in the grass in the backyard. Due to the ladder and equipment on the rooftop, it appears he was working on the roof of his two-story home. No one witnessed the event.

   How would you best manage this patient?

   _____

   _____

   _____

Skill Drills

Skill Drill A-1: Positioning the Patient

Test your knowledge of this skill drill by placing the photos below in the correct order. Number the first step with a "1," the second step with a "2," etc.

_____

Move the patient to a supine position with legs straight and arms at the sides.

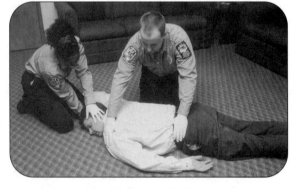

_____

Grasp the patient, stabilizing the cervical spine if needed.

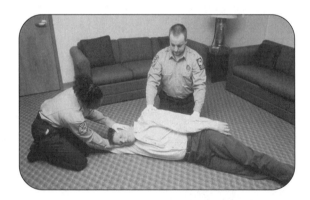

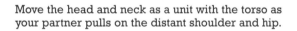
_____

Move the head and neck as a unit with the torso as your partner pulls on the distant shoulder and hip.

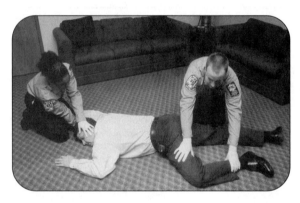

_____

Kneel beside the patient, leaving room to roll the patient toward you.

Skill Drill A-2: Performing Chest Compressions
Test your knowledge of this skill drill by filling in the correct words in the photo captions.

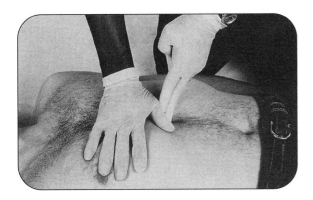

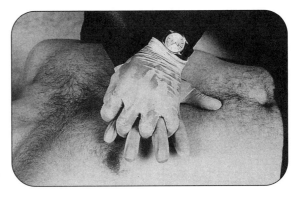

1. Place the _____ of one hand on the sternum, between the nipples.

2. Place the heel of your other hand over the _____ .

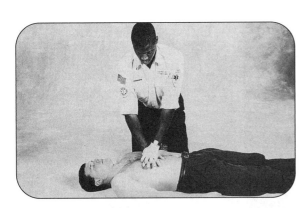

3. With your arms straight, _____ your elbows, and position your shoulders directly _____ your hands. Depress the sternum _____ inch(es) to _____ inch(es) using a direct downward movement.

Skill Drill A-3: Performing One-rescuer Adult CPR

Test your knowledge of this skill drill by placing the photos below in the correct order. Number the first step with a "1," the second step with a "2," etc.

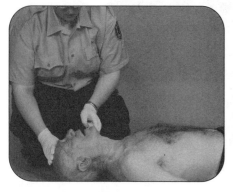

Open the airway.

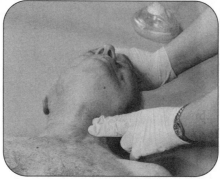

Check for a carotid pulse.

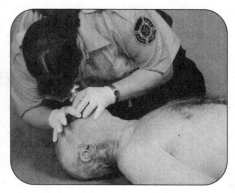

If not breathing, give two breaths of 1 second each.

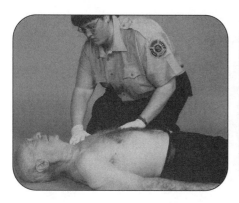

Establish unresponsiveness, and call for help.

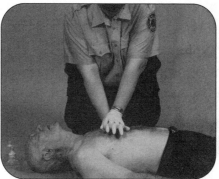

If no pulse is found, apply your AED. If there is no AED, place your hands in the proper position for chest compressions. Give 30 compressions at a rate of about 100/min.
Open the airway, and give two ventilations of 1 second each.
Perform five cycles of compressions and ventilations.
Stop CPR, and check for return of the carotid pulse.
Depending on patient condition, continue CPR, continue rescue breathing only, or place the patient in the recovery position and monitor breathing and pulse.

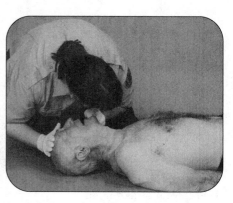

Look, listen, and feel for breathing. If breathing is adequate, place the patient in the recovery position and monitor.

Skill Drill A-4: Performing Two-rescuer Adult CPR
Test your knowledge of this skill drill by filling in the correct words in the photo captions.

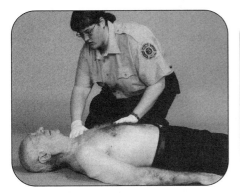

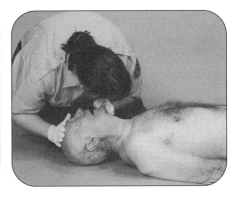

1. Establish _____, and take positions.

2. _____ the airway.

3. Look, listen, and feel for breathing. If breathing is adequate, place the patient in the _____ position and _____.

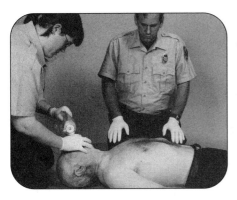

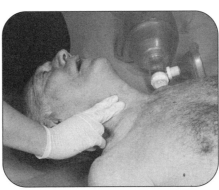

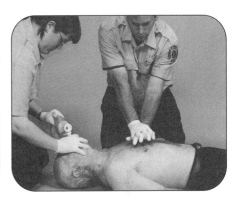

4. If not breathing, give _____ breaths of _____ second each.

5. Check for a _____ pulse.

6. If there is no pulse but an AED is available, apply it now. If no AED is available, begin _____ _____ at about _____/min (_____ compressions to _____ ventilations). After every _____ cycles, switch rescuer positions in order to minimize fatigue. Keep switch time to _____ seconds. Depending on patient _____, continue CPR, continue _____ only, or place in the recovery position and monitor.

Chapter 1: Introduction to Emergency Medical Care

Matching

1. H (page 8)
2. F (page 5)
3. M (page 4)
4. G (page 4)
5. A (page 4)
6. K (page 12)
7. B (page 13)
8. C (page 4)
9. L (page 13)
10. E (page 13)
11. I (page 12)
12. D (page 12)
13. J (page 6)

Multiple Choice

1. B (page 6)
2. A (page 12)
3. D (page 13)
4. C (page 13)
5. A (page 17)
6. C (page 12)
7. D (page 12)
8. D (page 10)
9. A (page 8)
10. B (page 9)

Fill-in

1. the U.S. DOT 1994 EMT-B National Standard Curriculum (page 5)
2. life-threatening, non-life threatening (page 5)
3. medications (page 10)
4. 9-1-1 (page 10)
5. medical control (page 12)

True/False

1. F (page 4)
2. T (page 10)
3. F (page 13)
4. T (page 16)
5. T (page 17)
6. T (page 18)

Short Answer

1. The EMT-B is one of the five levels of prehospital care. The EMT-B provides basic life support skills and may also provide some ALS skills, such as advanced airways and defibrillation. (page 10)
2. The Department of Transportation (DOT) has developed a series of guidelines, curriculum, funding sources, and assessment tools, all designed to develop and improve EMS in the United States. (page 9)
3. –Ensuring your own safety and the safety of your fellow EMT-Bs, the patient, and others at the scene
   –Locating and safely driving to the scene
   –Sizing up the scene and situation
   –Rapidly assessing the patient's gross neurologic, respiratory, and circulatory status
   –Providing any essential immediate intervention

–Performing a thorough, accurate patient assessment

–Obtaining an expanded SAMPLE history

–Reaching a clinical impression and providing prompt, efficient, prioritized patient care based on your assessment

–Communicating effectively with and advising the patient of any procedures you will perform

–Properly interacting and communicating with fire, rescue, and law enforcement responders at the scene

–Identifying patients who require rapid packaging and initiating transport without delay

–Identifying patients who do not need emergency care and will benefit from further detailed assessment and care before they are moved and transported

–Properly packaging the patients

–Safely lifting and moving the patient to the ambulance and loading the patient into it

–Providing safe appropriate transport to the hospital emergency department or other ordered facility

–Giving the necessary radio report to the medical control center or receiving hospital emergency department

–Providing any additional assessment or treatment while en route

–Monitoring the patient and checking vital signs while en route

–Documenting all findings and care on the run report

–Unloading the patient safely and, after giving a proper verbal report, transferring the patient's care to the emergency department staff

–Safeguarding the patient's rights (page 6)

4. On-line medical direction is provided through radio or telephone connections between the EMT-B and the medical control facility. Off-line medical direction is provided through written protocols, procedures, and standing orders. (page 12)

Word Fun

Ambulance Calls

1. You cannot release a minor to their own care unless they are considered an "emancipated minor." You should examine this patient thoroughly, treat him using rules of implied consent and make every attempt to contact his parents or legal guardians. If you cannot immediately contact his parents, you must release him to local law enforcement.

2. Call the dispatcher to send police if the environment appears hostile. Wait for them to arrive and then proceed to care for the patient. It would also be appropriate for the boy to come to you. Ask someone close by to bring the boy to the ambulance so that you can examine his injury. If in any doubt, wait for law enforcement.

3. Leaving any patient for any reason (without releasing him to the care of equally trained providers or the hospital) is legally viewed as abandonment. This scenario presents a dilemma. Obviously you are close to the scene, so you must make a quick decision based on the following questions. Is another ambulance en route, and if so, how long will their response time be? Can you split your crew? Can you place your patient in the back of the ambulance, respond to the other call, and continue caring for the patient while your partner assesses the unconscious assault victim? You will run into real world scenarios that require quick thinking, may not be "cut and dried," and may require you to think outside the box. You must however, always obey local protocols.

Chapter 2: The Well-Being of the EMT-B

Matching

1. D (page 64)
2. A (page 36)
3. C (page 43)
4. F (page 33)
5. B (page 44)
6. E (page 42)
7. M (page 42)
8. G (page 53)
9. L (page 52)
10. I (page 51)
11. J (page 44)
12. N (page 55)
13. H (page 52)
14. K (page 51)

Multiple Choice

1. C (page 25)
2. C (page 26)
3. A (page 26)
4. D (page 26)
5. B (page 26)
6. A (page 27)
7. D (page 27)
8. C (page 27)
9. B (page 27)
10. D (page 28)
11. D (page 28)
12. D (page 29)
13. D (page 30)
14. A (page 31)
15. B (page 32)
16. D (page 32)
17. D (page 33)
18. B (page 33)
19. D (page 33)
20. C (page 31)
21. D (page 34)
22. B (page 33)
23. D (page 36)
24. A (page 41)
25. A (page 34)
26. D (page 35)
27. D (page 36)
28. B (page 37)
29. C (page 35)
30. D (page 35)
31. D (page 33)
32. D (page 39)
33. B (page 40)
34. C (page 41)
35. B (page 44)

36. D (page 44)

37. A (page 47)

38. C (page 45)

39. D (page 49)

40. D (page 54)

41. B (page 56)

42. C (page 57)

43. D (page 65)

Fill-in

1. well-being (page 24)

2. emotional stress (page 24)

3. heart disease (page 25)

4. physician (page 25)

5. warm (page 21)

6. Fear (page 26)

7. depression (page 26)

8. minor (page 32)

9. high-stress (page 32)

True/False

1. T (page 24)

2. T (page 51)

3. T (page 26)

4. F (pages 42-43)

5. F (page 32)

6. T (page 34)

Short Answer

1. An infection control practice that assumes all body fluids are potentially infectious. (page 44)

2. **1.** Denial

   2. Anger/hostility

   3. Bargaining

   4. Depression

   5. Acceptance

   (page 26)

3. –Irritability toward coworkers, family and friends

   –Inability to concentrate

   –Difficulty sleeping, increased sleeping, or nightmares

   –Anxiety; indecisiveness; guilt

   –Loss of appetite (gastrointestinal disturbances)

   –Loss of interest in sexual activities

   –Isolation

   –Loss of interest in work

   –Increased use of alcohol

   –Recreational drug use (page 36)

4. –Change or eliminate stressors

   –Change partners to avoid a negative or hostile personality

   –Change work hours

   –Cut back on overtime

   –Change your attitude about the stressor

   –Stop wasting your energy complaining or worrying about things you cannot change

   –Try to adopt a more relaxed, philosophical outlook

   –Expand your social support system apart from your coworkers

   –Sustain friends and interests outside emergency services

   –Minimize the physical response to stress by employing various techniques (page 35)

5. **1.** Use soap and water.

   2. Rub hands together for at least 10 to 15 seconds to work up a lather.

   3. Rinse hands and dry with a paper towel.

   4. Use paper towel to turn off faucet. (page 44)

6. (page 57)

Level	Hazard	Protection Needed
0	Little to no hazard	None
1	Slightly hazardous	SCBA only (level C suit)
2	Slightly hazardous	SCBA only (level C suit)
3	Extremely hazardous	Full protection, with no exposed skin (level A or B suit)
4	Minimal exposure causes death	Special HazMat gear (level A suit)

7. **1.** Thin inner layer

   2. Thermal middle layer

   3. Outer layer

   (page 60)

8. **1.** Past history

   2. Posture

   3. Vocal activity

   4. Physical activity

   (pages 64-65)

Word Fun

```
            ¹C O V E R
             O
    ²C I S M M
             M
             U           ³B
    ⁴B       N           U
     S       I           R
    ⁵D I R E C T C O N T A C T ⁶
             A           O     I
        ⁷T U B E R C U L O S I S
    ⁸O       L           T     D
   ⁹P T S D  E
    P    H   D
  ¹⁰H E P A T I T I S
             S
  ¹¹P A T ¹²H O G E N
         I
         V           A
                     S
                     E
```

Ambulance Calls

1. Although family members understand that their loved one is terminally ill, it is common for a sudden feeling of panic to occur when the patient begins to deteriorate and you may be called to respond. This will be an emotionally charged event that will only heighten with the arrival of EMS personnel. Do not resuscitate orders apply to persons in cardiorespiratory arrest, and this patient is breathing and has a pulse. You will need to review the advance directives (living will documents), and this paperwork should be available along with the DNR orders. If the hospice nurse is available, you should speak directly with that person. If that person is not available, explain your concern for the patient and family members and explain your legal duties/requirements and notify medical control of the situation.

2. You must triage your patients. You may find this situation difficult, especially given that the crowd of onlookers will not understand any delay in patient care. You must treat the critical patients who will benefit from your care the most. This will likely prevent you from engaging in any sort of care for the patient who has the open head injury. You must regularly practice triaging patients, so that it will become an automatic response in the field. Patients like the one in this scenario can and do 'suck in' even the most seasoned providers. If you had plenty of resources, you could assist this patient not because it would change the patient's outcome but for other reasons (harvesting organs/PR issues). Always follow local protocols.

3. –Continue to treat your patient appropriately, including c-spine stabilization and transport.
   –Allow your cut to bleed as long as it is minimal and it will help to wash/clean it out.
   –Clean your wound with an alcohol gel if available.
   –Once patient care has been transferred at the receiving facility, immediately wash thoroughly with soap and water and report to your supervisor.
   –Follow up with prompt medical attention.

Skill Drills

Skill Drill 2-1: Proper Glove Removal Technique (page 46)

1. Partially remove the first glove by pinching at the wrist. Be careful to touch only the outside of the glove.
2. Remove the second glove by pinching the exterior with the partially gloved hand.
3. Pull the second glove inside-out toward the fingertips.
4. Grasp both gloves with your free hand touching only the clean, interior surfaces.

Chapter 3: Medical, Legal, and Ethical Issues

Matching

1. H (page 75)
2. I (page 75)
3. G (page 79)
4. E (page 75)
5. L (page 74)
6. A (page 75)
7. D (page 75)
8. F (page 74)
9. B (page 75)
10. C (page 76)
11. M (page 75)
12. N (page 76)
13. J (page 74)
14. K (page 73)

Multiple Choice

1. C (page 72)
2. A (page 72)
3. D (page 74)
4. D (page 74)
5. C (page 75)
6. B (page 78)
7. D (page 80)
8. D (page 81)
9. B (page 75)
10. A (page 75)
11. A (page 75)
12. C (page 76)
13. B (page 74)
14. D (page 83)
15. A (page 83)
16. D (page 84)
17. B (page 81)

Fill-in

1. scope of practice (page 72)
2. standard of care (page 73)
3. duty to act (page 74)
4. negligence (page 74)
5. termination (page 75)
6. expressed, implied (page 75)
7. assault, battery (page 75)
8. advance directive, DNR order (page 79)
9. refuse treatment (page 77)
10. special reporting (page 81)

True/False

1. T (page 74)
2. F (page 75)
3. T (page 75)
4. T (page 76)
5. T (page 75)

Short Answer

1. If the minor is emancipated, married, or pregnant (page 76)
2. You must continue to care for the patient until the patient is transferred to another medical professional of equal or higher skill level, or another medical facility. (page 75)

3. **1.** Obtain refusing party's signature on an official medical release form that acknowledges refusal.

   2. Obtain a signature from a witness of the refusal.

   3. Keep the refusal form with the incident report.

   4. Note the refusal on the incident report.

   5. Keep a department copy of the records for future reference. (page 78)

4. **1.** If it wasn't documented, it did not happen.

   2. Incomplete or disorderly records equate to incomplete or inexpert medical care. (page 81)

5. **1.** Inform medical control.

   2. Treat the patient as you would any patient.

   3. Take any steps necessary to preserve life.

   4. If saving the patient is not possible, take steps to make sure the organ remains viable. (page 84)

Word Fun

A crossword puzzle with the following answers: SCOPE OF PRACTICE, MEDICOLEGAL, NEGLIGENCE, BATTERY, EXPRESSED CONSENT, DUTY TO ACT, along with CERTIFICATION, ABANDONMENT, COMPETENT, ASSAULT, IMPLIED CONSENT, CONSENT, INFORMED CONSENT, ACTIONS.

Ambulance Calls

1. This patient needs to be evaluated at the hospital. Her parents will likely feel a right to be informed of their child's medical conditions and medical care. Laws regarding reproductive rights of minors vary from state to state. Some states allow minors to make decisions regarding birth control, prenatal care, or pregnancy termination without consenting parents, while others do not. You must know your local laws. You will have to provide information regarding the pregnancy to other health care providers directly involved in her care, and you should explain that fact and the necessity of such. Be tactful. Don't unnecessarily break your patient's trust by immediately sharing this knowledge with her parents. Document carefully and consult medical control.

2. You have a duty to act regardless of your current off-duty status (some states/provider levels/paid status varies) and you did the right thing by stopping to help the child. Once you have initiated care, you must ensure the child's parent(s) or legal guardian is notified. Although the grandfather is home, you now have another dilemma. The condition of the house/capability of the grandfather to care for the child while the mother is away is such that the question of neglect arises. You should speak with the grandfather and attempt to contact the mother of the child. If you believe that neglect or abuse of a child is occurring, you are legally required to intervene. You should document the condition of the house and notify medical control and/or child protective serves in accordance with your local protocols.

3. Assess the patient's mental status. If he is intoxicated, or has an altered mental status, he is treated under implied consent. If he is alert and oriented, you may attempt to talk him into being treated by explaining what you feel is necessary and what may happen if he does not receive care. If he has an altered mental status, orders from medical control may be obtained to restrain the patient with the help of law enforcement and transport him to the hospital.

Chapter 4: The Human Body

Matching

1. F (page 92)		**20.** B (page 106)	
2. K (page 123)		**21.** A (page 107)	
3. D (page 92)		**22.** A (page 107)	
4. G (page 93)		**23.** A (page 110)	
5. E (page 92)		**24.** B (page 111)	
6. L (page 123)		**25.** B (page 111)	
7. A (page 93)		**26.** A (page 110)	
8. C (page 93)		**27.** A (page 110)	
9. M (page 123)		**28.** C (page 111)	
10. J (page 93)		**29.** B (page 111)	
11. O (page 123)		**30.** C (page 111)	
12. B (page 93)		**31.** A (page 109)	
13. H (page 93)		**32.** C (page 111)	
14. N (page 93)		**33.** A (page 99)	
15. I (page 92)		**34.** C (page 126)	
16. B (page 100)		**35.** E (page 128)	
17. B (page 100)		**36.** B (page 128)	
18. A (page 100)		**37.** F (page 126)	
19. B (page 106)		**38.** D (page 128)	

Multiple Choice

1. B (page 92)
2. C (page 93)
3. B (page 113)
4. B (page 131)
5. D (page 119)
6. C (page 96)
7. A (page 99)
8. A (page 103)
9. D (page 95)
10. B (page 95)

Labeling

1. Directional Terms (page 93)

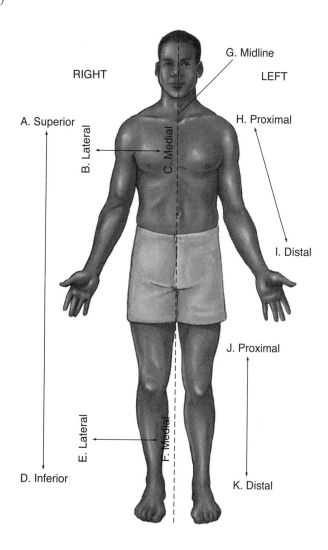

2. Anatomic Positions (page 95)

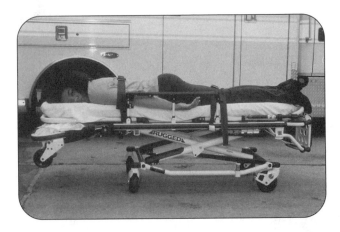

Prone

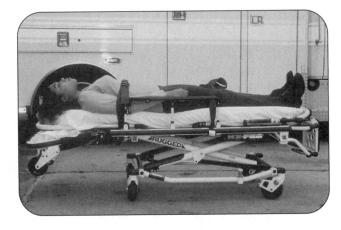

Supine

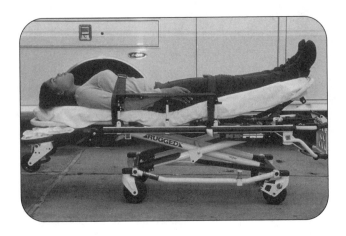

Shock position (modified Trendelenburg's position)

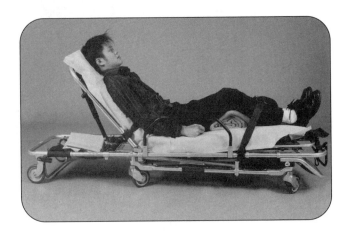

Fowler's position

3. Skeletal System (page 96)

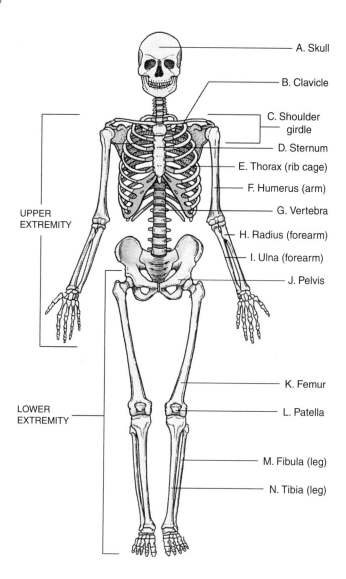

A. Skull

B. Clavicle

C. Shoulder girdle

D. Sternum

E. Thorax (rib cage)

F. Humerus (arm)

G. Vertebra

H. Radius (forearm)

I. Ulna (forearm)

J. Pelvis

K. Femur

L. Patella

M. Fibula (leg)

N. Tibia (leg)

UPPER EXTREMITY

LOWER EXTREMITY

4. The Skull (page 97)

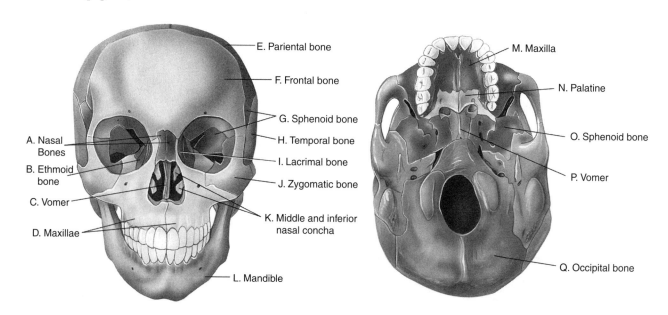

E. Pariental bone

F. Frontal bone

G. Sphenoid bone

H. Temporal bone

I. Lacrimal bone

J. Zygomatic bone

K. Middle and inferior nasal concha

A. Nasal Bones

B. Ethmoid bone

C. Vomer

D. Maxillae

L. Mandible

M. Maxilla

N. Palatine

O. Sphenoid bone

P. Vomer

Q. Occipital bone

5. The Spinal Column (page 99)

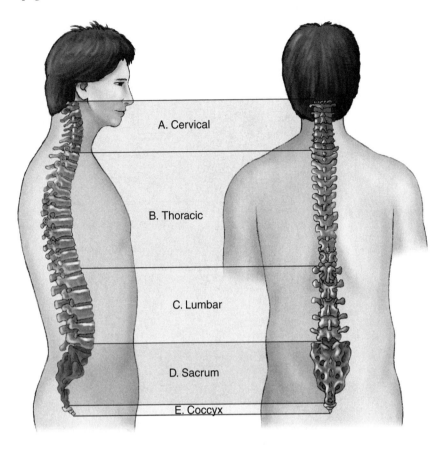

A. Cervical

B. Thoracic

C. Lumbar

D. Sacrum

E. Coccyx

6. The Thorax (page 101)

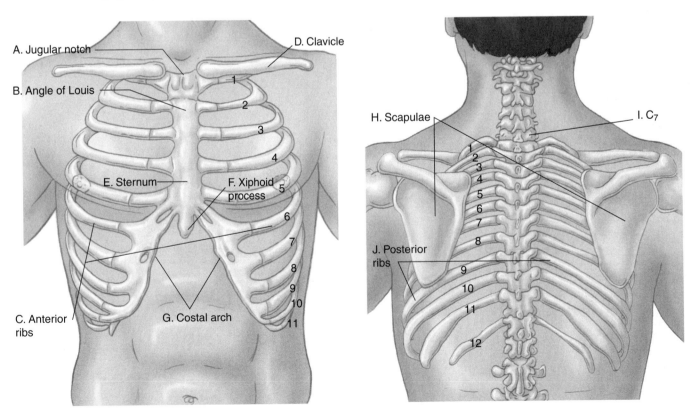

A. Jugular notch

B. Angle of Louis

C. Anterior ribs

D. Clavicle

E. Sternum

F. Xiphoid process

G. Costal arch

H. Scapulae

I. C$_7$

J. Posterior ribs

7. The Pelvis (page 104)

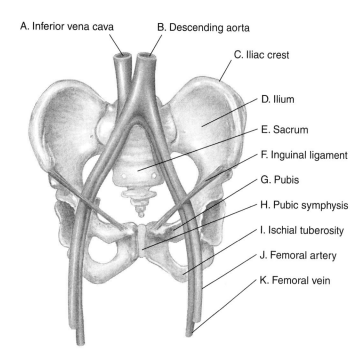

A. Inferior vena cava
B. Descending aorta
C. Iliac crest
D. Ilium
E. Sacrum
F. Inguinal ligament
G. Pubis
H. Pubic symphysis
I. Ischial tuberosity
J. Femoral artery
K. Femoral vein

8. The Lower Extremity (page 106)

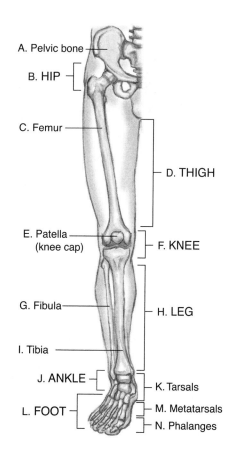

A. Pelvic bone
B. HIP
C. Femur
D. THIGH
E. Patella (knee cap)
F. KNEE
G. Fibula
H. LEG
I. Tibia
J. ANKLE
K. Tarsals
L. FOOT
M. Metatarsals
N. Phalanges

9. The Shoulder Girdle (page 107)

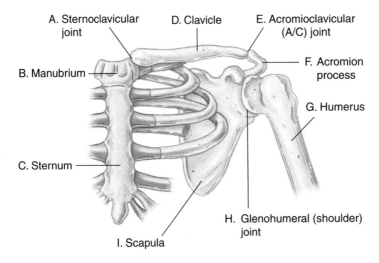

A. Sternoclavicular joint
D. Clavicle
E. Acromioclavicular (A/C) joint
B. Manubrium
F. Acromion process
G. Humerus
C. Sternum
H. Glenohumeral (shoulder) joint
I. Scapula

10. The Upper Extremity (page 107)

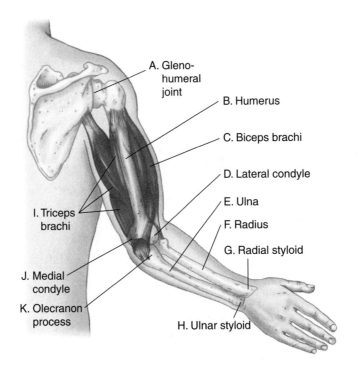

A. Gleno-humeral joint
B. Humerus
C. Biceps brachi
D. Lateral condyle
E. Ulna
F. Radius
G. Radial styloid
I. Triceps brachi
J. Medial condyle
K. Olecranon process
H. Ulnar styloid

11. Wrist and Hand (page 108)

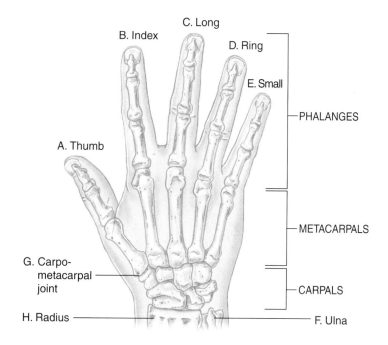

12. The Respiratory System (page 112)

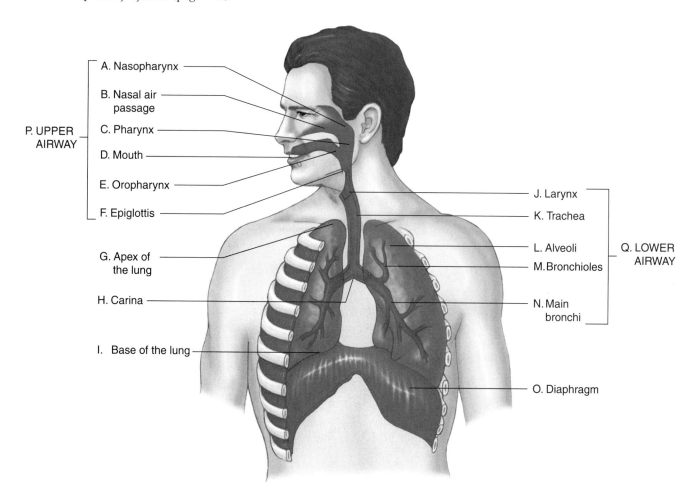

13. The Circulatory System (page 119)

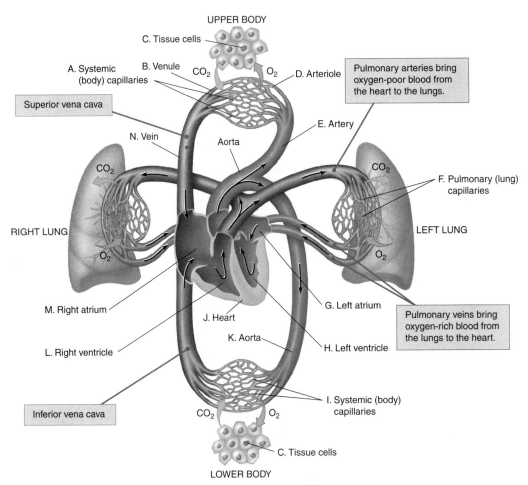

UPPER BODY

C. Tissue cells

A. Systemic (body) capillaries   B. Venule   CO₂   O₂   D. Arteriole

Pulmonary arteries bring oxygen-poor blood from the heart to the lungs.

Superior vena cava

E. Artery

N. Vein   Aorta

CO₂   CO₂

RIGHT LUNG   LEFT LUNG

O₂   O₂

F. Pulmonary (lung) capillaries

M. Right atrium   J. Heart   G. Left atrium

Pulmonary veins bring oxygen-rich blood from the lungs to the heart.

L. Right ventricle   K. Aorta   H. Left ventricle

Inferior vena cava

CO₂   O₂

I. Systemic (body) capillaries

C. Tissue cells

LOWER BODY

14. Electrical Conduction (page 121)

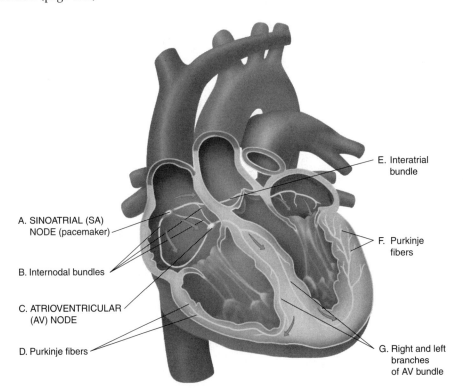

E. Interatrial bundle

A. SINOATRIAL (SA) NODE (pacemaker)

B. Internodal bundles

F. Purkinje fibers

C. ATRIOVENTRICULAR (AV) NODE

D. Purkinje fibers

G. Right and left branches of AV bundle

15. Central and Peripheral Pulses (page 124)

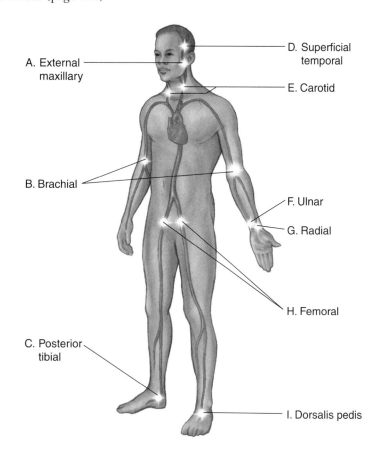

A. External maxillary
D. Superficial temporal
E. Carotid
B. Brachial
F. Ulnar
G. Radial
H. Femoral
C. Posterior tibial
I. Dorsalis pedis

16. Brain (page 126)

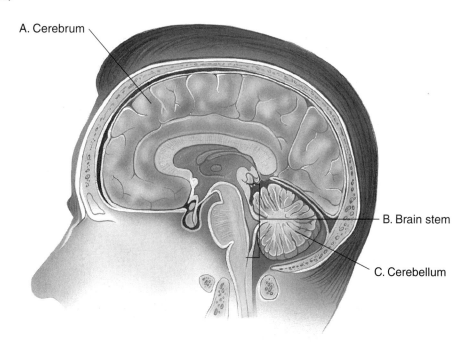

A. Cerebrum
B. Brain stem
C. Cerebellum

17. Anatomy of the Skin (page 130)

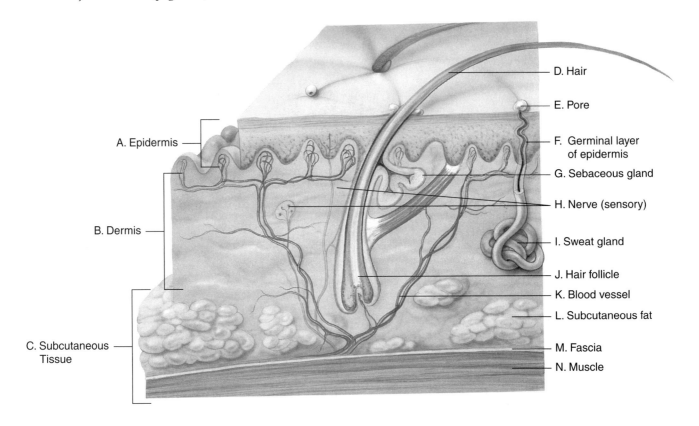

A. Epidermis
B. Dermis
C. Subcutaneous Tissue
D. Hair
E. Pore
F. Germinal layer of epidermis
G. Sebaceous gland
H. Nerve (sensory)
I. Sweat gland
J. Hair follicle
K. Blood vessel
L. Subcutaneous fat
M. Fascia
N. Muscle

18. Male Reproductive System (page 136)

FRONT VIEW

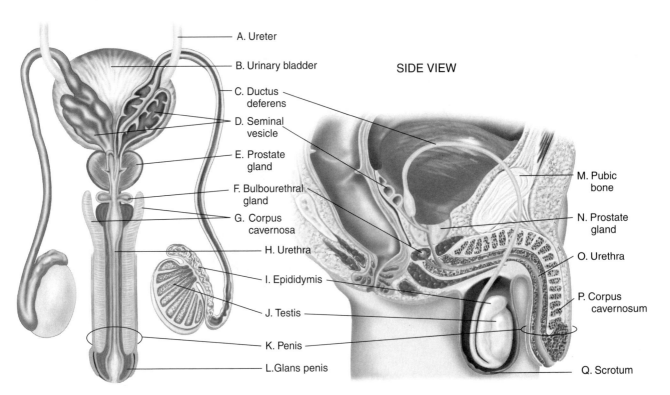

SIDE VIEW

A. Ureter
B. Urinary bladder
C. Ductus deferens
D. Seminal vesicle
E. Prostate gland
F. Bulbourethral gland
G. Corpus cavernosa
H. Urethra
I. Epididymis
J. Testis
K. Penis
L. Glans penis
M. Pubic bone
N. Prostate gland
O. Urethra
P. Corpus cavernosum
Q. Scrotum

19. Female Reproductive System (page 136)

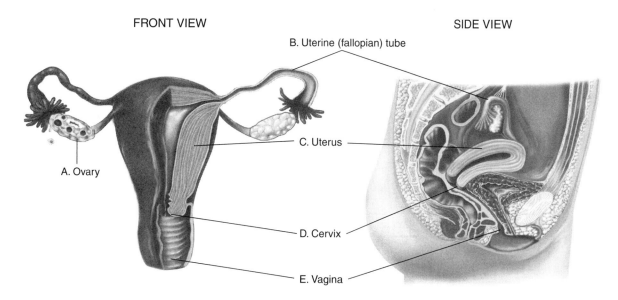

FRONT VIEW   SIDE VIEW

B. Uterine (fallopian) tube

C. Uterus

A. Ovary

D. Cervix

E. Vagina

Fill-in

1. 7 (page 99)
2. mandible (page 97)
3. 5 (page 113)
4. 12 (page 100)

5. 33 (page 100)
6. talus (page 106)
7. floating ribs (page 100)

True/False

1. F (page 122)
2. T (page 108)
3. T (pages 106-108)
4. F (page 120)
5. F (page 100)

Short Answer

1. Plasma – is a sticky, yellow fluid that carries the blood cells and nutrients
Red blood cells – give blood its red color and carry oxygen
White blood cells – play a role in the body's immune defense mechanism against infection
Platelets – essential in the formation of blood clots (pages 123 to 124)

2. Cervical spine – 7
Thoracic spine – 12
Lumbar spine – 5
Sacrum – 5
Coccyx – 4 (page 100)

3. RUQ – liver, gallbladder, portion of the colon
LUQ – stomach, spleen, portion of the colon
RLQ – large intestine, small intestine, appendix, ascending colon
LLQ – large intestine, small intestine, descending and the sigmoid portions of the colon (page 103)

4. **1.** superior and inferior vena cava

2. right atrium

3. right ventricle

4. pulmonary artery

5. lungs

6. pulmonary vein

7. left atrium

8. left ventricle

9. aorta (page 120)

Word Fun

```
D E R M I S                           M
I   E                       P         A
S   T                       E         N         A
T   R           P       B   R         D         T
A   O           O     I N F E R I O R           R
L   P           S       L   U       B           I
    E           T       A   S       L           U
T   R       V E N T R I C L E       M
R   I       R       E       O   I       F
A N T E R I O R       R     N   V       O
C   O       O       A           E       W
H   N       R     L A T E R A L         L
E   E                                   E
A   A G O N A L                         R
    L                 R A D I U S
```

Ambulance Calls

1. Lacerated liver, gall bladder, small intestine, large intestine, pancreas, diaphragm, right lung if the pathway is up, right kidney depending on length of knife. You could also have involvement of the other four quadrants based on the direction of travel of the blade.

The description would be a puncture wound or stab wound.

2. There is significant mechanism of injury in this scenario. Not only did the patient's vehicle impact a solid, stationary object but also with enough force to deform the steering wheel. This patient was not restrained and the abrupt deceleration caused driver's chest to impact the steering wheel with significant force. He could have numerous internal injuries including: pneumo and/or hemothorax, cardiac or pulmonary contusions, fractured ribs/sternum (not to mention spinal and head injuries). Immediate transport with full spinal precautions and high-flow oxygen is required.

3. Depending upon the condition of the track, the rider's use of protective gear, and the speed and height of the fall, he may or may not have sustained significant injuries. Given his self-splinting, it can be assumed that he has (at minimum) fractures to the left humerus, radius and/or ulna and clavicle. Treat and transport according to local protocols.

Chapter 5: Baseline Vital Signs and SAMPLE History

Matching

1. N (page 152)
2. E (page 157)
3. D (page 158)
4. L (page 154)
5. A (page 156)
6. H (page 148)
7. K (page 157)
8. F (page 155)
9. M (page 154)
10. O (page 154)
11. B (page 155)
12. G (page 158)
13. I (page 158)
14. C (page 155)
15. J (page 154)

Multiple Choice

1. C (page 147)
2. A (page 147)
3. B (page 149)
4. D (page 149)
5. A (page 149)
6. D (page 149)
7. D (page 150)
8. B (page 151)
9. C (page 151)
10. B (page 151)
11. C (pages 152-153)
12. C (page 153)
13. A (page 153)
14. D (page 154)
15. D (page 154)
16. D (page 154)
17. D (page 154)
18. B (page 155)
19. D (page 155)
20. D (page 155)
21. B (page 156)
22. D (page 156)
23. B (page 156)
24. A (page 157)
25. C (page 158)
26. A (page 160)
27. B (page 160)
28. D (page 160)
29. D (page 162)
30. C (page 149)
31. A (page 147)
32. B (page 158)
33. C (page 161)
34. C (page 162)
35. B (page 154)

Fill-in

1. Tidal volume (page 152)
2. conjunctiva (page 155)
3. deductive (page 146)
4. symptom (page 147)
5. spontaneous respirations (page 149)

6. Vital signs (page 149)
7. quality (page 150)
8. labored breathing (page 151)
9. fluid (page 151)
10. perfusion (page 154)

True/False

1. T (page 153)
2. T (page 158)
3. F (page 158)
4. T (page 151)
5. T (page 150)
6. F (page 153)
7. F (page 154)

8. T (page 154)
9. T (page 154)
10. T (page 155)
11. F (page 155)
12. F (page 161)
13. F (page 154)

Short Answer

1. 1. Pulse
    2. Respirations
    3. Pupils
    4. Blood pressure
    5. Skin
    6. Level of consciousness
    7. Capillary refill (page 149)
2. 1. Flushed (red)
    2. Pale (white, ashen, or graying)
    3. Jaundice (yellow)
    4. Cyanotic (blue-gray) (pages 154-155)
3. 1. Rate
    2. Rhythm
    3. Quality
    4. Depth (tidal volume) (pages 150-152)
4. Pressure exerted against the walls of the artery when the left ventricle contracts (page 157)
5. Pressure remaining against the walls of the artery when the left ventricle is at rest (page 157)
6. 1. Rate
    2. Strength
    3. Regularity (pages 153-154)
7. 1. Color
    2. Temperature
    3. Moisture (pages 154-155)
8. Gently compress the fingertip until it blanches. Release the fingertip, and count until it returns to its normal pink color. (page 155)
9. Pupils Equal And Round, Regular in size, react to Light (page 162)
10. A sign is a condition that can be seen, heard, felt, smelled, or measured. A symptom is something that the patient reports to you as a problem or feeling. (page 147)

Word Fun

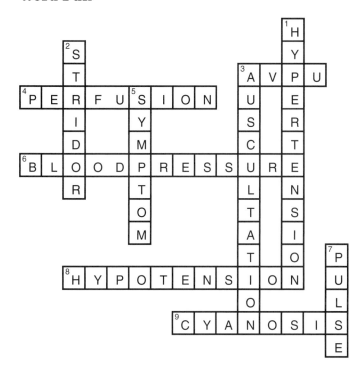

Ambulance Calls

1. You must closely monitor this patient's airway as its patency could change at any time. You should not take any actions to clear her airway unless it becomes blocked again (or partially blocked to the point of ineffective airway exchange). Obviously, no attempt should be made to pull the foreign body out of her airway. You must gently transport this patient in a position she finds most comfortable and be prepared to administer abdominal thrusts at any point should it become necessary.

2. Shoveling snow is very physical work (especially wet, heavy snow). The first snowfall often brings 9-1-1 calls for chest pain. People often don't fully understand how strenuous this activity is, and normally sedentary individuals will attempt to shovel snow without thought. This man isn't simply having angina, he is having a heart attack. He likely has heart disease, and although he reports his pain is "better," his chest heaviness continues despite rest. He should be transported immediately according to local protocols.

3. –Open and assess the airway.
   –Apply high-flow oxygen via nonrebreathing mask or BVM.
   –Assess carotid pulse.
   –If no pulse, initiate chest compressions.
   –If patient has a pulse, treat for shock and rapid transport.
   –Place patient in Trendelenburg position and cover to keep warm.
   –Continue assessment and obtain SAMPLE history en route.
   –Monitor vital signs and take a complete set every 5 minutes.

Skill Drills

Skill Drill 5-1: Obtaining a Blood Pressure by Auscultation or Palpation (page 159)

1. Apply the cuff snugly.
2. Palpate the brachial artery.
3. Place the stethoscope over the brachial artery, and grasp the ball-pump and turn-valve.
4. Close the valve and pump to 20 mm Hg above the point at which you stop hearing pulse sounds. Note the systolic and diastolic pressures as you let air escape slowly.
5. Open the valve, and quickly release remaining air.
6. When using the palpitation method, you should place your fingertips on the radial artery so that you feel the radial pulse.

Chapter 6: Lifting and Moving Patients

Matching

1. C (page 191)
2. G (page 200)
3. H (page 197)
4. E (page 202)
5. A (page 202)
6. I (page 197)
7. F (page 190)
8. B (page 200)
9. D (page 197)

Multiple Choice

1. D (page 204)
2. D (page 184)
3. D (page 171)
4. D (page 171)
5. D (page 173)
6. A (page 173)
7. B (page 176)
8. B (page 176)
9. D (page 179)
10. A (page 179)
11. D (page 182)
12. D (page 181)
13. D (page 182)
14. C (page 182)
15. D (page 183)
16. B (page 184)
17. A (page 185)
18. D (page 184)
19. C (page 185)
20. D (page 186)
21. D (page 185)
22. C (page 193)
23. A (page 202)
24. D (page 202)
25. D (page 197)
26. B (page 184)
27. A (page 174)
28. D (page 193)
29. B (page 193)
30. B (page 196)

Fill-in

1. body mechanics (page 170)
2. upright (page 171)
3. power lift (page 172)
4. palm (page 173)
5. locked-in (page 176)
6. 250 (page 180)
7. sideways (page 182)
8. locked (page 197)
9. overhead (page 198)
10. less strain (page 184)
11. spine movement (page 186)
12. direct ground lift (page 190)
13. extremity lift (page 191)
14. fluid resistant (page 198)
15. scoop stretcher (page 202)

True/False

1. T (page 200)
2. T (page 172)
3. F (page 180)
4. F (page 193)
5. T (page 184)
6. F (page 202)
7. F (page 197)
8. F (page 186)
9. F (page 190)
10. T (page 196)

Short Answer

1. –Front cradle
   –Firefighter's drag
   –One-person walking assist
   –Firefighter's carry
   –Pack strap (page 187)

2. –The vehicle or scene is unsafe.
   –The patient cannot be properly assessed before being removed from the car.
   –The patient needs immediate intervention that requires a supine position.
   –The patient's condition requires immediate transport to the hospital.
   –The patient blocks the EMT-B's access to another seriously injured patient. (page 186)

3. –Make sure there is sufficient lifting power.
   –Follow the manufacturer's directions for safe and proper use of the cot.
   –Make sure that all cots and patients are fully secured before you move the ambulance. (page 199)

4. –Be sure that you know or can find out the weight to be lifted and the limitations of the team's abilities.
   –Coordinate your movements with those of the other team members while constantly communicating with them.
   –Do not twist your body as you are carrying the patient.
   –Keep the weight that you are carrying as close to your body as possible while keeping your back in a locked-in position.
   –Be sure to flex at the hips, not at the waist, and bend at the knees, while making sure that you do not hyperextend your back by leaning back from your waist. (page 175)

5. Always keep your back in a straight, upright position and lift without twisting. (page 171)

Word Fun

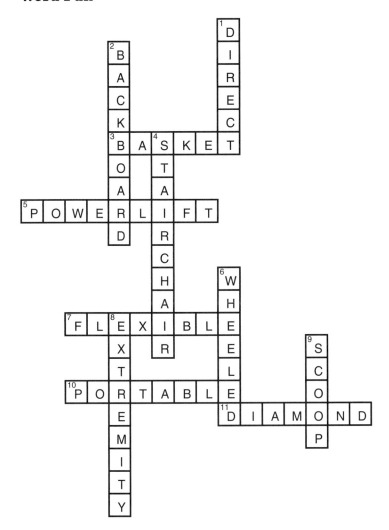

Ambulance Calls

1. –Immobilize the patient on a long spine board and apply high-flow oxygen.
   –Use four persons to carry the board back up the ledge.
   –Plan the route and brief your helpers before moving the patient.
   –Clarify whether you will move on "three" or count to three then move.
   –Coordinate the move until the patient is loaded into the ambulance.

2. It is highly unlikely that you will be able to move this patient, especially without significantly hurting yourself or your partner. You should immediately request additional personnel. You can attempt to move the patient, but if you cannot successfully do so, you will have to open his airway, etc. in his current position. Do the best you can until the patient can be moved.

3. Patients whose conditions will be exacerbated by physical activity should not walk to the gurney or ambulance. If a patient's condition is not such that it is medically necessary that you carry them, it is safer for them to walk on their own power to the ambulance. This patient should not/cannot walk. This produces a safety issue for you and your partner especially given the fact there is no elevator. Fortunately, this patient can sit upright and the use of a stair chair would be appropriate in this situation. Regardless, you should ask for more personnel given the patient's large size. Back injuries are very common in EMS. In order for providers to avoid these injuries, correct lifting techniques should be used and assistance should be requested whenever the patient is large or in a position not conducive to correct lifting procedures.

Skill Drills

Skill Drill 6-1: Performing the Power Lift (page 174)

1. Lock your back into an **upright**, inward curve. **Spread** and bend your legs. Grasp the backboard, palms up and just in front of you. **Balance** and center the weight between your arms.
2. Position your feet, **straddle** the object, and **distribute** weight.
3. **Straighten** your legs and lift, keeping your back locked in.

Skill Drill 6-2: Performing the Diamond Carry (page 176)

1. Position yourselves facing the patient.
2. After the patient has been lifted, the EMT-B at the foot turns to face forward.
3. EMT-Bs at the side each turn the head-end hand palm down and release the other hand.
4. EMT-Bs at the side turn toward the foot end.

Skill Drill 6-3: Performing the One-Handed Carrying Technique (page 177)

1. **Face** each other and use both **hands**.
2. Lift the backboard to **carrying height**.
3. **Turn** in the direction you will walk and **switch** to using one hand.

Skill Drill 6-4: Carrying a Patient on Stairs (page 178)

1. **Strap** the patient securely. Make sure one strap is tight across the **upper torso**, under the arms, and secured to the handles to prevent the patient from **sliding**.
2. Carry a patient down stairs with the **foot** end first, **head** elevated.
3. Carry the **head** end first going up stairs, always keeping the head elevated.

Skill Drill 6-5: Using a Stair Chair (page 181)

1. Position and secure the patient on the chair with **straps**.
2. Take your places at the **head** and **foot** of the chair.
3. A third **rescuer** "backs up" the rescuer carrying the **foot**.
4. **Lower** the chair to roll on landings, or for transfer to the cot.

Skill Drill 6-6: Performing the Rapid Extrication Technique (page 188)

1. First EMT-B provides in-line manual support of the head and cervical spine.
2. Second EMT-B gives commands, applies a cervical collar, and performs the initial assessment.
3. Second EMT-B supports the torso.
   Third EMT-B frees the patient's legs from the pedals and moves the legs together, without moving the pelvis or spine.
4. Second and Third EMT-Bs rotate the patient as a unit in several short, coordinated moves.
   First EMT-B (relieved by Fourth EMT-B or bystander as needed) supports the head and neck during rotation (and later steps).
5. First (or Fourth) EMT-B places the backboard on the seat against patient's buttocks.
6. Third EMT-B moves to an effective position for sliding the patient.
   Second and Third EMT-Bs slide the patient along the backboard in coordinated, 8" to 12" moves until the hips rest on the backboard.
7. Third EMT-B exits the vehicle, moves to the backboard opposite Second EMT-B, and they continue to slide the patient until patient is fully on the board.
8. First (or Fourth) EMT-B continues to stabilize the head and neck while Second and Third EMT-Bs carry the patient away from the vehicle.

Skill Drill 6-7: Extremity Lift (page 192)

1. Patient's hands are **crossed** over the chest.
First EMT-B grasps patient's wrists or **forearms** and pulls patient to a **sitting** position.

2. When the patient is sitting, First EMT-B passes his or her arms through patient's **armpits** and grasps the patient's opposite (or his or her own) **forearms** or **wrists**.
Second EMT-B kneels between the **legs**, facing in the same direction as the patient, and places his or her hands under the **knees**.

3. Both EMT-Bs rise to **crouching**.
On **command**, both lift and begin to move.

Skill Drill 6-8: Using a Scoop Stretcher (page 194)

1. Adjust stretcher **length**.
2. **Lift** patient slightly and **slide** stretcher into place, one side at a time.
3. **Lock** the stretcher ends together, avoiding **pinching**.
4. **Secure** the patient to the scoop stretcher and **transfer** it to the cot.

Skill Drill 6-9: Loading a Cot into an Ambulance (page 199)

1. Tilt the head of the cot **upward**, and place it into the patient compartment with the wheels on the floor.
2. Second rescuer on the side of the cot releases the **undercarriage** lock and lifts the undercarriage.
3. **Roll** the cot into the back of the ambulance.
4. Secure the cot to the **brackets** mounted in the ambulance.

Chapter 7: Airway

Matching

1. C (page 216)
2. I (page 216)
3. H (page 216)
4. G (page 216)
5. K (page 220)
6. E (page 216)
7. A (page 216)
8. F (page 216)
9. L (page 214)
10. D (page 214)
11. J (page 219)
12. B (page 193)

Multiple Choice

1. D (page 218)
2. B (pages 223-225)
3. B (page 221)
4. D (page 220)
5. A (page 220)
6. A (page 222)
7. C (page 227)
8. D (page 245)
9. C (page 246)
10. C (page 232)
11. B (page 225)
12. D (page 231)
13. C (page 226)
14. D (page 246)
15. A (page 242)
16. A (page 242)
17. C (page 218)
18. D (page 221)
19. B (page 222)
20. C (page 234)

Labeling

1. Upper and Lower Airways (page 215)

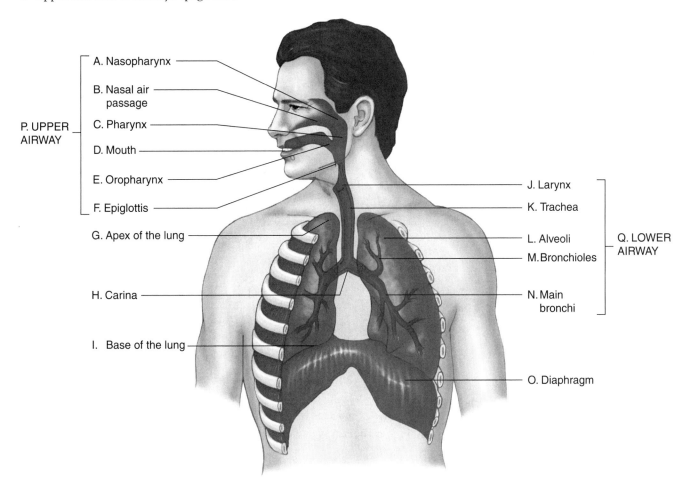

P. UPPER AIRWAY

A. Nasopharynx
B. Nasal air passage
C. Pharynx
D. Mouth
E. Oropharynx
F. Epiglottis
G. Apex of the lung
H. Carina
I. Base of the lung

J. Larynx
K. Trachea
L. Alveoli
M. Bronchioles
N. Main bronchi
O. Diaphragm

Q. LOWER AIRWAY

2. Thoracic Cage (page 215)

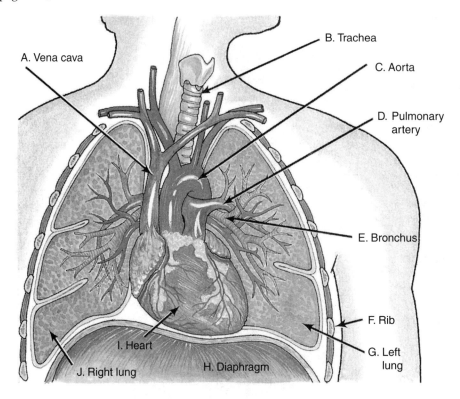

A. Vena cava

B. Trachea

C. Aorta

D. Pulmonary artery

E. Bronchus

F. Rib

G. Left lung

I. Heart

J. Right lung

H. Diaphragm

Fill-in

1. trachea (page 216)
2. higher (page 216)
3. 21, 78 (page 218)
4. carbon dioxide (page 219)
5. diaphragm, intercostal muscles (page 216)
6. low oxygen, high carbon dioxide (pages 218-219)
7. hypoxia (page 219)

True/False

1. F (page 226)
2. T (page 241)
3. F (page 227)
4. F (page 237)
5. F (page 236)

Short Answer

1. Nervousness, tachycardia, irritability, fear, apprehension. (Other signs: Mental status changes, use of accessory muscles for breathing, breathing difficulty, possible chest pain.) (page 222)
2. Adults: 12 to 20 breaths/min
   Children: 15 to 30 breaths/min
   Infants: 25 to 50 breaths/min (page 221)
3. Give slow, gentle breaths. (page 248)
4. **1.** Kneel above the patient's head.
   2. Extend the patient's neck unless you suspect a cervical spine injury.
   3. Open the mouth and suction as needed. Insert an airway adjunct as needed.
   4. Select a proper-sized mask.
   5. Position the mask on the patient's face.
   6. Use the C-clamp technique to hold the mask, then squeeze the bag every 5–6 seconds for adults, every 3–5 seconds for children and infants. (page 245)

5. **1.** Respiratory rate of less than 12 breaths/min or greater than 20 breaths/min

   2. Accessory muscle use

   3. Skin pulling in around the ribs during inspiration

   4. Pale, cyanotic, or cool (clammy) skin

   5. Irregular pattern of inhalation and exhalation

   6. Lung sounds that are decreased, unequal, or "wet"

   7. Labored breathing

   8. Shallow and/or uneven chest movement

   9. Two- or three-word sentences spoken (page 222)

6. They are the secondary muscles of respiration. They are not used in normal breathing. They include:

   1. Sternocleidomastoid (neck)

   2. Pectoralis major (chest)

   3. Abdominal muscles (page 222)

7. When the patient has severe trauma to the head or face. (page 227)

8. **1.** Select the proper-size airway and apply a water-soluble lubricant.

   2. Place the airway in the larger nostril with the curvature following the curve of the floor of the nose.

   3. Advance the airway gently.

   4. Continue until the flange rests against the skin. (page 227)

9. Tonsil tips are best because they have a large diameter and do not collapse. In addition, they are curved, which allows easy, rapid placement. (page 231)

10. 15 seconds. (page 232)

Word Fun

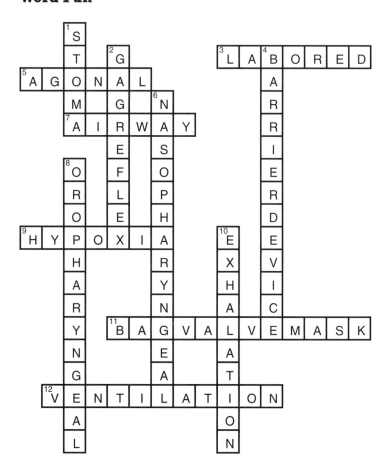

Ambulance Calls

1. Maintain cervical spine stabilization.
   Immediately open the airway with a modified jaw thrust.
   Suction to remove obstruction.
   Assess the airway for breathing (rate, rhythm, quality) and provide oxygen via nonrebreathing mask or BVM.
   Continue initial assessment, rapid extrication, and rapid transport.

2. You should reposition the head and attempt to re-ventilate the patient. If you are unsuccessful in your second attempt, you must take action to clear her airway. If you fail to clear her airway, she will like go into cardiac arrest as well. Choking victims have been known to walk away from others without indicating that they are choking and when this occurs in a restaurant setting, many of these people will go into a bathroom. If you find a patient in a restroom who cannot be ventilated, they likely have a foreign body airway obstruction (FBAO) most often as a result of ingested food.

3. The most common cause of airway obstruction in an unconscious patient is the tongue. The patient's husband told you that he helped her to the ground without injury, so it is safe to place her in the recovery position. If you were unsure of the presence of trauma, this position would not be used, instead using the jaw-thrust maneuver to manage her airway. In either case, you must continually monitor her condition and be prepared for vomitus.

Skill Drills

Skill Drill 7-1: Positioning the Unconscious Patient (page 224)

1. Support the head while your partner straightens the patient's legs.
2. Have your partner place his or her hand on the patient's far shoulder and hip.
3. Roll the patient as a unit with the person at the head calling the count to begin the move.
4. Open and assess the patient's airway and breathing status.

Chapter 8: Patient Assessment

Matching

1. P (page 268)
2. D (page 278)
3. M (page 289)
4. N (page 289)
5. B (page 278)
6. H (page 289)
7. L (page 276)
8. A (page 276)
9. O (page 272)
10. J (page 279)
11. G (page 278)
12. E (page 274)
13. F (page 283)
14. I (page 277)
15. K (page 274)
16. C (page 289)

Multiple Choice

1. C (pages 265-269)
2. D (page 226)
3. B (page 267)
4. C (page 267)
5. D (page 267)
6. A (page 267)
7. D (page 268)
8. B (page 269)
9. B (page 270)
10. C (page 274)
11. D (page 275)
12. A (page 276)
13. D (page 278)
14. A (page 277)
15. A (page 278)
16. C (page 278)
17. D (page 278)
18. D (page 278)
19. D (page 278)
20. C (page 279)
21. D (page 283)
22. B (page 268)
23. B (page 283)
24. C (page 289)
25. A (page 290)
26. D (page 263)
27. D (page 302)
28. B (page 307)
29. D (page 308)
30. A (page 264)
31. D (page 264)
32. D (page 270)
33. B (page 282)
34. B (page 294)

Fill-in

1. decisions (page 263)
2. body substance isolation (page 265)
3. victim (page 266)
4. safety (page 266)
5. Triage (page 268)
6. general impression (page 271)
7. life-threatening (page 271)
8. patency (page 275)
9. reevaluate (page 276)
10. cool (page 278)
11. initial assessment (page 279)

True/False

1. F (page 284)
2. F (page 272)
3. T (page 272)
4. T (page 277)
5. T (page 279)

Short Answer

1. To identify and initiate treatment of immediate or potential life threats. (page 271)
2. Immediate assessment of the environment, the patient's presenting signs and symptoms, and the patient's chief complaint. (page 271)
3. A-Airway
   B-Breathing
   C-Circulation (page 271)
4. Orientation to person, place, time, and event. Person (name) evaluates long-term memory. Place and time evaluate intermediate-term memory. Event evaluates short-term memory. (page 274)
5. 1. Identify the patient's chief complaint.
   2. Understand the circumstances surrounding the chief complaint.
   3. Direct further physical examination. (page 274)
6. Deformities, Contusions, Abrasions, Punctures/penetrations, Burns, Tenderness, Lacerations, Swelling (page 284)
7. 1. Ejection from a vehicle
   2. Death in the passenger compartment
   3. Fall greater than 15 to 20 feet
   4. Vehicle rollover
   5. High-speed vehicle collision
   6. Vehicle-pedestrian collision
   7. Motorcycle crash
   8. Unresponsiveness or altered mental status following trauma
   9. Penetrating head, chest, or abdominal trauma (page 267)

Word Fun

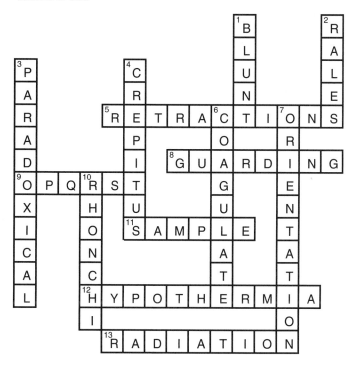

Ambulance Calls

1. –Maintain cervical spine control
   –Immediately manage the airway by suction and oxygen
   –Rapid survey and transport

   This patient is a load-and-go based on:
   • Mechanism of injury
   • Level of consciousness
   • Airway compromise

   Damage to vehicle indicates possible occult injuries.

2. This patient is having a very serious asthma attack. Accessory muscle use (nasal flaring, tracheal tugging, suprasternal and intercostal muscle retractions), work of breathing, wheezing, and one- to two-word responses all point to the seriousness of his attack. You should transport immediately, apply high-flow oxygen, and assist with MDI administration according to your local protocols.

3. Mechanism of injury is significant for this patient. Not only did he fall down a flight of wooden stairs, but he landed on a cement floor. His head injury is likely more significant than the bruising and laceration you can see. He could also have skull fractures, contusions, or intracranial bleeding. With any significant trauma to the head comes the likelihood for cervical spine fractures. You should also question the mechanism of his fall as medical conditions can sometimes precipitate injuries. Full c-spine precautions must be taken along with application of high-flow oxygen and prompt transport.

Skill Drills

Skill Drill 8-1: Performing a Rapid Physical Exam (pages 285 to 286)

1. Assess the head. Have your partner maintain in-line stabilization.
2. Assess the neck.
3. Apply a cervical spinal immobilization device on trauma patients.
4. Assess the chest. Listen to breath sounds on both sides of the chest.
5. Assess the abdomen.
6. Assess the pelvis. If there is no pain, gently compress the pelvis downward and inward to look for tenderness or instability.

7. Assess all four extremities. Assess pulse, motor, and sensory function.

8. Assess the back. In trauma patients, roll the patient in one motion.

Skill Drill 8-3: Performing the Detailed Physical Exam (pages 301 to 303)

1. Observe the face.

2. Inspect the area around the eyes and eyelids.

3. Examine the eyes for redness, contact lenses. Check pupil function.

4. Look behind the ear for Battle's sign.

5. Check the ears for drainage or blood.

6. Observe and palpate the head.

7. Palpate the zygomas.

8. Palpate the maxillae.

9. Palpate the mandible.

10. Assess the mouth and nose.

11. Check for unusual breath odors.

12. Inspect the neck.

13. Palpate the front and back of the neck.

14. Observe for jugular vein distention.

15. Inspect the chest and observe breathing motion.

16. Gently palpate over the ribs.

17. Listen to anterior breath sounds (midaxillary, midclavicular).

18. Listen to posterior breath sounds (bases, apices).

19. Observe the abdomen and pelvis.

20. Gently palpate the abdomen.

21. Gently compress the pelvis from the sides.

22. Gently press the iliac crests.

23. Inspect the extremities; assess distal circulation and motor sensory function.

24. Log roll the patient and inspect the back.

Chapter 9: Communications and Documentation

Matching

1. M (page 316)
2. G (page 317)
3. J (page 317)
4. K (page 317)
5. H (page 318)
6. L (page 317)
7. I (page 317)
8. C (page 318)
9. A (page 317)
10. F (page 319)
11. E (page 318)
12. D (page 317)
13. B (page 327)

Multiple Choice

1. D (pages 316-317)
2. A (page 317)
3. D (pages 316-317)
4. B (page 317)
5. B (page 318)
6. D (page 318)
7. C (page 319)
8. D (pages 319-320)
9. B (page 321)
10. D (page 322)
11. A (page 322)
12. B (page 323)
13. D (page 324)
14. D (page 324)
15. D (page 322)
16. A (page 323)
17. A (page 323)
18. A (page 325)
19. C (page 325)
20. D (page 324)
21. B (page 326)
22. D (page 326)
23. D (pages 326-327)
24. D (page 327)
25. C (pages 330-331)
26. D (page 330)
27. C (page 331)
28. A (pages 331-332)
29. B (page 332)
30. D (page 333)
31. D (page 333)
32. A (page 333)
33. D (page 336)
34. C (page 336)
35. D (page 334)
36. A (page 335)
37. D (page 335)
38. C (page 336)

Fill-in

1. patient care report (page 316)
2. transmitter, receiver (pages 316-317)
3. dedicated line (page 317)
4. telemetry (page 318)
5. cell phones (page 318)
6. Pagers (page 321)
7. importance (pages 320-321)
8. medical control (page 322)
9. slander (page 323)
10. medical control (page 323)
11. repeat (page 324)
12. standing orders (page 326)
13. eye contact (page 327)
14. honest (page 330)
15. interpreter (page 332)
16. minimum data set (page 332)
17. Competent (page 336)

True/False

1. T (pages 316-317)
2. T (pages 316-317)
3. T (page 318)
4. F (page 317)
5. T (page 317)
6. T (pages 332-333)
7. F (page 318)
8. F (page 319)
9. F (page 332)
10. T (page 329)

Short Answer

1. –Allocating specific radio frequencies for use by EMS providers
   –Licensing base stations and assigning appropriate radio call signs for those stations
   –Establishing licensing standards and operating specifications for radio equipment used by EMS providers
   –Establishing limitations for transmitter power output
   –Monitoring radio operations (pages 319-320)

2. –Monitor the channel before transmitting.
   –Plan your message.
   –Press the push-to-talk (PTT) button.
   –Hold the microphone 2 to 3 inches from your mouth and speak clearly.
   –Identify the person or unit you are calling, then identify your unit as the sender.
   –Acknowledge a transmission as soon as you can.
   –Use plain English.
   –Keep your message brief.
   –Avoid voicing negative emotions.
   –When transmitting a number with two or more digits, say the entire number first, then each digit separately.
   –Do not use profanity on the radio.
   –Use EMS frequencies only for EMS communications.
   –Reduce background noise.
   –Be sure other radios on the same frequency are turned down. (page 325)

3. –Continuity of care
   –Legal documentation
   –Education
   –Administrative
   –Research
   –Evaluation and continuous quality improvement (page 333)

4. –Traditional, written form with check boxes and a narrative section
   –Computerized, using electronic clipboard or similar device (page 335)

Word Fun

Ambulance Calls

1. –Go ahead and dispatch the closest ambulance for an emergency response.
   –Call for assistance from the fire department and local law enforcement.
   –Try to calm the caller down to obtain additional information.
   –If the caller is still of no help, ask her to get someone else to the phone.
   –Relay any additional information to the responding units.

2. You should determine if anyone on scene can translate for you (children will often speak both English and Spanish). If no one on scene can translate, you should contact a department translator. Every department should have a group of translators for different languages common to your community. (Ideally, you should speak languages commonly heard in your area.) If neither of these options is available, you should attempt as much nonverbal communication as possible, obtain baseline vital signs, and perform a physical exam to determine the nature of the problem. If the patient appears to refuse your help, you are in a tough situation. You cannot leave without knowing what the medical emergency is (if any), and you cannot obtain an informed refusal if clear communication does not occur.

3. Attempt to locate any identification that he may have on his person. If this is not possible or you cannot locate any forms of identification, you should notify law enforcement officers. The child should undergo a medical examination to ensure no injuries or other medical emergencies are present. Take care to communicate in a non-intimidating manner (taking care in level/tone of voice and posture) and attempt to establish trust with the patient.

Chapter 10: General Pharmacology

Matching

1. I (page 345)

2. J (page 344)

3. F (page 344)

4. G (page 350)

5. D (page 344)

6. H (page 344)

7. B (page 344)

8. C (page 344)

9. E (page 346)

10. A (page 348)

Multiple Choice

1. C (page 344)

2. A (page 355)

3. A (page 345)

4. A (page 345)

5. D (page 346)

6. D (page 347)

7. B (page 349)

8. D (page 351)

9. B (page 351)

10. A (pages 354-355)

11. A, B, C (pages 354-355)

12. C (page 349)

13. C (page 351)

14. A (page 344)

15. B (page 344)

16. D (page 350)

Fill-in

1. Glucose (page 350)

2. Epinephrine (page 351)

3. sublingually (page 355)

4. intravenous injection (page 345)

5. solutions (page 347)

6. medication (page 344)

True/False

1. F (page 349)

2. F (page 350)

3. F (page 351)

4. T (page 354)

5. F (page 345)

6. T (page 356)

7. F (page 353)

8. F (page 354)

Short Answer

1. Intravenous   Intramuscular   Transcutaneous
   Oral   Intraosseous   Inhalation
   Sublingual   Subcutaneous   Per rectum (pages 345-346)

2. 1. Obtain an order from medical control.

   2. Verify the proper medication and prescription.

   3. Verify the form, dose, and route.

   4. Check the expiration date and condition of medication.

   5. Reassess vital signs, especially heart rate and blood pressure, at least every 5 minutes or as the patient's condition changes.

   6. Document. (pages 355-356)

3. Many drugs adsorb (stick to) activated charcoal, preventing the drugs from being absorbed by the body. It needs to be shaken because it is a suspension and should be given in a covered container with a straw. (page 350)

4. 1. Dilates lung passages

   2. Constricts blood vessels

   3. Increases heart rate and blood pressure (page 351)

5. By pressing it into the skin (page 351)

6. **1.** Relaxes coronary arteries and veins

   2. Less blood is returned to the heart

   3. Decreases blood pressure

   4. Relaxes veins throughout the body

   5. May cause headaches (page 355)

7. To aim the spray properly (page 354)

Word Fun

Across:
3. TRANSCUTANEOUS
5. ABSORPTION
6. INTRAVENOUS
7. EPINEPHRINE
8. INTRAOSSEOUS
9. NITROGLYCERIN

Down:
1. INDICATION
2. SUBLINGUAL
4. ADSORPTION

Ambulance Calls

1. This patient is in serious trouble. The history of the events combined with his level of consciousness, stridor, and hypotension are obvious signs of an anaphylactic reaction. You must immediately administer epinephrine in order to counteract the effects of the insect stings, i.e. histamine release. If available, ALS providers should be requested. If not available, transport to the nearest appropriate facility should occur without delay. With the presence of multiple stings, this patient will likely need repeat doses of epinephrine as well as the administration of antihistamines, breathing treatments, and possibly advanced airway maneuvers. Your partner should apply high-flow oxygen and remove any remaining stingers in the neck or face by scraping them from the skin. Your prompt action is essential to patient survival.

2. This patient is suffering from hypoglycemia. Although this patient is confused, he is able to talk and swallow. Some states allow EMT-B's to perform blood glucose tests, and this would provide information regarding this patient's blood glucose level. If your local protocols do not allow for this skill, you can gather much information about this patient through his physical signs and medical history (all indicative of low blood sugar). You should administer (at least) one tube of oral glucose and reassess his mentation and vital signs. Provide treatment and transport according to local protocols.

3. –Place patient in position of comfort.

   –Give 100% oxygen via nonrebreathing mask.

   –Check blood pressure!

   –Check expiration date on nitroglycerin.

   –Contact medical control for permission to assist patient with 1 nitroglycerin tablet SL.

   –Monitor vital signs.

   –Rapid transport.

Chapter 11: Respiratory Emergencies

Matching

1. B (page 373)
2. D (page 369)
3. H (page 370)
4. J (page 372)
5. L (page 374)
6. C (page 373)
7. G (page 368)

8. E (page 370)
9. M (page 389)
10. A (page 370)
11. K (page 376)
12. F (page 373)
13. I (page 375)

Multiple Choice

1. D (page 384)
2. C (page 366)
3. D (page 368)
4. C (page 366)
5. B (page 367)
6. D (page 368)
7. B (page 368)
8. A (page 389)
9. B (page 370)
10. A (page 369)
11. D (page 371)
12. A (page 372)
13. D (pages 372-373)
14. B (pages 372-373)
15. C (page 374)
16. C (page 373)
17. B (page 373)
18. D (pages 389-390)
19. C (page 374)
20. A (page 375)
21. B (page 376)
22. B (page 368)
23. C (page 377)
24. A (page 366)
25. A (page 384)
26. B (page 379)
27. D (page 392)
28. D (page 380)
29. C (page 385)
30. D (page 385)
31. D (page 385)
32. C (page 390)

Labeling

Obstruction, scarring, and dilation of the alveolar sac (page 372)

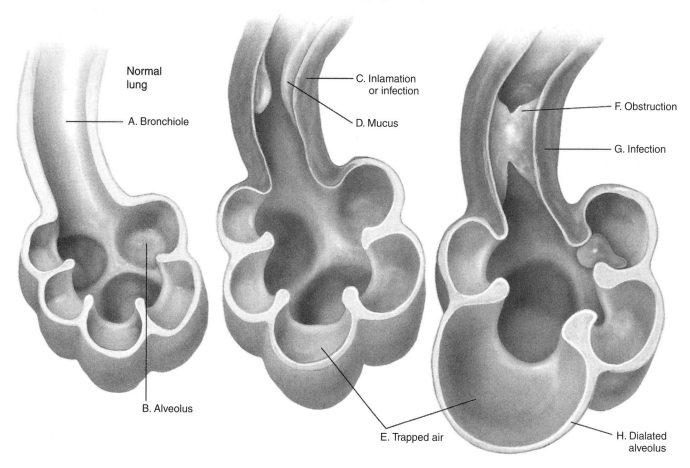

Normal lung

A. Bronchiole

B. Alveolus

C. Inlamation or infection

D. Mucus

E. Trapped air

F. Obstruction

G. Infection

H. Dialated alveolus

Fill-in

1. carbon dioxide (page 366)
2. oxygen (page 367)
3. oxygen (page 368)
4. alveoli (page 366)
5. trachea (page 366)
6. 12, 20 (page 367)
7. carbon dioxide (page 366)
8. breathing status (page 379)
9. transport decision (page 379)
10. SAMPLE, OPQRST (page 380)

True/False

1. F (page 370)
2. T (page 373)
3. F (page 373)
4. F (page 373)
5. T (page 375)
6. T (page 376)
7. F (page 376)
8. T (page 389)
9. F (pages 373 & 379)
10. T (page 370)
11. T (page 378)
12. T (page 382)

Short Answer

1. 1. Normal rate and depth
   2. Regular pattern of inhalation and exhalation
   3. Good audible breath sounds on both sides of the chest
   4. Regular rise and fall on both sides of the chest
   5. Pink, warm, dry skin (page 367)

2. 1. Pulmonary vessels are obstructed from absorbing oxygen and releasing carbon dioxide by fluid, infection, or collapsed air spaces.
   2. Damaged alveoli
   3. Air passages obstructed by muscle spasm, mucus, weakened airway walls
   4. Blood flow to the lungs obstructed
   5. Pleural space is filled with air or excess fluid (page 368)

3. 1. Patient is unable to coordinate administration and inhalation.
   2. Inhaler is not prescribed for patient.
   3. You did not obtain permission from medical control or local protocol.
   4. Patient has already met maximum prescribed dose before your arrival. (page 385)

4. An ongoing irritation of the respiratory tract; excess mucus production obstructs small airways and alveoli. Protective mechanisms are impaired. Repeated episodes of irritation and pneumonia can cause scarring and alveolar damage, leading to COPD. (page 370)

5. 1. Respiratory rate of slower than 12 breaths/min or faster than 20 breaths/min
   2. Muscle retractions above the clavicles between ribs, below rib cage, especially in children
   3. Pale or cyanotic skin
   4. Cool, damp (clammy) skin
   5. Shallow or irregular respirations
   6. Pursed lips
   7. Nasal flaring (pages 379)

6. A condition characterized by a chronically high blood level of carbon dioxide in which the respiratory center no longer responds to high blood levels of carbon dioxide. In these patients, low blood oxygen causes the respiratory center to respond and stimulate respiration. If the arterial level of oxygen is then raised, as happens when the patient is given additional oxygen, there is no longer any stimulus to breathe; both the high carbon dioxide and low oxygen drives are lost. (page 368)

7. 1. Is the air going in?
   2. Does the chest expand with each breath?
   3. Does the chest fall with each breath?
   4. Is the rate adequate for the age of your patient? (page 378)

Word Fun

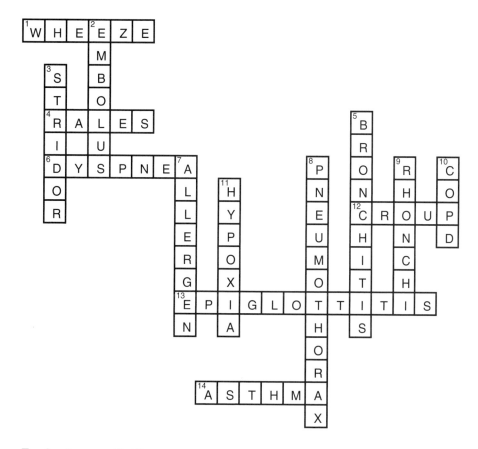

Ambulance Calls

1. This child has all the classics signs of epiglottitis. You should do nothing to excite or frighten the child as doing so will likely cause his airway to spasm and close. Remember to use non-threatening body language (place yourself below his or her eye level), give the child distance (until you establish trust) and perform any exam using the toe-to-head method. This is a true emergency that requires immediate transport to the hospital, but you must do so tactfully. Use parents to assist with patient care efforts such as applying humidified oxygen (blow-by or mask).

2. This patient's chief complaint, age, weight, smoking and birth control use all place her at risk of pulmonary embolism. PE patients most often experience a sudden onset of shortness of breath that they describe as sharp and worsening with inspiration. You should provide high flow oxygen, obtain vital signs, perform a secondary exam en route to the hospital (including auscultation of lung sounds) and prompt transport to the nearest appropriate facility.

3. –Place patient in position of comfort.
   –Provide high-flow oxygen via nonrebreathing mask.
   –Monitor vital signs.
   –Provide rapid transport.

Skill Drills

Skill Drill 11-1: Assisting a Patient With a Metered-Dose Inhaler (page 388)

1. Ensure inhaler is at room temperature or **warmer**.
2. Remove oxygen mask. Hand inhaler to patient. Instruct about breathing and **lip seal**.
3. Instruct patient to press inhaler and inhale. Instruct about **breath holding**.
4. Reapply **oxygen**. After a few **breaths**, have patient repeat **dose** if order/protocol allows.

Assessment Review

1. D (page 392)
2. B (page 393)
3. A (page 392)
4. B (page 393)
5. D (page 393)

Emergency Care Summary

Respiratory Distress
nonrebreathing
10 to 15
oropharyngeal
nasopharyngeal

Asthma
metered-dose
expiration date
exhale

Infection of Upper or Lower Airway
humidified

Acute Pulmonary Edema
ventilatory support

Chronic Obstructive Pulmonary Disease
full-flow
15 L

Spontaneous Pneumothorax
supplemental

Pleural Effusions
high-flow

Obstruction of the Upper Airway
BLS

Pulmonary Embolism
cardiac arrest

Hyperventilation
respirations
(page 393)

Chapter 12: Cardiovascular Emergencies

Matching

1. L (page 402)
2. C (page 403)
3. O (page 403)
4. G (page 403)
5. N (page 403)
6. K (page 404)
7. H (page 402)
8. M (page 402)
9. B (page 405)
10. D (page 409)
11. F (page 404)
12. I (page 407)
13. J (page 410)
14. A (page 410)
15. E (page 410)

Multiple Choice

1. D (page 402)
2. D (page 402)
3. A (page 402)
4. B (page 402)
5. A (page 403)
6. B (page 404)
7. C (page 404)
8. A (page 404)
9. C (page 404)
10. A (page 405)
11. B (page 407)
12. B (page 407)
13. D (page 408)
14. B (page 408)
15. D (page 408)
16. D (page 408)
17. C (page 408)
18. C (page 408)
19. C (page 408)
20. C (page 409)
21. D (page 409)
22. D (page 409)
23. A (page 409)
24. A (page 412)
25. D (page 411)
26. B (page 412)
27. C (page 409)
28. C (page 413)
29. B (page 414)
30. C (page 415)
31. D (page 415)
32. D (page 415)
33. C (page 415)
34. B (page 419)
35. D (page 419)
36. C (page 421)
37. C (page 421)
38. B (page 421)
39. C (page 422)
40. D (page 421)
41. D (page 423)
42. A (page 432)
43. D (page 426)
44. C (page 427)
45. C (page 429)

Labeling

1. The Right and Left Sides of Heart (page 403)

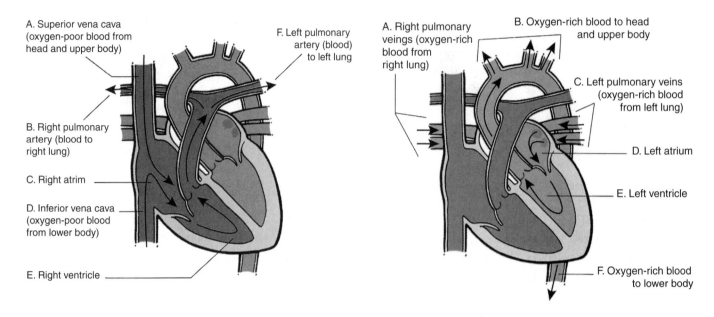

A. Superior vena cava (oxygen-poor blood from head and upper body)

F. Left pulmonary artery (blood) to left lung

B. Right pulmonary artery (blood to right lung)

C. Right atrim

D. Inferior vena cava (oxygen-poor blood from lower body)

E. Right ventricle

A. Right pulmonary veings (oxygen-rich blood from right lung)

B. Oxygen-rich blood to head and upper body

C. Left pulmonary veins (oxygen-rich blood from left lung)

D. Left atrium

E. Left ventricle

F. Oxygen-rich blood to lower body

2. Electrical Conduction (page 403)

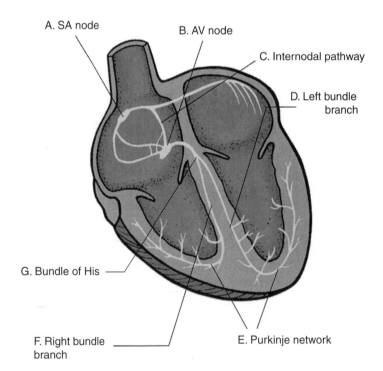

A. SA node

B. AV node

C. Internodal pathway

D. Left bundle branch

G. Bundle of His

F. Right bundle branch

E. Purkinje network

3. Pulse Points (page 406)

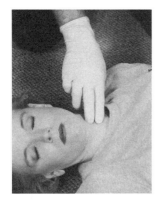

A. Carotid

B. Femoral

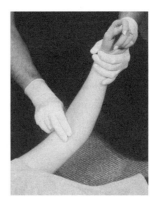

C. Brachial

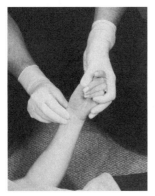

D. Radial

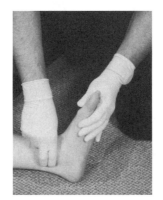

E. Posterior tibial

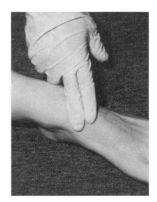

F. Dorsalis pedis

Fill-in

1. septum (page 402)

2. aorta (page 402)

3. right (page 402)

4. AV (page 403)

5. dilation (page 403)

6. Red blood (page 404)

7. Diastolic (page 404)

8. four (page 402)

9. left (page 402)

True/False

1. F (page 402)

2. F (page 403)

3. T (page 405)

4. F (page 407)

5. T (page 408)

6. F (page 408)

7. F (page 415)

8. F (page 424)

9. T (page 408)

10. F (page 404)

11. T (page 420)

12. T (page 420)

Short Answer

1. Automated: Operator needs only to apply pads and turn on the machine. It performs all functions for analyzing and shocking. This type of defibrillator often has a computer voice synthesizer to advise the EMT which steps to take.

Semi-automated: Operator applies pads, turns on the machine, and pushes button to shock. (pages 420-421)

2. **1.** Not having a charged battery

2. Applying the AED to a patient who is moving

3. Applying the AED to a responsive patient with a rapid heart rate (page 421)

3. **1.** If the patient regains a pulse

   2. After six to nine shocks have been delivered

   3. If the machine gives three consecutive "no shock" messages (page 427)

4. **1.** Place pads correctly.

   2. Make sure no one is touching the patient.

   3. Do not defibrillate a patient who is in pooled water.

   4. Dry the chest before defibrillating a wet patient.

   5. Do not defibrillate a patient who is touching metal that others are touching.

   6. Remove nitroglycerin patches and wipe the area with a dry towel before defibrillation. (pages 426-427)

5. **1.** Obtain an order from medical direction.

   2. Take the patient's blood pressure; continue with administration only if the systolic blood pressure is greater than 100 mm Hg.

   3. Check that you have the right medication, right patient, and right delivery route.

   4. Check the expiration date of the nitroglycerin.

   5. Question the patient about the last dose he or she took and its effects.

   6. Be prepared to have the patient lie down.

   7. Give the medication sublingually.

   8. Advise the patient to keep his or her mouth closed to allow the medication to dissolve.

   9. Recheck blood pressure within 5 minutes.

   10. Record each medication and the time of administration.

   11. Perform continued assessment. (page 417)

6. **1.** It may or may not be caused by exertion, but can occur at any time.

   2. It does not resolve in a few minutes.

   3. It may or may not be relieved by rest or nitroglycerin. (page 408)

7. **1.** Sudden death

   2. Cardiogenic shock

   3. Congestive heart failure (page 409)

8. **1.** Sudden onset of weakness, nausea, or sweating without an obvious cause

   2. Chest pain/discomfort that does not change with each breath

   3. Pain in lower jaw, arms, or neck

   4. Sudden arrhythmia with syncope

   5. Pulmonary edema

   6. Sudden death

   7. Increased and/or irregular pulse

   8. Normal, increased, or decreased blood pressure

   9. Normal or labored respirations

   10. Pale or gray skin

   11. Feelings of apprehension (pages 408-409)

9. Remove clothing from the patient's chest area. Apply the pads to the chest: one just to the right of the sternum, just below the clavicle, the other on the left chest with the top of the pad 2 to 3 inches below the armpit. Ensure that the pads are attached to the patient cables (and that they are attached to the AED in some models). (page 426)

Word Fun

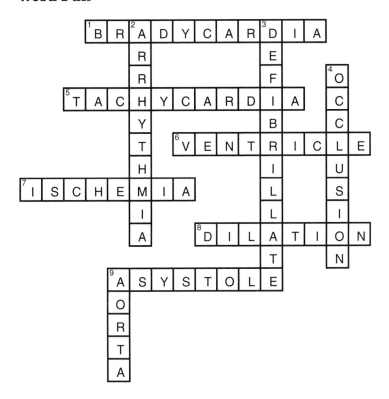

Ambulance Calls

1. –Place patient in the position of comfort.
   –Provide high-flow oxygen via nonrebreathing mask.
   –Monitor vital signs.
   –Provide normal transport.

2. Denial is one of the biggest indicators of heart attack. Although this patient is considered "younger" and otherwise healthy, he is having signs and symptoms of a myocardial infarction. It may take some convincing of the need for treatment and transport, but you must be clear on the potential consequences of his refusal of care (informed refusal). If he initially refuses, explain what physical signs you see that lead you to believe he is likely having a heart attack, express genuine concern for his well-being and speak directly with medical control. More often than not, when patients hear the same or similar information from a physician, it makes a different impact. If he allows you to examine, treat and transport him, apply high-flow oxygen, transport promptly and follow local protocols.

3. This patient could be having a heart attack. A common symptom that women (especially post-menopausal women) experience when having a heart attack is the sudden onset of generalized weakness. Although chest pain is a common indicator of heart attack, if a patient is not experiencing pain or pressure it does not necessarily mean they are not experiencing a cardiac event. It is important to note that heart disease is the number one killer of women in the United States, taking more lives than cancer and killing more women than men every year. If your patient exhibits any combination of the "associated symptoms" of heart attack such as nausea, vomiting, shortness of breath, pain or numbness in the neck, jaw, back, arm(s) or cool, pale, sweaty skin, you should suspect the possibility of a heart attack. You should apply high-flow oxygen, obtain vital signs, allow the patient to maintain a position of comfort, and provide immediate transport to the nearest appropriate facility.

Skill Drills

Skill Drill 12-2: AED and CPR (pages 428-429)

1. Stop CPR if in progress.
   Assess responsiveness.
   Check breathing and pulse.
   If unresponsive and not breathing adequately, give two slow ventilations.

2. If pulseless, begin CPR.
Prepare the AED pads.
Turn on the AED; begin narrative if needed.

3. Apply AED pads.
Stop CPR.

4. Verbally and visually clear the patient.
Push the Analyze button if there is one.
Wait for the AED to analyze rhythm.
If no shock advised, perform CPR for 2 minutes.
If shock advised, recheck that all are clear and push the Shock button.
Immediately initiate 5 cycles (approximately 2 minutes) of CPR, beginning with chest compressions. Reanalyze rhythm.
Press Shock if advised (second shock).
Push the Analyze button, if needed, to analyze rhythm again.
Press Shock if advised (third shock).

5. Check pulse.
If pulse is present, check breathing.
Gather additional information on the arrest event.

6. If breathing adequately, give oxygen and transport.
If not, open airway, ventilate, and transport.
If no pulse, perform 5 cycles of CPR (approximately 2 minutes).
Clear the patient and analyze again.
If necessary, repeat one cycle of up to three shocks.
Transport and call medical control.
Continue to support breathing or perform CPR, as needed.

Assessment Review

1. D (page 432) **4.** D (page 432)
2. A (page 432) **5.** D (page 432)
3. D (page 432)

Emergency Care Summary

Chest Pain   **Cardiac Arrest**

Nitroglycerin   Defibrillation
systolic   pulseless
100   apnea
dose   two
fainting   pocket mask
underneath   pads
5   breastbone
100   collarbone
analyze
rhythm
nonrebreathing
arrest event
(page 433)

Chapter 13: Neurologic Emergencies

Matching

1. D (page 440)
2. G (page 441)
3. E (page 441)
4. B (page 440)
5. I (page 441)
6. F (page 441)
7. J (page 442)
8. C (page 442)
9. M (page 451)
10. H (page 451)
11. L (page 444)
12. K (page 442)
13. A (page 444)

Multiple Choice

1. D (page 440)
2. A (page 440)
3. A (page 441)
4. C (page 442)
5. C (page 442)
6. A (page 442)
7. B (page 442)
8. B (page 443)
9. D (page 444)
10. A (page 451)
11. D (page 451)
12. D (page 453)
13. D (page 451)
14. C (page 440)
15. B (page 457)
16. C (page 457)
17. D (page 457)
18. D (pages 457-458)
19. B (page 444)
20. D (page 444)
21. A (pages 444-445)
22. B (page 445)
23. C (page 428)
24. C (page 445)
25. A (page 445)
26. D (page 448)
27. D (pages 442-443)
28. A (page 443)
29. D (page 451)

Labeling

1. Brain (page 441)

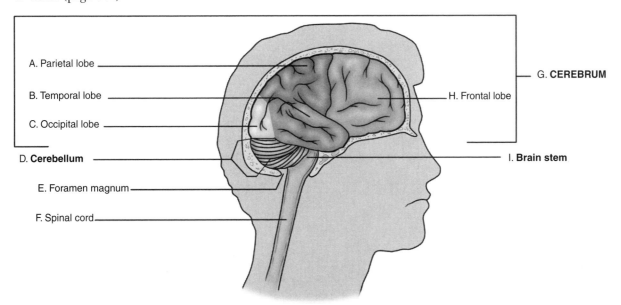

A. Parietal lobe

B. Temporal lobe

C. Occipital lobe

D. **Cerebellum**

E. Foramen magnum

F. Spinal cord

G. **CEREBRUM**

H. Frontal lobe

I. **Brain stem**

2. Spinal Cord (page 441)

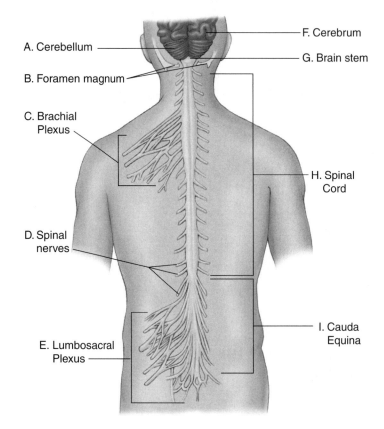

A. Cerebellum

B. Foramen magnum

C. Brachial Plexus

D. Spinal nerves

E. Lumbosacral Plexus

F. Cerebrum

G. Brain stem

H. Spinal Cord

I. Cauda Equina

Fill-in

1. twelve (page 441)

2. cerebellum (page 440)

3. emotion and thought (page 440)

4. head (page 441)

5. three (page 440)

6. nerves (page 441)

7. opposite, same (page 440)

8. cerebrum (page 440)

9. Incontinence (page 452)

10. brain (page 440)

11. epidural (page 445)

12. hemiparesis (page 452)

13. altered mental status (page 458)

True/False

1. F (page 444)

2. F (page 451)

3. T (page 451)

4. F (page 451)

5. T (page 452)

6. F (page 444)

Short Answer

1. **1.** Facial droop—Ask patient to show teeth or smile.
  2. Arm drift—Ask patient to close eyes and hold arms out with palms up.
  3. Speech—Ask patient to say, "The sky is blue in Cincinnati." (page 448)

2. Newer clot-busting therapies may be helpful in reversing damage in certain kinds of strokes, but treatment must be started within 3 hours after onset of the event. (page 450)

3. A period of time after a seizure, generally lasting from 5 to 30 minutes, that is characterized by some degree of altered mental status and labored respirations (page 452)

4. Infarcted cells are dead. Ischemic cells are still alive, although they are not functioning properly because of hypoxia. (page 442)

5. **1.** Hypoglycemia
  2. Postictal state
  3. Subdural or epidural bleeding (pages 444-445)

Word Fun

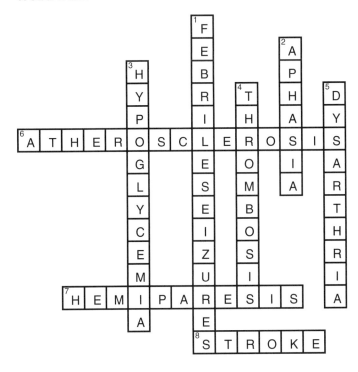

Ambulance Calls

1. Given these signs, your patient is likely experiencing a left hemispheric stroke. Any problems related to his ability to understand or use language will be frustrating for both you and the patient. After performing your initial assessment, the Cincinnati Stoke Scale can provide valuable information regarding the presence of a stroke. This assessment measures abnormalities in speech and the presence of facial droop and arm drift. Stroke is a true emergency that requires prompt transport. You should apply high-flow oxygen and take care during transport to prevent injury of affected body parts as the patient will not be able to protect them on their own. It is helpful to place the patient on the affected side and elevate their heads approximately 6" to facilitate swallowing. Be sure to relate your positive findings of stroke to the hospital to avoid any unnecessary delays in patient care upon arrival to the emergency department.

2. Maintain the airway—high-flow oxygen.
  Suction if necessary or position lateral recumbent to clear secretions.
  Check glucose level.
  Provide rapid transport.

3. This patient's signs and symptoms cause you to suspect the presence of a hemorrhagic stroke. You should apply high-flow oxygen and provide immediate transport. This patient will be more likely to experience seizure activity than patients suffering from ischemic stroke. There is no way for you to determine the type or extent of her stroke, as this can be accomplished only in the hospital. Your job is to recognize the seriousness of the situation, provide supportive measures within your scope of practice, and provide prompt transport and notification to the receiving facility.

Assessment Review

1. D (page 459)
2. B (page 459)
3. D (page 459)
4. D (page 459)
5. C (page 459)

Emergency Care Summary

Stroke

teeth

arm drift

Cincinnati

Seizure

Spontaneous

Inappropriate words

Withdraws to pain

Glasgow Coma Scale score

Altered Mental Status

Cincinnati Stroke Scale

Glasgow Coma Scale

(page 460)

Chapter 14: The Acute Abdomen

Matching

1. I (page 469)
2. D (page 466)
3. E (page 466)
4. M (page 467)
5. K (page 469)
6. A (page 470)
7. C (page 471)
8. J (page 470)
9. F (page 470)
10. B (page 466)
11. G (page 469)
12. N (page 468)
13. L (page 470)
14. O (page 466)
15. H (page 466)

Multiple Choice

1. C (page 470)
2. A (page 470)
3. A (page 469)
4. A (page 466)
5. D (page 473)
6. A (page 469)
7. B (page 466)
8. C (page 469)
9. B (page 466)
10. D (page 466)
11. C (page 471)
12. D (pages 468-469)
13. B (page 471)
14. C (page 469)
15. A (page 474)
16. C (page 476)

Labeling

1. Solid Organs (page 467)

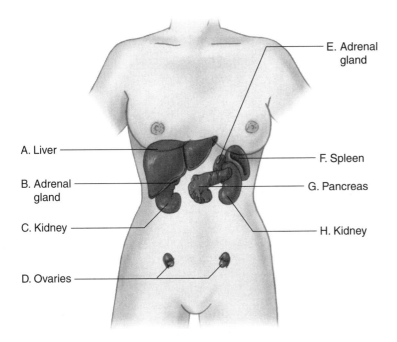

2. Hollow Organs (page 467)

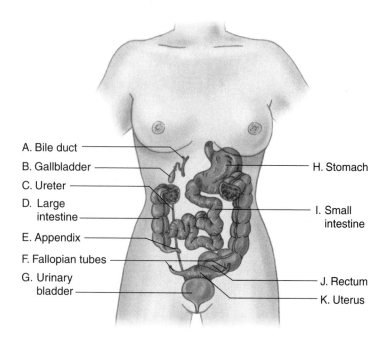

A. Bile duct
B. Gallbladder
C. Ureter
D. Large intestine
E. Appendix
F. Fallopian tubes
G. Urinary bladder
H. Stomach
I. Small intestine
J. Rectum
K. Uterus

3. Retroperitoneal Organs (page 467)

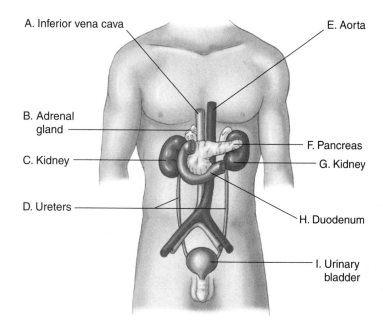

A. Inferior vena cava
B. Adrenal gland
C. Kidney
D. Ureters
E. Aorta
F. Pancreas
G. Kidney
H. Duodenum
I. Urinary bladder

True/False

1. F (page 466)
2. F (pages 468-469)
3. F (page 468)
4. F (page 466)
5. T (page 466)
6. F (page 471)
7. F (page 472)
8. T (page 469)
9. T (page 469)

Short Answer

1. Occurs because of connections between the body's two nervous systems. The abdominal organs are supplied by autonomic nerves, which, when irritated, stimulate close-lying sensory (somatic) nerves. (page 466)

2. No. It is too complex and treatment is the same. (page 466)

3. Paralysis of muscular contractions in the bowel results in retained gas and feces. Nothing can pass through. (page 470)

4. Bleeding and fluid shifts. (page 470)

5. 1. Explain to the patient what you are about to do.

   2. Place the patient in a supine position with the legs drawn up and flexed at the knees unless trauma is suspected.

   3. Determine whether the patient is restless or quiet, whether motion causes pain, or whether any characteristic position, distention, or obvious abnormality is present.

   4. Palpate the four quadrants of the abdomen.

   5. Determine whether the patient can relax the abdominal wall on command.

   6. Determine whether the abdomen is tender when palpated. (page 476)

Word Fun

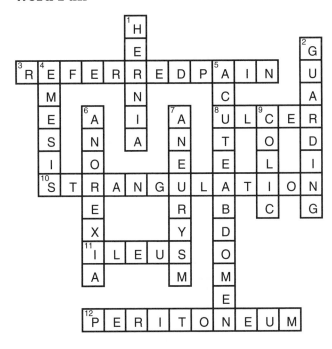

Ambulance Calls

1. Possible appendicitis.
   Place patient in the position of comfort.
   Apply high-flow oxygen.
   Keep patient warm.
   Rapid transport.
   Obtain a SAMPLE history.
   Document OPQRST.
   Monitor patient closely.

2. This patient has a bowel obstruction (fecal impaction) that is causing his pain and tenderness. If left untreated, his condition will deteriorate. When assessing his abdomen, explain what you will do, place him with knees slightly toward his abdomen and gently palpate all four quadrants to determine the presence of rigidity or masses. If a patient points to a specific location of pain, palpate that area last. You must take great care in moving and transporting him, as it will be particularly painful if he is bumped or jostled. Apply high-flow oxygen, allow him to find his position of comfort, and move him gently. Do not delay transport to attempt to determine the cause of abdominal pain.

3. Individuals who have experienced a kidney stone(s) will tell you that it is an extremely painful experience. Provide prompt, gentle transport. Monitor his airway, breathing and circulation, be prepared for continued vomiting, apply oxygen (which can ease nausea), obtain vital signs, do not give anything by mouth, and keep the patient as comfortable as possible. Be sure to thoroughly document all information regarding their signs and symptoms as well as any treatment you provide. Always follow local protocols.

Assessment Review

1. D (page 476)
2. A (page 476)
3. B (page 476)
4. C (page 470)
5. D (page 476)

Emergency Care Summary

supine

flexed

trauma

distention

abnormality

Palpate

quadrants

abdominal wall

palpated

(page 476)

Chapter 15: Diabetic Emergencies

Matching

1. H (page 482)
2. F (page 483)
3. B (page 482)
4. K (page 484)
5. L (page 482)
6. N (page 485)
7. E (page 483)
8. G (page 483)
9. C (page 483)
10. A (page 485)
11. M (page 482)
12. D (page 485)
13. J (page 484)
14. I (page 482)

Multiple Choice

1. A (page 483)
2. A (page 484)
3. D (page 482)
4. C (page 482)
5. B (page 482)
6. A (page 484)
7. C (page 483)
8. D (page 482)
9. D (page 485)
10. B (page 483)
11. B (page 486)
12. B (page 483)
13. D (page 483)
14. A (page 482)
15. A (page 482)
16. A (page 483)
17. C (page 483)
18. B (page 484)
19. B (page 484)
20. D (page 483)
21. C (page 485)
22. B (page 484)
23. A (page 485)
24. B (pages 484-485)
25. C (page 487)
26. D (page 485)
27. A (page 493)
28. A (page 490)
29. D (page 490)
30. D (page 408)
31. B (page 485)
32. C (page 485)
33. A (page 487)
34. D (page 485)
35. A (page 485)
36. C (page 492)
37. C (page 485)
38. D (page 483)
39. C (page 488)
40. D (page 494)
41. A (page 488)
42. D (page 488)

Fill-in

1. diabetes mellitus (page 482)
2. autoimmune (page 483)
3. ineffective (page 483)
4. diabetic coma (page 485)
5. sugar, insulin (pages 488-489)

True/False

1. T (page 483)
2. F (page 483)
3. T (page 484)
4. T (page 485)
5. T (page 485)
6. T (page 483)
7. F (page 482)
8. F (page 482)
9. T (page 492)
10. T (page 482)
11. T (page 483)

Short Answer

1. Insulin is a hormone that enables glucose to enter body cells. (page 482)
2. **1.** Glucose
   2. Insta-Glucose (page 488)
3. A patient who is unconscious or not able to swallow should not be given oral glucose. (page 490)
4. **1.** Diabinase
   2. Orinase
   3. Micronase
   4. Glucotrol (page 483)
5. **1.** Ketoacidosis
   2. Dehydration
   3. Hyperglycemia (page 485)
6. Kussmaul respirations; dehydration; fruity odor on breath; rapid, weak pulse; normal or slightly low blood pressure; varying degrees of unresponsiveness. (page 485)
7. Insulin shock; it develops rapidly as opposed to diabetic coma, which takes longer to develop. (page 484)
8. It will immediately benefit the patient in insulin shock and is unlikely to worsen the condition of the patient in a diabetic coma. (page 490)

Word Fun

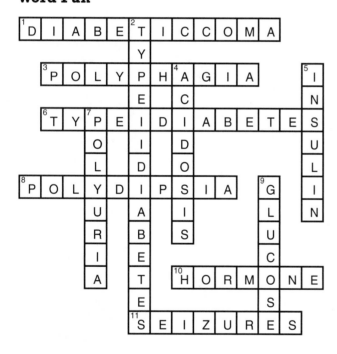

Ambulance Calls

1. Turn the patient on her side immediately or use suction to clear the airway.
   Insert an oral or nasal airway and apply high-flow oxygen.
   Attempt to obtain a blood glucose level.
   Transport patient rapidly, since you should never give anything by mouth to an unresponsive patient.
   Monitor the patient closely.
2. When experiencing hypoglycemia, diabetics can exhibit very strange behavior. Some family members can be quite surprised to see or hear their loved ones doing and saying things quite out of their normal behavior. When patients are experiencing low blood sugar, they may act as if they are intoxicated or mentally unstable. Perform a physical exam and look for clues regarding potential medical conditions. Some EMT-Bs can perform blood glucose testing, and this should be done for any patients with decreased levels of consciousness. If this patient is alert enough to swallow, administer oral glucose and provide prompt transport. Continue to monitor their ABCs for any changes.

3. You cannot give this patient anything by mouth as he is unconscious and therefore unable to protect his own airway. If available, you should request ALS providers, as they can administer intravenous dextrose (glucose). If you have no emergency providers available with this scope of practice within your system, you must transport this patient immediately. You may encounter family members who do not understand why you cannot "just give him some sugar." Hopefully, there will be no delays in explaining his need for transport or the seriousness of his condition. Perform a thorough assessment (including blood glucose testing if permitted), provide high-flow oxygen, monitor his ABCs (using airway adjuncts, positive pressure ventilations and suctioning as needed), and provide prompt transport.

Skill Drills

Skill Drill 15-1: Administering Glucose (page 491)

1. Make sure that the tube of glucose is intact and has not **expired**.
2. Squeeze a generous amount of oral glucose onto the **bottom third** of a **bite stick** or tongue depressor.
3. Open the patient's **mouth**. Place the tongue depressor on the **mucous membranes** between the cheek and the gum with the **gel side** next to the cheek. Repeat until the entire tube has been used.

Comparison Table

	Hyperglycemia	**Hypoglycemia**
History		
Food intake	Excessive	Insufficient
Insulin dosage	Insufficient	Excessive
Onset	Gradual (hours to days)	Rapid, within minutes
Skin	Warm and dry	Pale and moist
Infection	Common	Uncommon
Gastrointestinal Tract		
Thirst	Intense	Absent
Hunger	Absent	Intense
Vomiting	Common	Uncommon
Respiratory System		
Breathing	Rapid, deep (Kussmaul respirations)	Normal or rapid
Odor of breath	Sweet, fruity	Normal
Cardiovascular System		
Blood pressure	Normal to low	Low
Pulse	Normal or rapid and full	Rapid, weak
Nervous System		
Consciousness	Restless merging to coma	Irritability, confusion, seizure, or coma
Urine		
Sugar	Present	Absent
Acetone	Present	Absent
Treatment		
Response	Gradual, within 6 to 12 hours following medical treatment	Immediately after administration of glucose

Assessment Review

1. B (page 494)
2. B (page 494)
3. C (page 494)
4. D (page 494)
5. D (page 494)

Emergency Care Summary

expiration date

generous

third

mucous membranes

cheek

(page 494)

Chapter 16: Allergic Reactions and Envenomations

Matching

1. D (page 500)
2. A (page 500)
3. H (page 500)
4. E (page 500)
5. B (page 505)
6. G (page 500)
7. C (page 502)
8. F (page 502)

Multiple Choice

1. A (page 508)
2. D (page 500)
3. A (page 500)
4. D (page 502)
5. C (pages 505-506)
6. B (page 503)
7. D (page 505)
8. C (page 508)
9. D (page 507)
10. C (page 502)
11. A (page 503)
12. D (page 513)
13. C (page 501)
14. C (page 500)
15. A (page 500)
16. C (page 500)
17. B (page 500)
18. D (page 508)
19. D (page 508)
20. D (page 508)

Fill-in

1. bronchial passages (page 500)
2. urticaria (page 500)
3. barbed (page 501)
4. anaphylactic (page 500)
5. hypoperfusion (page 505)
6. bronchioles (page 507)
7. imminent death (page 500)

True/False

1. T (page 500)
2. T (page 500)
3. T (page 500)
4. F (page 504)

Short Answer

1. Increased blood pressure, tachycardia, pallor, dizziness, chest pain, headache, nausea, vomiting (page 512)

2.
   1. Insect bites/stings
   2. Medications
   3. Plants
   4. Food
   5. Chemicals (pages 500-501)

3.
   1. Obtain order from medical control.
   2. Follow BSI techniques.
   3. Make sure medication was prescribed for that patient.
   4. Check for discoloration or expiration of medications.
   5. Remove cap.
   6. Wipe thigh with alcohol if possible.
   7. Place tip against lateral midthigh.
   8. Push firmly until activation.
   9. Hold in place until medication is injected.
   10. Remove and dispose.
   11. Record the time and dose.
   12. Reassess and record patient's vital signs. (page 508)

4. Respiratory: Sneezing or itchy, runny nose; chest or throat tightness; dry cough; hoarseness; rapid, noisy, or labored respirations; wheezing and/or stridor
   Circulatory: Decreased blood pressure; increased pulse (initially); pale skin and dizziness; loss of consciousness and coma (page 506)

Word Fun

Ambulance Calls

1. Scene safety is not an issue in this case as it was a police dog that caused the patient's injuries. Rabies is a non-issue as well, because this dog has been inoculated. Because there is still the chance for infection and the need for wound care/sutures, the patient should be evaluated by a physician. You should cover the wound with sterile dressings and provide transport to the emergency department.

2. Based on the information you have, there is no way to determine if the child fell out of the tree or climbed down. You should assume that the child fell, which will require the use of full spinal precautions. You also have the issue of scene safety, as the damaged hive is now lying on the ground next to the patient. You must take care not to receive multiple stings yourself, so proceed with caution. If the child is having a severe allergic reaction, hopefully you will have access to an EpiPen Junior. Provide high flow oxygen and prompt transport according to local protocols.

3. –Check the EpiPen for clarity, expiration date, etc.
   –Obtain a physician's order to administer the EpiPen to the patient.
   –Administer the EpiPen and promptly dispose of auto-injector.
   –Apply high-flow oxygen.
   –Rapid transport.
   –Monitor the patient and assess vital signs frequently.

Skill Drills

Skill Drill 16-1: Using an Auto-injector (page 509)

1. Remove the auto-injector's **safety cap**, and quickly wipe the thigh with **antiseptic**.
2. Place the tip of the auto-injector against the **lateral** part of the thigh.
3. Push the **injector** firmly against the **thigh**, and hold it in place until all the medication is injected.

Skill Drill 16-2: Using an AnaKit (pages 510-511)

1. Prepare the injection site with antiseptic, and remove the needle cover.
2. Hold the syringe upright, and carefully use the plunger to remove air.
3. Turn the plunger one-quarter turn.
4. Quickly insert the needle into the muscle.
5. Hold the syringe steady, and push the plunger until it stops.
6. Have the patient chew and swallow the Chlo-Amine antihistamine tablets provided in the kit.
7. If available, apply a cold pack to the sting site.

Assessment Review

1. A (page 513)
2. D (page 513)
3. D (page 513)
4. C (page 513)
5. D (page 513)

Emergency Care Summary
Using an Auto-injector
lateral

10

Using an AnaKit
upright

one-quarter

muscle

antihistamine

cold pack

(page 513)

Chapter 17: Substance Abuse and Poisoning

Matching

1. F (page 518)
2. H (page 518)
3. G (page 519)
4. D (page 528)
5. J (page 533)
6. I (page 518)
7. K (page 529)
8. E (page 531)
9. B (page 518)
10. A (page 528)
11. C (page 532)

Multiple Choice

1. B (page 523)
2. B (page 518)
3. D (page 536)
4. C (page 533)
5. D (pages 518-519)
6. C (page 520)
7. C (page 527)
8. B (pages 528-529)
9. C (page 529)
10. D (page 530)
11. A (page 532)
12. C (page 531)
13. D (page 533)
14. B (page 535)
15. A (page 521)
16. D (page 531)
17. D (page 533)
18. D (page 532)
19. B (page 531)
20. D (page 521)
21. C (page 521)
22. B (page 521)
23. C (page 523)
24. D (page 522)
25. A (page 524)
26. D (page 528)
27. D (page 529)
28. A (page 521)
29. D (page 521)
30. D (page 522)
31. D (page 523)
32. A (page 520)
33. C (page 520)
34. C (page 538)
35. D (page 527)
36. B (page 527)
37. C (page 527)
38. D (page 527)

Fill-in

1. ABCs (pages 521-522)
2. alcohol (page 528)
3. adsorbing (page 527)
4. 5 to 10 minutes, 15 to 20 minutes (page 522)
5. respiratory depression (page 529)
6. hypoglycemia (page 528)
7. recognize (page 520)
8. 1 gram, kilogram (page 527)
9. outward (page 522)
10. ingestion (page 522)
11. delirium tremens (DTs) (page 529)
12. ignite (page 522)
13. addiction (page 522)
14. Hypovolemia (page 529)

True/False

1. T (page 527)
2. F (pages 520-523)
3. F (page 527)
4. F (page 521)
5. F (page 523)

6. T (page 529)
7. T (page 533)
8. F (page 528)
9. T (page 529)
10 T (page 532)

Short Answer

1. Activated charcoal adsorbs (binds to) the toxin and keeps it from being absorbed in the gastrointestinal tract. (page 523)
2. 1. Ingestion
   2. Inhalation
   3. Injection
   4. Absorption (page 521)
3. Hypertension, tachycardia, dilated pupils, and agitation/seizures (page 531)
4. 1. The organism itself causes the disease.
   2. The organism produces toxins that cause disease. (page 534)
5. Symptoms of acetaminophen overdose do not appear until the damage is irreversible, up to a week later. Finding evidence at the scene can save a patient's life. (page 533)
6. They describe patient presentation in cholinergic poisoning (organophosphate insecticides, wild mushrooms).
   DUMBELS: Defecation, urination, miosis, bronchorrhea, emesis, lacrimation, salivation
   SLUDGE: Salivation, lacrimation, urination, defecation, gastrointestinal irritation, eye constriction/emesis (page 533)
7. 1. Opioid analgesics
   2. Sedative-hypnotics
   3. Inhalants
   4. Sympathomimetics
   5. Hallucinogens
   6. Anticholinergic agents
   7. Cholinergic agents (pages 528-533)
8. 1. What substance did you take?
   2. When did you take it or become exposed to it?
   3. How much did you ingest?
   4. What actions have been taken?
   5. How much do you weigh? (page 518)
9. Because they ignite when they come into contact with water (page 522)

Word Fun

Ambulance Calls

1. You should attempt to identify the substance. Some rat poisons are actually "blood thinning agents" or anticoagulants, such as Coumadin. You should collect the substance and call the poison control center (**1-800-222-1222**) and/or the hospital emergency department for patient care instructions. Some substances require the administration of activated charcoal, while others do not. Perform your initial assessment, and provide high flow oxygen (blow by) and prompt transport. Know your local protocols.

2. This patient obviously abuses alcohol and illegal substances. However, you cannot automatically assume that his decrease in mentation is directly related to alcohol intoxication or influence of other substances. He may have other medical conditions, which may mimic intoxication or even be obscured by it. You should perform a thorough assessment including vital signs, monitor his ABCs (as these could change at any time), and transport to the nearest appropriate medical facility for evaluation. It is also important to be aware of the possibility of used needles when performing assessments and/or removal of clothing when visualizing any potential injuries. Protect yourself.

3. Maintain the airway with an adjunct and high-flow oxygen via BVM device or nonrebreathing mask with 100% oxygen.
Monitor vital signs and provide supportive measures.
Provide rapid transport.
Take the pill bottle along to the emergency department.
Be alert for possible vomiting.
Monitor patient closely and be prepared for the possible need for CPR.

Assessment Review

1. B (page 538)
2. D (page 538)
3. C (page 538)
4. B (page 538)
5. A (page 538)

Emergency Care Summary

suctioning

high-flow oxygen

SAMPLE

vital signs

ALS

hallucinogens

cholinergic

plant

food

activated charcoal

altered mental status

swallow

12.5

50

Chapter 18: Environmental Emergencies

Matching

1. J (page 545)
2. H (page 564)
3. K (page 545)
4. F (page 554)
5. A (page 562)
6. N (page 546)
7. I (page 545)
8. C (page 560)
9. B (page 554)
10. E (page 545)
11. G (page 545)
12. M (page 553)
13. L (page 554)
14. D (page 560)

Multiple Choice

1. D (page 545)
2. A (page 545)
3. D (page 545)
4. C (page 546)
5. C (page 547)
6. D (page 546)
7. A (page 546)
8. C (page 547)
9. D (page 551)
10. D (pages 551-552)
11. A (page 554)
12. D (page 554)
13. D (page 554)
14. A (page 554)
15. D (page 555)
16. A (page 557)
17. A (page 557)
18. D (page 557)
19. B (page 560)
20. D (page 566)
21. C (page 562)
22. D (page 566)
23. B (page 562)
24. B (page 564)
25. A (page 564)
26. D (page 564)
27. A (page 564)
28. D (page 569)
29. A (page 573)
30. A (page 574)
31. D (page 572)
32. C (page 574)
33. D (page 569)
34. D (page 572)
35. D (page 555)
36. B (page 546)
37. A (page 545)
38. B (page 546)
39. D (page 551)

Fill-in

1. moderate to severe (pages 549-550)
2. ascent (page 564)
3. rewarming (page 552)
4. self-protection (page 553)
5. Shivering (page 564)
6. diving reflex (page 567)

True/False

1. T (page 553)
2. T (page 547)
3. F (page 546)
4. T (page 553)
5. T (page 554)
6. F (page 555)
7. T (page 567)
8. F (page 572)
9. F (page 572)
10. T (page 572)
11. F (page 574)
12. F (page 574)

Short Answer

1. **1.** Increase heat production: shiver, jump, walk around, etc.
    2. Move to another area where heat loss decreases: out of wind, into sun, etc.
    3. Wear insulated clothing: layer with wool, down, synthetics, etc. (pages 545-546)

2. **1.** Move the patient out of the hot environment and into the ambulance.
    2. Set air conditioning to maximum cooling.
    3. Remove patient's clothing.
    4. Administer high-flow oxygen.
    5. Apply cool packs to patient's neck, groin, and armpits.
    6. Cover patient with wet towels, or spray with cool water and fan.
    7. Keep fanning.
    8. Transport immediately.
    9. Notify the hospital. (pages 555-556)

3. An air embolism is a bubble of air in the blood vessels caused by breath-holding during rapid ascent. The resulting high pressure in the lungs causes alveolar rupture. (page 564)

4. Treatment of air embolism and decompression sickness (page 565)

5. **1.** Remove the patient from the cold.
    2. Handle injured part gently and protect from further injury.
    3. Administer oxygen.
    4. Remove wet or restricting clothing. (page 552)

6. **1.** Do not break blisters.
    2. Do not rub or massage area.
    3. Do not apply heat or rewarm unless instructed by medical control.
    4. Do not allow patient to stand or walk on a frostbitten foot. (page 552)

7. **1.** Have the patient lie flat and stay quiet.
    2. Wash the bite area with soapy water.
    3. Splint the extremity.
    4. Mark the skin with a pen to monitor advancing swelling. (page 572)

8. Black widow: Bite has a systemic effect (venom is neurotoxic).
    Brown recluse: Bite destroys tissue locally (venom is cytotoxic). (pages 569-570)

Word Fun

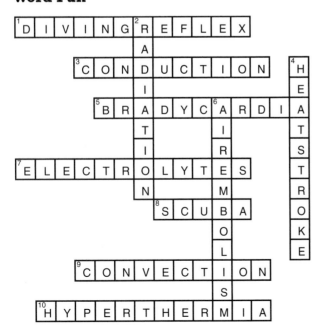

Ambulance Calls

1. Provide BLS.
   Administer oxygen.
   Transport the patient in the left lateral recumbent position with the head down.
   Transport to a facility with hyperbaric chamber access.

2. Your main concern for this patient is hypothermia. Although this patient could also likely have localized cold injuries such as frostbite or frostnip, hypothermia can be fatal. You must handle this patient carefully, remove them from the cold environment, remove any wet clothing, and prevent further heat loss. Assessing the extent of the hypothermia through mentation will be difficult in this case, as your patient is likely confused. Take note of the presence of shivering (this protective mechanism stops at core temperatures <90° F), and if possible take their temperature (rectally). Assess airway, breathing and circulation; provide warm, humidified oxygen (if possible) and passive rewarming measures, such as increasing the heat in the patient compartment and prompt transport.

3. This patient has severe dehydration. Her use of diuretics and laxatives combined with strenuous exercise and a "body wrap," have depleted her body's volume. There is also the possibility that she is suffering from hyperthermia, as her lack of fluids will affect her body's thermoregulatory mechanisms. She needs fluid replacement therapy and assessment of her core temperature. You should remove excess layers of clothing (in this case, the body wrap), give high flow oxygen, encourage oral fluid replacement if fully alert, and provide non-aggressive cooling measures.

Skill Drills

Skill Drill 18-1: Treating for Heat Exhaustion (page 556)

1. Remove **extra clothing**.

2. Move the patient to a **cooler environment**.
   Give **oxygen**.
   Place the patient in a **supine** position, elevate the legs, and **fan** the patient.

3. If the patient is **fully alert**, give water by mouth.

4. If nausea develops, **transport** on the side.

Skill Drill 18-2: Stabilizing a Suspected Spinal Injury in the Water (page 563)

1. Turn the patient to a supine position by rotating the entire upper half of the body as a single unit.

2. As soon as the patient is turned, begin artificial ventilation using the mouth-to-mouth method or a pocket mask.

3. Float a buoyant backboard under the patient.

4. Secure the patient to the backboard.

5. Remove the patient from the water.

6. Cover the patient with a blanket and apply oxygen if breathing. Begin CPR if breathing and pulse are absent.

Assessment Review

1. B (page 576)

2. A (page 576)

3. B (page 576)

4. B (page 576)

5. C (page 578)

Emergency Care Summary
Cold Injuries
Insulate

convection

cold environment

Heat Injuries
oxygen

supine

elevated

fan

liter

nausea

left-lateral

Drowning and Diving Injuries
supine

pocket-mask

backboard

oxygen

Lightning Injuries
conscious

cardiac arrest

cardiac arrest

Spider Bites
BLS

rigidity

spider

Snake Bites
plunger

one-quarter

muscle

antihistamine

cold pack

Injuries From Marine Animals
discharge

reduce

vinegar

credit card

(page 578)

Chapter 19: Behavioral Emergencies

Matching

1. C (page 584)
2. F (page 588)
3. B (page 586)
4. D (page 585)
5. E (page 586)
6. A (page 584)

Multiple Choice

1. D (page 584)
2. B (page 584)
3. D (page 584)
4. B (page 585)
5. A (page 585)
6. B (page 585)
7. A (page 584)
8. D (page 585)
9. C (page 586)
10. D (page 586)
11. A (page 586)
12. B (page 590)
13. D (page 586)
14. B (page 587)
15. D (page 588)
16. C (page 590)
17. A (page 590)
18. D (page 592)
19. D (page 592)
20. C (page 593)
21. B (page 593)
22. D (page 586)
23. A (page 585)
24. D (page 586)
25. D (page 592)
26. B (page 593)

Fill-in

1. Behavior (page 584)
2. behavioral crisis (page 584)
3. depression (page 585)
4. Organic brain syndrome (page 586)
5. suicide (page 590)

True/False

1. T (page 585)
2. F (page 586)
3. F (page 585)
4. F (page 593)
5. F (page 586)
6. F (page 589)
7. F (page 590)
8. T (page 592)
9. F (page 584)
10. F (page 590)
11. T (page 593)

Short Answer

1. A behavioral crisis is a temporary change in behavior that interferes with ADL or that is unacceptable to the patient or others. A mental health problem is this kind of behavioral change recurring on a regular basis. (page 584-585)

2.
   1. Improper functioning of the central nervous system.
   2. Drugs or alcohol
   3. Psychogenic circumstances (page 586)

3.
   1. The degree of force necessary to keep the patient from injuring self or others
   2. Patient's gender, size, strength, and mental status
   3. The type of abnormal behavior the patient is exhibiting (page 593)

4.
   1. Be prepared to spend extra time.
   2. Have a definite plan of action.
   3. Identify yourself calmly.
   4. Be direct.
   5. Assess the scene.
   6. Stay with the patient.
   7. Encourage purposeful movement.
   8. Express interest in the patient's story.
   9. Do not get too close to the patient.
   10. Avoid fighting with the patient.
   11. Be honest and reassuring.
   12. Do not judge. (page 586)

5.
   1. Depression at any age
   2. Previous suicide attempt
   3. Current expression of wanting to commit suicide or sense of hopelessness
   4. Family history of suicide
   5. Age older than 40 years, particularly for single, widowed, divorced, alcoholic, or depressed individuals
   6. Recent loss of spouse, significant other, family member, or support system
   7. Chronic debilitating illness or recent diagnosis of serious illness
   8. Holidays
   9. Financial setback, loss of job, police arrest, imprisonment, or some sort of social embarrassment
   10. Substance abuse, particularly with increasing usage
   11. Children of an alcoholic parent
   12. Severe mental illness
   13. Anniversary of death of loved one, job loss, marriage, etc.
   14. Unusual gathering or new acquisition of things that can cause death, such as purchase of a gun, a large volume of pills, or increased use of alcohol (page 590)

Word Fun

A crossword puzzle with the following filled-in answers:

- Down 1: ALTERBDMENTAL... (Column starting with A-L-T-E-R, continuing B-D-M-E-N-T-A-L)
- Vertical word 1: A L T E R (then) B D M E N T A L P S T A T U S
- Vertical word 2: F U N C T I O N A L D I S O R D E R
- 3 Across: B E H A V I O R A L C R I S I S
- 5 Across: M E N T A L D I S O R D E R
- 6 Across: P S Y C H O G E N I C
- 7 Across: D E P R E S S I O N

Ambulance Calls

1. Removing restraints (especially when the patient has a known history of recent violence) is ill advised and potentially against local protocols. Restraints, although uncomfortable, afford you and your patient a safe environment. You should continually monitor the restraints to ensure they are not too tight and that no manner of restraints (whether applied initially by you in the field or by hospital personnel for an interfacility transport) affects the patient's ability to breathe. Know your local laws regarding the use of restraints and your local protocols for appropriate use and discontinuation.

2. Although he tells you that he is fine, you cannot simply walk away. Those individuals who contemplate suicide often tell people they are "fine." With information passed along by his sister, you must take action. Because there is no evidence that the patient has tried to harm himself and he did not express directly to you that he intends to, you must use persuasive techniques to gain consent to treatment and transport.

3. –Be understanding and listen.
   –Explain to the patient that she needs medical care.
   –Monitor vital signs and reassure patient en route.

Assessment Review

1. D (page 595)   4. D (page 595)
2. A (page 595)   5. D (page 595)
3. D (page 595)

Emergency Care Summary

control
safely transport
tension
emotional distress
violence
(page 595)

Chapter 20: Obstetric and Gynecologic Emergencies

Matching

1. D (page 600)
2. C (page 600)
3. L (page 600)
4. B (page 601)
5. N (page 600)
6. H (page 600)
7. M (page 600)
8. E (page 601)
9. F (page 600)
10. O (page 616)
11. K (page 618)
12. I (page 602)
13. A (page 611)
14. G (page 616)
15. J (page 603)

Multiple Choice

1. D (page 611)
2. B (page 605)
3. D (page 612)
4. B (page 613)
5. A (page 614)
6. C (page 615)
7. D (page 615)
8. D (page 616)
9. A (page 618)
10. D (page 618)
11. B (page 620)
12. D (page 602)
13. B (page 602)
14. D (page 602)
15. A (page 602)
16. C (page 602)
17. B (page 603)
18. B (page 603)
19. D (page 602)
20. C (page 602)
21. D (page 603)
22. D (page 618)
23. C (page 603)
24. B (page 603)
25. A (page 603)
26. C (page 620)
27. D (page 602)
28. C (page 608)
29. C (page 602)
30. C (page 611)
31. D (page 613)

Tables

Apgar Scoring System

Area of Activity	Score		
	2	1	0
Appearance	Entire infant is pink.	Body is pink, but hands and feet remain blue.	Entire infant is blue and pale.
Pulse	More than 100 beats/min	Fewer than 100 beats/min	Absent pulse.
Grimace or Irritability	Infant cries and tries to move foot away from finger snapped against its sole.	Infant gives a weak cry in response to stimulus.	Infant does not cry or react to stimulus.
Activity or Muscle Tone	Infant resists attempts to straighten out hips and knees.	Infant makes weak attempts to resist straightening.	Infant is completely limp, with no muscle tone.
Respiration	Rapid respirations	Slow respirations	Absent respirations

Labeling

1. Anatomic Structures of the Pregnant Woman (page 600)

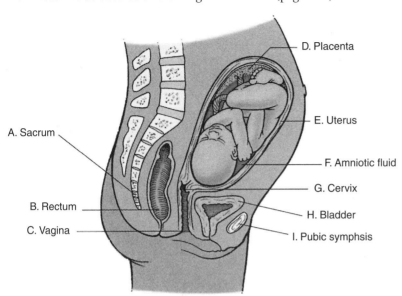

A. Sacrum

B. Rectum

C. Vagina

D. Placenta

E. Uterus

F. Amniotic fluid

G. Cervix

H. Bladder

I. Pubic symphsis

Fill-in

1. placenta (page 613)
2. arteries, vein (page 601)
3. 500 to 1000 mL (page 601)
4. 36, 40 (page 602)
5. trimesters (page 602)
6. body fluids (page 603)
7. ectopic pregnancy (page 603)
8. resuscitate (page 604)
9. fontanels (page 611)

True/False

1. T (page 600)
2. F (page 602)
3. F (page 602)
4. F (page 602)
5. F (page 607)
6. T (page 616)
7. T (page 612)
8. F (page 613)
9. F (page 613)
10. T (page 618)
11. T (page 620)

Short Answer

1. Early: spontaneous abortion (miscarriage) or ectopic pregnancy
   Later: Placenta previa or placenta abruptio (page 603)
2. On the left side, to prevent supine hypotensive syndrome (low blood pressure occurring from the weight of the fetus compressing the inferior vena cava) (page 603)

3. **1.** Uterine contractions

   2. Bloody show

   3. Rupture of amniotic sac (page 602)

4. **1.** When delivery can be expected in a few minutes.

   2. When some natural disaster or catastrophe makes it impossible to reach the hospital.

   3. When no transportation is available. (page 607)

5. Exert gentle pressure on the head as it emerges to prevent rapid expulsion with a strong contraction. (page 607)

6. The brain is covered by only skin and membrane at the fontanels. (page 611)

7. Exerting gentle pressure horizontally across the perineum with a sterile gauze pad may reduce the risk of perineal tearing. (page 611)

8. **1.** During a breech delivery to protect the infant's airway

   2. When the umbilical cord is prolapsed (pages 617-618)

9. **1.** Prematurity

   2. Low birth weight

   3. Severe respiratory depression (page 620)

Word Fun

Ambulance Calls

1. This patient has classic signs of pre-eclampsia. If she does not receive medical care soon to control her hypertension, she will likely experience seizure activity. Provide high flow oxygen, obtain a set of vital signs (especially blood pressure), and provide prompt transport with the patient on her left side.

2. Recent laws have been enacted to provide protection for new mothers who do not wish to keep their babies. They can release their newborns and infants to fire stations and hospitals without fear of criminal charges (as the young woman in this scenario chose to do). You should notify other providers in your station as you begin measures to dry, warm, stimulate, and suction the baby's airway. These measures are very effective in improving a newborn's oxygenation and perfusion status. Blow-by oxygen will help immensely, but if those measures fail to quickly improve the newborn's status, bag valve mask ventilations should be initiated. Provide chest compressions if the newborn's heart rate is <100 beats/min, and take note of the condition of the umbilical cord to ensure no blood loss occurs. Provide prompt transport and notify the hospital of the incoming patient.

3. –Position the mother for delivery and apply high-flow oxygen.
   –As crowning occurs, use a clamp to puncture the sac, away from the baby's face.
   –Push the ruptured sac away from the infant's face as the head is delivered.
   –Clear the baby's mouth and nose immediately.
   –Continue with the delivery as normal.

Skill Drills

Skill Drill 20-1: Delivering the Baby (page 610)

1. Support the **bony** parts of the head with your hands as it emerges. Suction fluid from the **mouth**, then **nostrils**.
2. As the **upper shoulder** appears, guide the head **down** slightly, if needed, to deliver the **shoulder**.
3. Support the head and upper body as the **lower shoulder** delivers, guiding the head **up** if needed.
4. Handle the slippery, delivered infant firmly but gently, keeping the neck in **neutral** position to **maintain** the airway.
5. Place the umbilical cord clamps **2"** to **4"** apart and **cut** between them.
6. Allow the **placenta** to deliver itself. Do not pull on the **cord** to speed delivery.

Skill Drill 20-2: Giving Chest Compressions to an Infant (page 617)

1. Find the proper position: just below the **nipple line**, middle or **lower third** of the sternum.
2. Wrap your hands around the body, with your **thumbs** resting at that position.
3. Press your thumbs gently against the sternum, compressing 1/2" to 3/4" deep.

Assessment Review

1. A (page 624)
2. A (page 624)
3. A (page 623)
4. D (page 623)
5. B (page 623)

Emergency Care Summary
Delivering the Baby
bony

mouth

nostrils

down

up

neutral

2, 4

Giving Chest Compressions to an Infant
middle

lower

1/2, 3/4

Premature Infant
90, 95

32.2, 35

bulb syringe

bleeding

oxygen

(page 624)

Chapter 21: Kinematics of Trauma

Matching

1. H (page 643)
2. G (page 638)
3. F (page 632)
4. E (page 632)

5. C (page 633)
6. A (page 633)
7. D (page 633)
8. B (page 632)

Multiple Choice

1. B (page 632)
2. C (page 632)
3. C (page 632)
4. B (page 633)
5. A (page 633)
6. D (pages 637-640)
7. C (pages 637-640)
8. C (page 633)
9. D (page 636)
10. A (page 636)
11. C (page 636)
12. B (page 636)
13. D (page 638)

14. A (page 638)
15. D (page 638)
16. B (page 638)
17. C (page 638)
18. B (page 640)
19. C (page 640)
20. B (page 641)
21. C (page 634)
22. A (page 634)
23. B (page 636)
24. C (page 635)
25. A (page 632)

Fill-in

1. Injuries (page 632)
2. patterns, events (page 632)
3. Traumatic injury (page 632)
4. $KE = 1/2 \, MV^2$ (page 634)
5. Polaroids (page 637)
6. deceleration (page 638)
7. submarining (page 638)

True/False

1. T (page 632)
2. F (page 632)
3. F (page 633)
4. T (page 639)
5. T (page 639)
6. T (page 633)
7. T (page 632)

Short Answer

1. Potential energy is the product of mass (weight), force of gravity, and height, and is mostly associated with the energy of falling objects. (page 633)

2. **1.** Collision of the car against another car or other object
   2. Collision of the passenger against the interior of the car
   3. Collision of the passenger's internal organs against the solid structures of the body (pages 637-639)

3. **1.** The height of the fall
   2. The surface struck
   3. The part of the body that hits first, followed by the path of energy displacement (page 642)

4. A bullet, because of its speed, creates pressure waves that emanate from its path, causing distant damage. (page 643)

5. The size (mass) and speed (velocity) of the projectile affect the potential damage. If the mass is doubled, the potential energy is doubled. If the velocity is doubled, the potential energy is quadrupled. (page 633)

Word Fun

Crossword puzzle solution:

Down 1: MECHANISM
Across 3: KINETIC ENERGY
Down 2: POTENTIAL ENERGY
Across 6: BLUNT TRAUMA
Down 3: KSOFINJUY
Down 4: TRAUMATIC INJUNCTION
Down 5: CAVITATION
Across 7: DECELERATION
Across 8: WORK

Ambulance Calls

1. Given the highway speeds and lack of a shoulder belt and airbag along with his complaints of head and neck pain, your index of suspicion for head and spinal injuries is very high. It is a positive sign that he is awake and able to communicate, however, this should not encourage you to spend any more time on scene than is necessary to extricate this patient and place him in full spinal precautions. His condition could change at any time, and his inability to remember the details of the event likely indicate the presence of a closed head injury. Provide high-flow oxygen and prompt transport.

2. Each story is 10 feet, so this patient fell approximately 20 feet from the ladder to the hard ground (assuming you live in an area with cold winter weather). He is now unconscious and lying in the snow. He not only has significant injuries, but now the high possibility of hypothermia, the severity of which will depend upon the length of time he has been exposed to the elements as well as local weather conditions. Assume he has significant head and spinal injuries; determine if he is responsive, and manage his airway, as he will be unable to protect it. Provide high-flow oxygen and prompt transport to the nearest appropriate facilities, taking care not to waste time in determining other injuries.

3. –Apply high-flow oxygen.
   –Stabilize object in place with bulky dressings.
   –Monitor vital signs.
   –Transport in a supine position.
   –Rapid transport due to abdominal penetration.

Chapter 22: Bleeding

Matching

1. J (page 650)	**8.** I (page 668)
2. E (page 650)	**9.** M (page 666)
3. K (page 650)	**10.** A (page 668)
4. F (page 651)	**11.** D (page 656)
5. C (page 650)	**12.** L (page 654)
6. H (page 650)	**13.** G (page 655)
7. B (page 655)	

Multiple Choice

1. D (page 652)	**20.** B (page 655)
2. D (page 650)	**21.** C (page 655)
3. C (page 650)	**22.** C (page 655)
4. C (page 650)	**23.** D (page 656)
5. B (page 650)	**24.** B (page 656)
6. A (page 650)	**25.** C (page 659)
7. D (page 650)	**26.** A (page 660)
8. B (page 651)	**27.** A (page 660)
9. D (page 652)	**28.** C (page 661)
10. C (page 652)	**29.** D (page 662)
11. B (page 653)	**30.** C (page 663)
12. C (page 653)	**31.** D (page 656)
13. C (page 653)	**32.** D (page 666)
14. A (page 653)	**33.** B (page 666)
15. D (page 654)	**34.** B (page 667)
16. B (page 654)	**35.** D (page 668)
17. B (page 655)	**36.** D (page 668)
18. D (page 655)	**37.** D (page 668)
19. A (page 655)	**38.** C (page 669)

Labeling

1. The left and right sides of the heart (page 651)

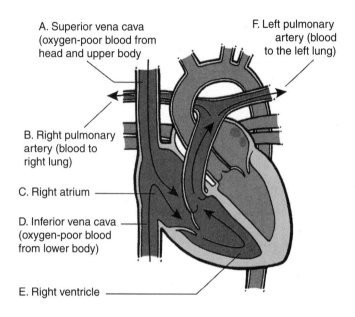

A. Superior vena cava (oxygen-poor blood from head and upper body

F. Left pulmonary artery (blood to the left lung)

B. Right pulmonary artery (blood to right lung)

C. Right atrium

D. Inferior vena cava (oxygen-poor blood from lower body)

E. Right ventricle

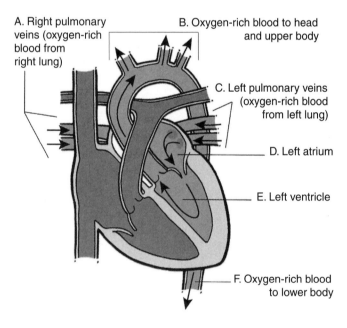

A. Right pulmonary veins (oxygen-rich blood from right lung)

B. Oxygen-rich blood to head and upper body

C. Left pulmonary veins (oxygen-rich blood from left lung)

D. Left atrium

E. Left ventricle

F. Oxygen-rich blood to lower body

2. Perfusion (page 653)

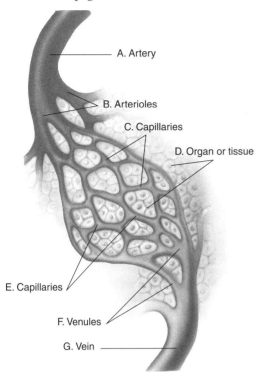

A. Artery
B. Arterioles
C. Capillaries
D. Organ or tissue
E. Capillaries
F. Venules
G. Vein

3. Arterial pressure points (page 662)

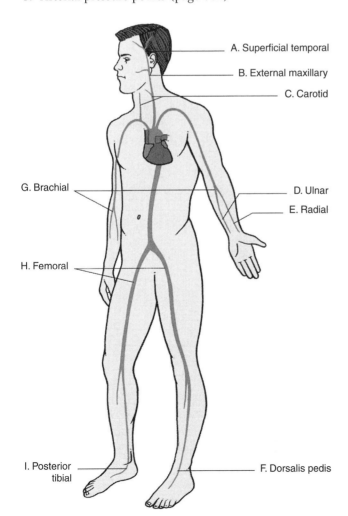

A. Superficial temporal
B. External maxillary
C. Carotid
G. Brachial
D. Ulnar
E. Radial
H. Femoral
I. Posterior tibial
F. Dorsalis pedis

Fill-in

1. perfusion (page 653)
2. involuntary (page 650)
3. lungs (page 653)
4. hypoperfusion (page 653)
5. inferior (page 651)
6. white cells, red cells, platelets, and plasma (page 652)
7. oxygenated (page 650)
8. heart, brain, lungs, and kidneys (page 653)
9. 4 to 6 (page 654)
10. rapid trauma assessment (page 658)

True/False

1. F (page 655)
2. F (page 655)
3. F (page 659)
4. T (page 659)
5. F (page 666)
6. T (page 666)
7. T (page 661)
8. F (page 660)
9. F (page 663)
10. T (page 662)

Short Answer

1. It redirects blood away from nonessential organs to the heart, brain, lungs, and kidneys. (page 653)
2. Artery: Bright red, spurting

   Vein: Dark color with steady flow

   Capillary: Darker color, oozes (page 655)
3.
   1. Direct pressure and elevation
   2. Pressure dressings
   3. Pressure points (for upper and lower extremities)
   4. Splints
   5. Air splints
   6. PASG
   7. Tourniquets (page 659)
4.
   1. Change in mental status (restlessness, anxiety, combativeness)
   2. Weakness, fainting, or dizziness on standing (early sign) or at rest (later sign)
   3. Tachycardia
   4. Thirst
   5. Nausea and vomiting
   6. Cold, moist (clammy) skin
   7. Shallow, rapid breathing
   8. Dull eyes
   9. Slightly dilated pupils, slow to respond to light
   10. Capillary refill in infants and children of more than 2 seconds
   11. Weak, rapid (thready) pulse
   12. Decreasing blood pressure
   13. Altered level of consciousness (page 669)
5.
   1. BSI
   2. Maintain the airway.
   3. Administer high-flow oxygen.
   4. Control all obvious external bleeding.

5. Apply splints to extremity, if a limb is involved.
6. Monitor and record the patient's vital signs.
7. Give the patient nothing by mouth.
8. Elevate the legs 6 to 12 inches in nontrauma patients.
9. Keep the patient warm.
10. Provide immediate transport. (page 670)

Word Fun

Ambulance Calls

1. Control bleeding with direct pressure, elevation, pressure point, and a tourniquet as a LAST resort.
   Apply high-flow oxygen.
   Place patient in the position of comfort.
   Monitor vital signs.
   Provide rapid transport.

2. You must control bleeding through direct pressure, elevation, pressure dressings and pressure points as needed. You must also transport the amputated portion of the limb with the patient to the hospital. Quickly attempt to determine how much blood has been lost and assess his skin and vital signs as these will provide accurate indicators of the significance of blood loss. Place the patient in the Trendelenburg position as needed; apply high-flow oxygen and provide prompt transport.

3. This is an isolated injury that, depending upon the severity of the laceration to the antecubital vein, can result in significant blood loss. Attempt to control bleeding through direct pressure, elevation, pressure dressings and pressure points as needed. Something as benign as a pen can cause significant damage in the hands of a determined person.

Skill Drills

Skill Drill 22-1: Controlling External Bleeding (page 661)

1. Apply **direct pressure** over the wound.
   Elevate the injury above the **level** of the **heart** if no **fracture** is suspected.
2. Apply a **pressure dressing**.
3. Apply pressure at the appropriate **pressure point** while continuing to hold **direct pressure**.

Skill Drill 22-2: Applying a Pneumatic Antishock Garment (PASG) (page 664)

1. Apply the garment so that the top is below the lowest rib.
2. Enclose both legs and the abdomen.
3. Open the stopcocks.
4. Inflate with the foot pump, and close the stopcocks when the patient's systolic blood pressure reaches 100 mm Hg or the Velcro crackles.
5. Check the patient's blood pressure again. Monitor the vital signs.

Assessment Review

1. C (page 672)
2. A (page 672)
3. B (page 672)
4. D (page 672)
5. C (page 672)

Emergency Care Summary

External Bleeding

cervical

Controlling External Bleeding

pressure

Using PASG for Control of Massive Soft-Tissue Bleeding in the Extremities

air splint

Applying a Tourniquet

twice

square

Treating Epistaxis

15

shock

Internal Bleeding
Steps to Caring for Patient with Internal Bleeding

splint

nothing

Using PASG for Treatment of Shock

abdominal

response

(page 673)

Chapter 23: Shock

Matching

1. B (page 678)
2. G (page 678)
3. H (page 678)
4. C (page 678)
5. E (page 681)
6. A (page 681)
7. F (page 682)
8. I (page 683)
9. D (page 684)

Multiple Choice

1. D (page 678)
2. B (page 678)
3. D (page 680)
4. C (page 680)
5. D (page 680)
6. D (page 680)
7. D (page 680)
8. C (page 683)
9. A (page 681)
10. B (page 681)
11. C (page 690)
12. A (page 682)
13. C (page 682)
14. D (page 682)
15. A (page 682)
16. B (page 682)
17. A (page 683)
18. A (page 683)
19. D (page 683)
20. B (page 684)
21. A (page 683)
22. B (page 688)
23. C (page 688)
24. B (page 689)
25. D (page 694)
26. B (page 691)

Fill-in

1. Hypoperfusion (page 678)
2. contraction (page 678)
3. nonessential, essential (page 678)
4. perfusion (page 678)
5. heart, vessels, blood (page 678)
6. shock (hypoperfusion) (page 678)
7. Sphincters, contract, dilate (page 678)
8. Diastolic, systolic (page 678)
9. Blood (page 678)
10. involuntary (page 680)

True/False

1. T (page 683) **6.** T (page 691)

2. T (page 682) **7.** T (page 682)

3. T (page 678) **8.** F (page 678)

4. F (page 678) **9.** F (page 680)

5. F (page 684) **10.** F (page 684)

Short Answer

1. Causes: allergic reaction (most severe form)
   Signs/Symptoms: Can develop within seconds; mild itching/rash; burning skin; vascular dilation; generalized edema; profound coma; rapid death
   Treatment: Manage airway. Assist ventilations. Administer high-flow oxygen. Determine cause. Assist with administration of epinephrine. Transport promptly.
   (page 689)

2. Causes: Inadequate heart function; disease of muscle tissue; impaired electrical system; disease or injury
   Signs/Symptoms: Chest pains; irregular pulse; weak pulse; low blood pressure; cyanosis (lips, under nails); anxiety
   Treatment: Position comfortably. Administer oxygen. Assist ventilations. Transport promptly.
   (page 689)

3. Causes: Loss of blood or fluid
   Signs/Symptoms: Rapid, weak pulse; low blood pressure; change in mental status; cyanosis (lips, under nails); cool, clammy skin; increased respiratory rate
   Treatment: Secure airway. Assist ventilations. Administer high-flow oxygen. Control external bleeding. Elevate legs. Keep warm. Transport promptly.
   (page 689)

4. Causes: Excessive loss of fluid and electrolytes due to vomiting, urination, or diarrhea
   Signs/Symptoms: Rapid, weak pulse; low blood pressure; change in mental status; cyanosis (lips, under nails); cool, clammy skin; increased respiratory rate
   Treatment: Secure airway. Assist ventilations. Administer high-flow oxygen. Determine illness. Transport promptly.
   (page 689)

5. Causes: Damaged cervical spine, which causes widespread blood vessel dilation
   Signs/Symptoms: Bradycardia (slow pulse); low blood pressure; signs of neck injury
   Treatment: Secure airway. Spinal immobilization. Assist ventilations. Administer high-flow oxygen. Transport promptly.
   (page 689)

6. Causes: Temporary, generalized vascular dilation; anxiety; bad news; sight of injury/blood; prospect of medical treatment; severe pain; illness; tiredness
   Signs/Symptoms: Rapid pulse; normal or low blood pressure
   Treatment: Determine duration of unconsciousness. Record initial vital signs and mental status. Suspect head injury if patient is confused or slow to regain consciousness. Transport promptly.
   (page 689)

7. Causes: Severe bacterial infection
   Signs/Symptoms: Warm skin; tachycardia; low blood pressure
   Treatment: Transport promptly. Administer oxygen en route. Provide full ventilatory support. Elevate legs. Keep patient warm. (page 689)

8. **1.** Poor pump function
   2. Blood or fluid loss from blood vessels
   3. Poor vessel function (blood vessels dilate) (page 681)

9. Falling blood pressure, labored or irregular breathing, ashen, mottled, or cyanotic skin, thready or absent peripheral pulses, dull eyes or dilated pupils, and poor urinary output. (page 684)

10. **1.** Remove the autoinjector's safety cap, and quickly wipe the thigh with antiseptic.
    2. Place the tip of the autoinjector against the lateral part of the thigh.
    3. Push the autoinjector firmly against the thigh, and hold it in place until all the medication is injected. (page 693)

Word Fun

Ambulance Calls

1. When assessing the victim of a fall, you must take into consideration not only the patient's complaints and obvious injuries, but also the mechanism of injury including the height of the fall, the surface on which he or she landed, the position in which he or she landed, and any medical and/or previous traumatic injuries that could exacerbate the injuries. This patient landed on a hard surface, most likely a wooden floor, and now presents with numbness and tingling of his lower body. These are all indicators of spinal cord injury. Assume spinal fractures are present; ensure that you assess pulse, motor, and sensation of all extremities prior to placing him in full spinal precautions and reassess again after immobilized. Document your findings and continue this assessment en route to the hospital.

2. This patient is in shock and it seems to be related to an infectious organism (septic shock). In this case, although he has not lost blood volume through hemorrhage, his vessels or 'container' has become too large, making his available blood volume inadequate. Immediate transport is required. The patient should be given high-flow oxygen and placed in the Trendelenburg position to facilitate available perfusion to essential organs (heart, lungs, brain, and kidneys).

3. –Treat for anaphylactic shock.
   –Apply high-flow oxygen while inquiring if the patient has an EpiPen.
   –Obtain orders and administer EpiPen.
   –Monitor vital signs.
   –Rapid transport.

Skill Drills
Skill Drill 23-1: Treating Shock (page 687)
1. Keep the patient supine, open the airway, and check breathing and pulse.
2. Control obvious external bleeding.
3. Splint any broken bones or joint injuries.
4. Give high-flow oxygen if you have not already done so, and place blankets under and over the patient.
5. If no fractures are suspected, elevate the legs 6 to 12 inches.

Assessment Review
1. D (page 692)
2. A (page 692)
3. B (page 692)
4. B (page 692)
5. C (page 692)

Emergency Care Summary
General Shock
spinal

supine

circulation

pulse

Anaphylactic Shock
lateral

injected

(page 693)

Chapter 24: Soft-Tissue Injuries

Matching

1. G (page 698)
2. B (page 698)
3. D (page 698)
4. F (page 699)
5. H (page 698)
6. J (page 705)
7. E (page 705)
8. A (page 706)
9. C (page 705)
10. I (page 711)

Multiple Choice

1. C (page 698)
2. B (page 698)
3. B (page 698)
4. A (page 698)
5. B (page 699)
6. C (page 699)
7. B (page 700)
8. B (page 700)
9. D (page 700)
10. C (page 700)
11. C (page 702)
12. C (page 703)
13. B (page 700)
14. A (page 705)
15. D (page 705)
16. B (page 710)
17. D (page 707)
18. D (page 707)
19. A (page 704)
20. D (page 707)
21. D (page 711)
22. C (page 712)
23. C (page 714)
24. D (page 715)
25. D (page 715)
26. C (page 716)
27. B (page 716)
28. C (page 716)
29. D (page 716)
30. D (page 718)
31. B (page 723)
32. D (page 724)
33. A (page 726)
34. D (page 727)

Labeling

1. The Skin (page 699)

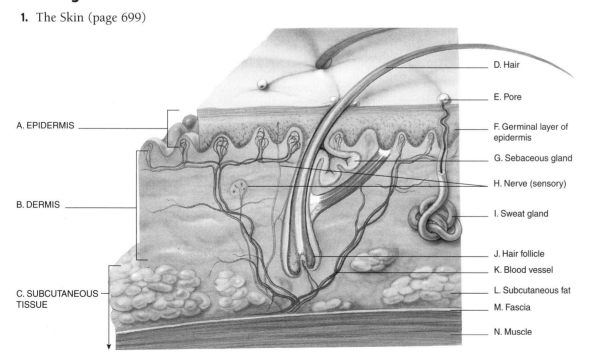

A. EPIDERMIS

B. DERMIS

C. SUBCUTANEOUS TISSUE

D. Hair

E. Pore

F. Germinal layer of epidermis

G. Sebaceous gland

H. Nerve (sensory)

I. Sweat gland

J. Hair follicle

K. Blood vessel

L. Subcutaneous fat

M. Fascia

N. Muscle

2. The Rule of Nines (page 718)

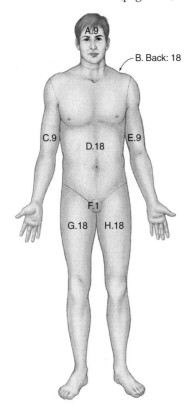

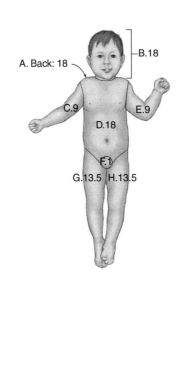

Fill-in

1. moist (page 699)

2. cool (page 698)

3. dermis (page 698)

4. subcutaneous (page 699)

5. constrict (page 699)

6. bacteria, water (page 699)

7. dermis (page 698)

8. temperature (page 699)

9. epidermis, dermis (page 698)

10. radiated (page 699)

True/False

1. T (page 716)

2. F (page 716)

3. T (page 716)

4. T (page 718)

5. T (page 715)

6. T (page 719)

7. F (page 720)

8. T (page 724)

9. T (page 726)

10. T (page 726)

11. F (page 726)

12. F (page 727)

13. T (page 727)

14. F (page 700)

15. F (page 705)

Short Answer

1. **1.** superficial

  2. partial-thickness

  3. full-thickness (page 716)

2. **1.** closed

  2. open

  3. burns (page 699)

3. R=rest

   I=ice

   C=compression

   E=elevation

   S=splinting (page 703)

4. –Full- or partial-thickness burns covering more than 20% of total body surface area

   –Burns involving the hands, feet, face, airway, or genitalia (page 717)

5. Brush off dry chemicals and/or remove clothing, then flush the burned area with large amounts of water. (page 722)

6. First, there may be deep tissue injury not visible on the outside. Second, there is a danger of cardiac arrest from the electrical shock. (page 724)

7. A wound caused by a penetrating object into the chest that causes air to enter the chest. The air enters the chest area through the wound, but remains in the pleural space and the lung does not expand. With exhalation, air passes back through the wound, making a "sucking" sound. (page 708)

8. **1.** Control bleeding

   2. Protect from further damage

   3. Prevent further contamination and infection (page 726)

9. **1.** Abrasions

   2. Lacerations

   3. Avulsions

   4. Penetrating (page 704)

10. **1.** Depth (superficial/partial/full)

    2. Extent (% of body burned)

    3. Involvement of critical areas (face, upper airway, hands, feet, genitalia)

    4. Pre-existing medical conditions or other injuries

    5. Age of younger than 5 years or older than 55 years (page 715)

Word Fun

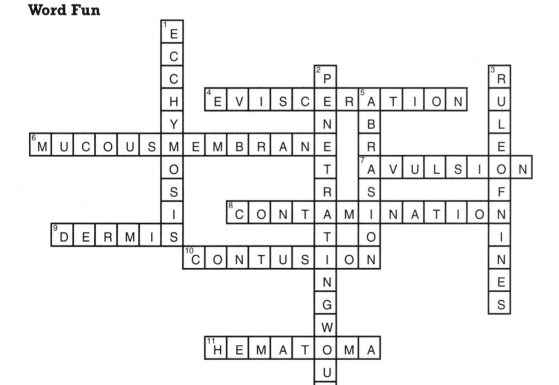

Ambulance Calls

1. –BSI precautions
   –Apply direct pressure.
   –Elevate the extremity and apply a pressure dressing.
   –Try a pressure point if bleeding is not controlled (tourniquet as a last resort).
   –Once bleeding is controlled, splint the arm to decrease movement.
   –Apply high-flow oxygen.
   –Transport in position of comfort, normal response.
   –Monitor vital signs.

2. Unfortunately, this scenario has occurred in households throughout the country. This is why it is so important to 'turn pot handles in' when cooking in the home of a small, inquisitive child. You must evaluate the child quickly to determine the extent and severity of the burns. Assess airway, breathing, and circulation, and quickly apply sterile dressings and high-flow oxygen. Promptly transport the patient according to local protocols.

3. Apply direct pressure to control any bleeding using sterile dressings. Have the patient lie down; as this injury will be quite painful and even the toughest of people can suddenly feel faint, especially if they look at the injury. Find the piece of avulsed tissue, wrap it in sterile dressings, and transport it with you to the hospital. Oxygen via nasal cannula can assist with nausea that the patient may experience.

Skill Drills

Skill Drill 24-1: Controlling Bleeding from a Soft-Tissue Injury (page 711)

1. Apply **direct pressure** with a **sterile** bandage.
2. Maintain pressure with a **roller** bandage.
3. If bleeding continues, apply a second **dressing** and **roller** bandage over the first.
4. **Splint** the extremity.

Skill Drill 24-2: Stabilizing an Impaled Object (page 713)

1. Do not attempt to **move** or **remove** the object. Stabilize the impaled body part.
2. Control **bleeding** and **stabilize** the object in place using **soft dressings**, **gauze**, and/or **tape**.
3. Tape a **rigid** item over the stabilized object to protect it from **movement** during transport.

Skill Drill 24-3: Caring for Burns (page 721)

1. Follow BSI precautions to help prevent infection.
   If it is safe to do so, remove the patient from the burning area; extinguish or remove hot clothing and jewelry as needed.
   If the wound(s) is still burning or hot, immerse the hot area in cool, sterile water, or cover with a wet, cool dressing.
2. Provide high-flow oxygen and continue to assess the airway.
3. Estimate the severity of the burn, then cover the area with a dry, sterile dressing or clean sheet.
   Assess and treat the patient for any other injuries.
4. Prepare for transport.
   Treat for shock.
5. Cover the patient with blankets to prevent loss of body heat.
   Transport promptly.

Assessment Review

1. C (page 728)
2. B (page 728)
3. D (page 728)
4. A (page 728)
5. D (page 730)

Emergency Care Summary

Closed Injuries
heart

Open Injuries
roller

roller

Burns
sterile

traumatic

Abdominal Injuries
abdomen

Impaled objects
gauze

Neck Wounds
occlusive

carotid

Chemical Burns
15

20

Electrical Burns
AED

Small Animal and Human Bites
antibiotic

(page 730)

Chapter 25: Eye Injuries

Matching

1. D (page 736)
2. C (page 737)
3. B (page 737)
4. E (page 737)
5. J (page 736)
6. G (page 736)
7. A (page 736)
8. F (page 736)
9. I (page 736)
10. H (page 737)

Multiple Choice

1. B (page 736)
2. D (page 736)
3. A (page 736)
4. A (page 737)
5. D (page 739)
6. C (page 738)
7. D (page 744)
8. D (page 745)
9. B (page 745)
10. D (page 745)
11. B (page 746)
12. C (page 746)
13. D (page 747)
14. A (page 747)
15. B (page 747)
16. A (page 748)
17. D (page 749)
18. C (page 750)

Fill-in

1. lacrimal glands (page 736)
2. optic nerve (page 737)
3. camera (page 737)
4. chemical burn (page 750)
5. orbit (page 736)
6. abnormalities (page 740)
7. orbit (page 741)
8. Conjunctivitis (page 741)
9. conjunctiva (page 736)
10. sclera (page 736)

True/False

1. F (page 742)
2. F (page 736)
3. T (page 736)
4. F (page 736)
5. F (page 750)
6. F (page 742)
7. F (page 741)
8. T (page 737)
9. F (page 741)
10. T (page 736)

Short Answer

1. A condition in which the retina is separated from its attachments at the back of the eye. Common findings include flashing lights, floaters, or a cloud or shade over the patient's vision. However, pain is not a common complaint. (page 749)

2. **1.** Never exert pressure on the eye.

   2. If part of the eyeball is exposed, gently apply a moist, sterile dressing to prevent drying.

   3. Cover the eye with a protective shield or sterile dressing. (page 747)

3. **1.** One pupil larger than the other

   2. Eyes not moving together or pointing in different directions

   3. Failure of the eyes to follow movement when instructed

   4. Bleeding under the conjunctiva

   5. Protrusion or bulging of the eye (page 749)

Word Fun

Ambulance Calls

1. You may find examination of the eye to be difficult as the child is very upset and frightened. To prevent any further injury to the eye, the other eye must be covered. This will also likely frighten the child. Anxiety will be lessened if the child is accompanied to the hospital with an adult who knows him. Request that the teacher ride with the child to the emergency department. Notify the receiving facility of the seriousness of the injury.

2. Position patient supine on stretcher.
   Set up IV bags to use as irrigation hoses.
   Connect IV line to a nasal cannula and place the prongs over the nose so that the fluid runs into the eyes.
   Monitor patient and flush eyes continuously en route to the hospital.
   Provide rapid transport.

3. This patient could have sustained facial fractures surrounding his eye (orbital fractures) as well as trauma to the eye itself. Note if the eye has held its shape or if you see any drainage from the eye, or swelling or deformity of the facial bones. Take spinal precautions as appropriate; cover both eyes to prevent further damage; apply oxygen and transport to the nearest appropriate facility.

Skill Drills

Skill Drill 25-1: Removing a Foreign Object From Under the Upper Eyelid (page 743)

1. Have the patient look **down**, grasp the **upper lashes**, and gently pull the lid away from the eye.

2. Place a **cotton-tipped applicator** on the outer surface of the upper lid.

3. Pull the lid **forward** and **up**, folding it back over the applicator.

4. Gently remove the foreign object from the eyelid with a **moistened, sterile** applicator.

Skill Drill 25-2: Stabilizing a Foreign Object Impaled in the Eye (page 745)

1. To prepare a doughnut ring, wrap a **2"** roll around your fingers and thumb **seven** or **eight** times. Adjust the diameter by **spreading** your fingers.

2. Wrap the remainder of the roll, . . .

3. . . . working around the ring.

4. Place the dressing over the **eye** to hold the impaled object in place, then **secure** it with a **gauze** dressing.

Assessment Review

1. A (page 751)

2. D (page 751)

3. C (page 751)

4. B (page 752)

5. B (page 752)

Emergency Care Summary

Foreign Objects

lashes

2" roll

Burns

5, 20

saline

Lacerations

moist

Blunt Trauma

metal shield

(page 752)

Chapter 26: Face and Throat Injuries

Matching

1. C (page 758)
2. B (page 758)
3. E (page 758)
4. F (page 758)
5. G (page 758)
6. D (page 760)
7. A (page 758)

Multiple Choice

1. D (page 758)
2. B (page 758)
3. C (page 758)
4. D (page 758)
5. A (page 758)
6. D (page 759)
7. C (page 759)
8. D (page 760)
9. A (page 765)
10. D (page 765)
11. B (page 766)
12. C (page 766)
13. A (page 767)
14. D (page 767)
15. D (page 768)
16. A (page 769)

Labeling

Face/Skull (page 759)

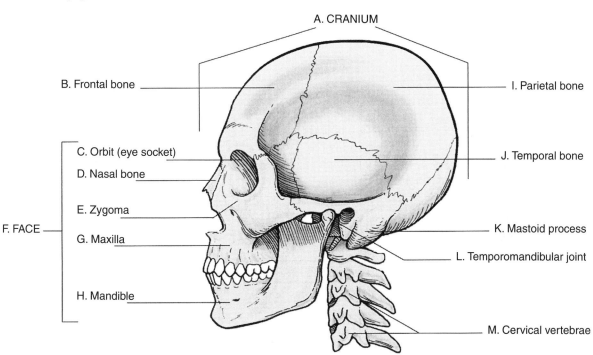

A. CRANIUM
B. Frontal bone
C. Orbit (eye socket)
D. Nasal bone
E. Zygoma
F. FACE
G. Maxilla
H. Mandible
I. Parietal bone
J. Temporal bone
K. Mastoid process
L. Temporomandibular joint
M. Cervical vertebrae

Fill-in

1. carotid (page 760)
2. cervical (page 758)
3. temporal (page 758)
4. trachea (page 759)
5. cartilage (page 759)

6. men, women (page 759)
7. foramen magnum (page 758)
8. maxilla (page 758)
9. parietal (page 758)
10. trachea (pages 759)

True/False

1. T (page 760)
2. T (page 761)
3. F (page 764)
4. F (page 767)
5. T (page 768)
6. T (page 758)
7. F (page 758)
8. F (page 760)
9. T (page 761)
10. F (page 765)

Short Answer

1. Direct manual pressure with a dry dressing. Use roller gauze around the circumference of the head to hold pressure dressing in place. (page 764)
2. Apply direct pressure above and below the injury. (page 769)

Word Fun

Ambulance Calls

1. Depending upon where the dog's teeth have punctured the skin, you may have a variety of soft-tissue injuries and swelling. If you notice the presence of subcutaneous emphysema, the dog punctured or perforated the child's trachea. You must also assume the presence of cervical spine injuries and must take appropriate precautions. Assess his level of consciousness, airway, breathing, and circulation. Control any bleeding and apply other dressings as needed after airway management is accomplished and while en route to the hospital. Always follow local protocols.

2. Apply direct pressure to the bleeding site using gloved fingertips and a sterile occlusive dressing. Secure the dressing in place and apply pressure if necessary. You may need to treat for shock. Provide prompt transport with the patient immobilized to a backboard and apply high-flow oxygen en route.

3. You should determine what objects were used to cause injury to this man's face. Baseball bats would be readily available and would increase your index of suspicion. You should determine the presence of head and neck pain. If the area of injury is limited to his nose, and the need for spinal precautions are not indicated, you can instruct the patient in controlling his bleeding by ensuring he is pushing on the cartilage of his nose and does not lean his head backwards. Swallowing blood will cause nausea. Do not allow the patient to blow his nose and use ice as needed to reduce swelling and pain. Transport according to local protocols.

Skill Drills

Skill Drill 26-1: Controlling Bleeding From a Neck Injury (page 770)

1. Apply **direct pressure** to control bleeding.
2. Use **roller gauze** to secure a dressing in place.
3. Wrap the bandage around and under the patient's **shoulder**.

Assessment Review

1. D (page 771)
2. B (page 771)
3. D (page 772)
4. A (page 772)
5. C (page 772)

Emergency Care Summary

Nose Injuries

forward

Ear Injuries

moist

Facial Fractures

bone

saliva

Neck Injuries

roller gauze

(page 772)

Chapter 27: Chest Injuries

Matching

1. B (page 778)   **6.** H (page 780)

2. D (page 779)   **7.** I (page 780)

3. C (page 779)   **8.** J (page 789)

4. A (page 779)   **9.** F (page 780)

5. E (page 779)   **10.** G (page 780)

Multiple Choice

1. B (page 779)   **11.** D (page 786)

2. A (page 779)   **12.** D (page 786)

3. C (page 779)   **13.** D (page 786)

4. D (page 780)   **14.** D (page 786)

5. D (page 780)   **15.** D (page 787)

6. D (page 781)   **16.** A (page 787)

7. B (page 787)   **17.** C (page 788)

8. C (page 787)   **18.** D (page 788)

9. D (page 784)   **19.** D (page 789)

10. C (page 785)   **20.** B (page 789)

Labeling

Anterior Aspect of the Chest (page 778)

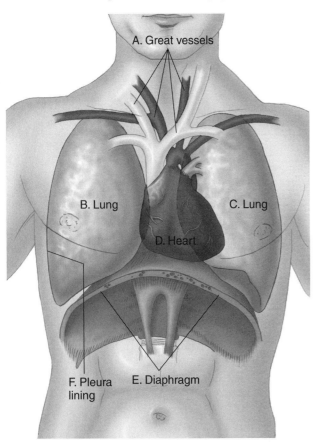

The Ribs (page 779)

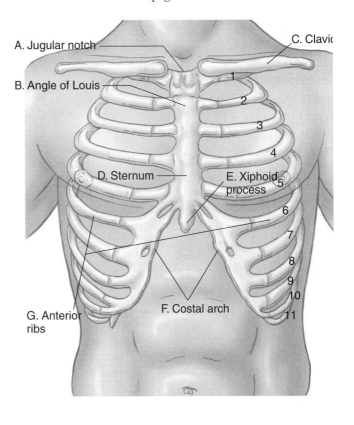

Fill-in

1. back (page 779)
2. decreases (page 779)
3. sternum (page 779)
4. bronchi (page 779)
5. phrenic (page 779)
6. ribs (page 779)
7. diaphragm (page 778)
8. Pleura (page 779)
9. aorta (page 779)
10. contracts (page 779)

True/False

1. T (page 780)
2. F (page 780)
3. T (page 786)
4. F (page 787)
5. F (page 786)
6. F (page 789)
7. F (page 778)
8. T (page 779)
9. F (page 780)
10. T (page 780)

Short Answer

1. –Pain at the site of injury
   –Pain localized at the site of injury that is aggravated by or increased with breathing
   –Dyspnea (difficulty breathing, shortness of breath)
   –Hemoptysis (coughing up blood)
   –Failure of one or both sides of the chest to expand normally with inspiration
   –Rapid, weak pulse and low blood pressure
   –Cyanosis around the lips or fingernails. (page 780)

2. **1.** Seal the wound with a large airtight dressing that seals all four sides.

   2. Seal the wound with a dressing that seals three sides with the fourth side as a flutter valve.

   Your local protocol will dictate the way you are to care for this injury. (page 785)

3. Tape a bulky pad against the segment of the chest. (page 788)

4. Sudden severe compression of the chest, causing a rapid increase of pressure within the chest. Characteristic signs include distended neck veins, facial and neck cyanosis, and hemorrhage in the sclera of the eye. (page 788)

Word Fun

Across and down crossword solution:

1. HEMOTHORAX
2. HEMOPTYSIS
3. TENSIONPNEUMOTHORAX
4. PUPLMONARY (PULMONARYCONTUSION)
5. PARADOXICALMOTION
6. FLUTTERVALVE
7. TACHYPNEA
8. FLAILCHEST
9. PERICARDIALTAMPONADE
10. DYSPNEA
11. PNEUMOTHORAX

Ambulance Calls

1. You should be concerned with the presence of rib and sternal fractures as well as pulmonary contusions and pneumothoracies. This patient may require assistance in breathing as it will be extremely painful for him to breathe in. If he requires assistance with a BVM device, you must be careful not to become too aggressive in your ventilations. Time your ventilations with the patient's respirations and be gentle. Provide high-flow oxygen and take spinal precautions according to your local protocols.

2. This patient's chest has been crushed with a significant amount of weight. This mechanism of injury indicates the high potential for rib, sternal, thoracic fractures and other soft-tissue injuries as well. Patients who experience sternal fractures will find it difficult to be placed supine, as this will likely increase their pain. There will be little you can do to ease this pain as they should be immobilized on a long backboard. Provide prompt transport, high-flow oxygen, and monitor the pulse, motor, and sensations particularly distal to the suspected spinal injury.

3. –Apply high-flow oxygen via nonrebreathing mask or BVM device.
   –Fully c-spine immobilized patient.
   –Rapid transport
   –Monitor vital signs en route.

Assessment Review

1. B (page 791)
2. D (page 792)
3. C (page 792)
4. D (page 792)
5. A (page 792)

Emergency Care Summary

Pneumothorax
occlusive

flutter valve

Hemothorax
shock

Rib Fracture
spinal injury

Flail Chest
high-flow

(page 792)

Chapter 28: Abdomen and Genitalia Injuries

Matching

1. G (page 798)
2. D (page 798)
3. C (page 798)
4. H (page 807)
5. E (page 807)
6. B (page 802)
7. A (page 809)
8. F (page 798)

Multiple Choice

1. D (page 798)
2. D (page 798)
3. C (page 798)
4. D (page 798)
5. B (page 798)
6. C (page 798)
7. A (page 798)
8. C (page 799)
9. B (page 799)
10. A (page 799)
11. D (page 799)
12. C (page 799)
13. B (page 799)
14. A (page 800)
15. B (page 800)
16. B (page 800)
17. B (page 804)
18. D (page 800)
19. D (page 800)
20. D (page 801)
21. C (page 801)
22. D (page 802)
23. D (page 802)
24. D (page 802)
25. A (page 807)
26. C (page 808)
27. D (page 809)
28. D (page 809)
29. A (page 809)
30. B (page 809)
31. D (page 810)
32. D (page 812)

Labeling

1. Hollow organs (page 798)

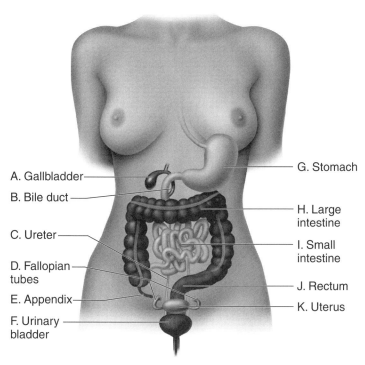

A. Gallbladder

B. Bile duct

C. Ureter

D. Fallopian tubes

E. Appendix

F. Urinary bladder

G. Stomach

H. Large intestine

I. Small intestine

J. Rectum

K. Uterus

2. Solid organs (page 799)

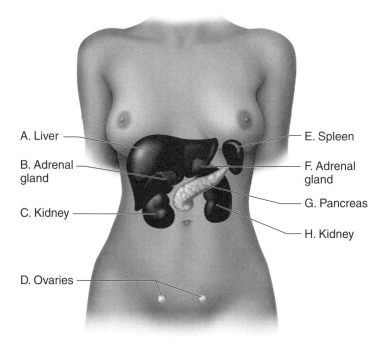

A. Liver

B. Adrenal gland

C. Kidney

D. Ovaries

E. Spleen

F. Adrenal gland

G. Pancreas

H. Kidney

Fill-in

1. solid (page 798)

2. urinary (page 807)

3. retroperitoneal (page 807)

4. external signs (page 809)

5. peritonitis (page 798)

6. inflammatory response (page 798)

7. blunt injuries (page 799)

8. penetrating injuries (page 799)

9. flank (page 800)

10. evisceration (page 802)

True/False

1. F (page 798)

2. T (page 800)

3. F (page 801)

4. T (page 809)

5. F (page 798)

6. F (page 799)

7. T (page 798)

8. F (page 799)

9. T (page 801)

10. T (page 802)

Short Answer

1. Stomach, intestines, ureters, bladder, gallbladder, and rectum (page 798)

2. Liver, spleen, pancreas, and kidneys (page 798)

3. Pain, shock signs, bruises, lacerations, bleeding, tenderness, guarding, and difficulty with movement because of pain (page 800)

4. –Inspect the patient's back and sides for exit wounds.
   –Apply a dry, sterile dressing to all open wounds.
   –If the penetrating object is still in place, apply a stabilizing bandage around it to control external bleeding and to minimize movement of the object. (page 802)

5. –Cover with moistened sterile dressing.
   –Secure dressing with bandage.
   –Secure bandage with tape. (page 802)

6. –An abrasion, laceration, or contusion in the flank
   –A penetrating wound in the region of the lower rib cage (the flank) or the upper abdomen
   –Fractures on either side of the lower rib cage or of the lower thoracic or upper lumbar vertebrae (page 809)

Word Fun

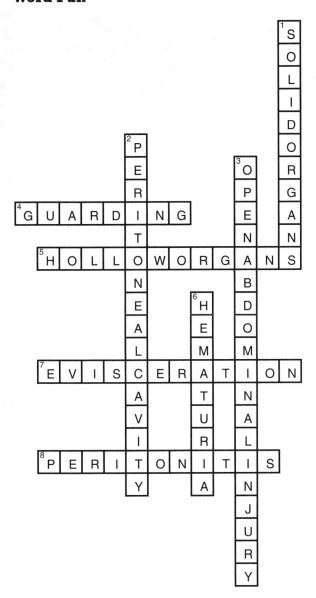

Ambulance Calls

1. Assess the ABCs.
   Apply high-flow oxygen.
   Control any bleeding.
   Stabilize the knife in place with bulky dressings—DO NOT REMOVE.
   Keep movement of patient to the bare minimum, so as not to create further injury. (Sliding the patient very carefully onto a backboard may help to minimize movement.)
   Monitor vital signs.
   Provide rapid transport.
   Bandage minor lacerations en route.

2. Quickly visualize the area to determine how badly he has cut himself, and if he has in fact amputated any portion of his penis. You will need to control bleeding, as blood loss in this area can be significant. Use pressure dressings/pressure points as needed to control bleeding; provide high-flow oxygen and prompt transport. Also, request the presence of a police officer during transport as this patient will likely need to be restrained and could be unpredictable during transport.

3. Cover the abdomen and the portion of the protruding bowel with a moistened, sterile dressing and/or an occlusive dressing. Secure these dressings with tape; allow the patient to draw his knees up as needed for comfort; apply high-flow oxygen; cover the patient to preserve warmth and promptly transport to the hospital.

Assessment Review

1. A (page 813)
2. B (page 813)
3. D (page 814)
4. D (page 814)
5. C (page 814)

Emergency Care Summary
Kidney and Urinary Bladder Injuries

vital signs

External Male Genitalia Injuries

Avulsion

saline

minutes

Amputation

ice

Lacerated Head of the Penis

sterile dressing

Urethral Injuries

urine

External Female Genitalia Injuries

compress

Internal Female Genitalia Injuries, Pregnancy

left

left

Rectal Bleeding

compresses

Sexual Assault

airway maintenance

major bleeding

paper

plastic

(page 814)

Chapter 29: Musculoskeletal Care

Matching

1. G (page 820)
2. J (page 820)
3. D (page 821)
4. I (page 822)
5. F (page 822)
6. B (page 824)
7. K (page 825)
8. A (page 825)
9. C (page 822)
10. E (page 824)
11. H (page 838)

Multiple Choice

1. A (page 856)
2. B (page 821)
3. A (page 821)
4. B (page 822)
5. C (page 822)
6. B (page 822)
7. C (page 823)
8. A (page 823)
9. D (page 824)
10. A (page 824)
11. B (page 824)
12. C (page 824)
13. D (page 824)
14. B (page 824)
15. C (page 825)
16. D (page 825)
17. A (page 825)
18. C (page 825)
19. D (page 825)
20. A (page 825)
21. C (page 825)
22. B (page 825)
23. A (page 826)
24. C (page 826)
25. D (page 828)
26. C (page 828)
27. D (page 830)
28. D (page 829)
29. D (page 832)
30. D (page 836)
31. C (page 838)
32. D (page 838)
33. B (page 841)
34. D (page 846)
35. B (page 848)
36. D (page 853)
37. D (page 857)

38. C (page 859)
39. B (page 859)
40. D (page 860)
41. C (page 860)
42. B (page 860)
43. A (page 861)
44. C (page 862)

Labeling

The Human Skeleton (page 822)

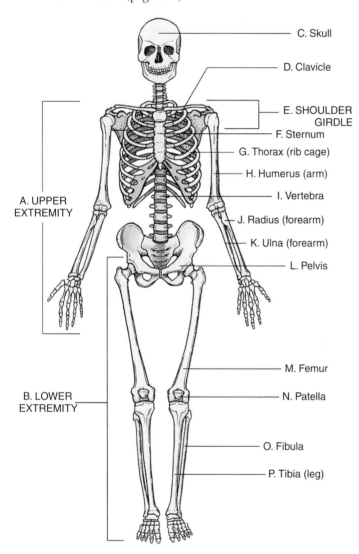

C. Skull

D. Clavicle

E. SHOULDER GIRDLE

F. Sternum

G. Thorax (rib cage)

H. Humerus (arm)

I. Vertebra

J. Radius (forearm)

K. Ulna (forearm)

L. Pelvis

A. UPPER EXTREMITY

M. Femur

N. Patella

B. LOWER EXTREMITY

O. Fibula

P. Tibia (leg)

Fill-in

1. wasting (page 821)
2. red (page 821)
3. hinge (page 822)
4. clavicle (page 848)
5. mechanism of injury (page 834)
6. open fracture (page 824)

7. sciatic nerve (page 857)
8. femur (page 856)
9. crepitus (page 827)
10. reduce (page 828)
11. neurovascular status (page 829)

True/False

1. T (page 836) **6.** F (page 837)

2. T (page 836) **7.** T (page 837)

3. F (page 838) **8.** T (page 837)

4. T (page 838) **9.** T (page 832)

5. T (page 837) **10.** F (page 832)

Short Answer

1. **1.** Direct

   2. Indirect

   3. Twisting

   4. High-energy (page 824)

2. Deformity
   Tenderness (point)
   Guarding
   Swelling
   Bruising
   Crepitus
   False motion
   Exposed fragments
   Pain
   Locked joint (page 825)

3. **1.** Pulse

   2. Capillary refill

   3. Sensation

   4. Motor function (page 832)

4. **1.** Remove clothing from the area.

   2. Note and record the patient's neurovascular status distal to the site of the injury.

   3. Cover all wounds with a dry, sterile dressing before splinting.

   4. Do not move the patient before splinting.

   5. For a suspected fracture, immobilize the joints above and below the fracture.

   6. For a joint injury, immobilize the bones above and below the injured joint.

   7. Pad all rigid splints.

   8. Maintain manual immobilization to minimize movement of the limb and to support the injury site.

   9. Use a constant, gentle manual traction to align the limb.

   10. If you encounter resistance to limb alignment, splint the limb in its deformed position.

   11. Immobilize all suspected spinal injuries in a neutral in-line position on a backboard.

   12. If the patient has signs of shock, align the limb in the normal anatomic position and provide transport.

   13. When in doubt, splint. (page 836)

5. **1.** Stabilize the fracture.

   2. Align the limb.

   3. Avoid potential neurovascular compromise. (page 838)

Word Fun

The completed crossword contains the following answers:

- 3 across: TRACTION
- 4 across / SCIATIC NERVE
- 6 across: POSITION OF FUNCTION
- 7 across: COMPARTMENT SYNDROME
- 8 across: HEMATURIA
- 11 across: CLAVICLE
- 12 across: SYMPHSIS PUBIS
- 13 across: ULNA
- 14 across: SKELETAL MUSCLE

Down answers:
- 1: GLENOID
- 2: OPEN FRACTURE
- 4: DISPLACED
- 5: SCIATIC NERVE (OF SOSA)
- 9: ECCHYMOSIS
- 10: CREPITUS

Ambulance Calls

1. This patient likely has a compression injury to his lumbar spine. The force exerted on his body from the landing will be transferred up from his feet through his legs to his pelvis and spine. You must take all spinal precautions, apply high-flow oxygen, and provide prompt transport to the nearest appropriate facility. Continue to monitor any changes in the pulse, motor, and sensation, specifically in his lower body.

2. –Splint the arm in position found, since circulation is intact.
   –Use a board splint for support with a sling and swathe.
   –Immobilize hand in the position of function.
   –Apply oxygen as needed.
   –Transport in the position of comfort.
   –Normal transport.
   –Monitor vital signs.

3. This man not only has probable injuries and swelling to his airway but significant likelihood for distraction injuries to his cervical spine. You must take c-spine precautions as you open his airway and determine the presence of breathing. The information provided does not include whether he has a pulse. Assess airway, breathing, and circulation; apply full spinal precautions, high-flow oxygen, positive pressure ventilations, and CPR as needed.

Skill Drills

Skill Drill 29-1: Assessing Neurovascular Status (pages 833-834)

1. Palpate the **radial** pulse in the upper extremity.
2. Palpate the **posterior tibial** pulse in the lower extremity.
3. Assess capillary refill by blanching a fingernail or **toenail**.
4. Assess sensation on the flesh near the **tip** of the **index** finger.
5. On the foot, first check sensation on the flesh near the **tip** of the **great toe**.
6. Also check foot sensation on the **lateral side**.
7. Evaluate motor function by asking the patient to **open** the hand. (Perform motor tests only if the hand or foot is not **injured**. **Stop** a test if it causes pain.)
8. Also ask the patient to **make a fist**.
9. To evaluate motor function in the foot, ask the patient to **extend** the foot.
10. Also have the patient **flex** the foot and **wiggle** the toes.

Skill Drill 29-2: Caring for Musculoskeletal Injuries (page 837)

1. Cover open wounds with a **dry**, **sterile** dressing, and **apply pressure** to control bleeding.
2. Apply a splint and elevate the extremity about 6" (slightly above the level of the **heart**).
3. Apply **cold packs** if there is swelling, but do not place them **directly** on the skin.
4. **Position** the patient for transport and **secure** the injured area.

Skill Drill 29-3: Applying a Rigid Splint (page 839)

1. Provide gentle **support** and **in-line traction** of the limb.
2. Second EMT-B places the splint **alongside** or **under** the limb.
   Pad between the limb and the splint as needed to ensure even pressure and contact.
3. Secure the splint to the limb with **bindings**.
4. Assess and record **distal neurovascular** functions.

Skill Drill 29-4: Applying a Zippered Air Splint (page 841)

1. Support the injured limb and apply gentle **traction** as your partner applies the open, deflated splint.
2. Zip up the splint, inflate it by **pump** or by **mouth**, and test the **pressure**. Check and record **distal neurovascular** function.

Skill Drill 29-5: Applying an Unzippered Air Splint (page 842)

1. **Support** the injured limb. Have your partner place his or her arm through the splint to grasp the patient's **hand** or **foot**.
2. Apply gentle **traction** while sliding the splint onto the injured limb.
3. **Inflate** the splint.

Skill Drill 29-6: Applying a Vacuum Splint (page 843)

1. **Stabilize** and **support** the injury.
2. Place the splint and **wrap** it around the limb.
3. **Draw** the air **out of** the splint through the suction valve, and then **seal** the valve.

Skill Drill 29-7: Applying a Hare Traction Splint (page 844)

1. Expose the injured limb and check pulse, motor, and sensory function.
   Place the splint beside the uninjured limb, adjust the splint to proper length, and prepare the straps.
2. Support the injured limb as your partner fastens the ankle hitch about the foot and ankle.
3. Continue to support the limb as your partner applies gentle in-line traction to the ankle hitch and foot.
4. Slide the splint into position under the injured limb.
5. Pad the groin and fasten the ischial strap.

6. Connect the loops of the ankle hitch to the end of the splint as your partner continues to maintain traction. Carefully tighten the ratchet to the point that the splint holds adequate traction.
7. Secure and check support straps.
   Assess pulse, motor, and sensory functions.
8. Secure the patient and splint to the backboard in a way that will prevent movement of the splint during patient movement and transport.

Skill Drill 29-8: Applying a Sager Traction Splint (page 846)

1. After exposing the injured area, check the patient's pulse and motor and sensory function.
   Adjust the thigh strap so that it lies anteriorly when secured.
2. Estimate the proper length of the splint by placing it next to the injured limb.
   Fit the ankle pads to the ankle.
3. Place the splint at the inner thigh, apply the thigh strap at the upper thigh, and secure snugly.
4. Tighten the ankle harness just above the malleoli.
   Snug the cable ring against the bottom of the foot.
5. Extend the splint's inner shaft to apply traction of about 10% of body weight.
6. Secure the splint with elasticized cravats.
7. Secure the patient to a long backboard.
   Check pulse, motor, and sensory functions.

Skill Drill 29-9: Splinting the Hand and Wrist (page 855)

1. Move the hand into the **position of function**. Place a soft **roller bandage** in the palm.
2. Apply a **padded board** splint on the **palmar** side with fingers **exposed**.
3. Secure the splint with a **roller bandage**.

Assessment Review

1. C (page 864)
2. B (page 864)
3. A (page 865)
4. B (page 865)
5. D (page 865)

Emergency Care Summary
Applying a Hare Traction Splint
sensory

in-line traction

ischial

support straps

Applying a Sager Traction Splint
anteriorly

inner

upper

malleoli

10%

(page 865)

Chapter 30: Head and Spine Injuries

Matching

1. D (page 872)
2. E (page 872)
3. H (page 875)
4. G (page 875)
5. A (page 876)
6. C (page 872)
7. B (page 880)
8. I (page 874)
9. J (page 877)
10. F (page 872)

Multiple Choice

1. D (page 872)
2. B (page 874)
3. C (page 872)
4. C (page 890)
5. A (page 872)
6. B (page 876)
7. D (page 874)
8. D (page 872)
9. C (page 875)
10. A (page 877)
11. D (page 882)
12. D (page 881)
13. B (page 877)
14. D (page 890)
15. B (page 889)
16. A (page 891)
17. A (page 886)
18. B (page 893)
19. D (page 886)
20. B (page 879)
21. D (page 879)
22. D (page 879)
23. B (page 880)
24. B (page 880)
25. D (page 880)
26. C (page 880)
27. C (page 882)
28. D (page 884)
29. C (page 890)
30. A (page 890)
31. D (page 898)
32. C (page 900)
33. A (page 905)

Labeling

1. The Brain (page 873)

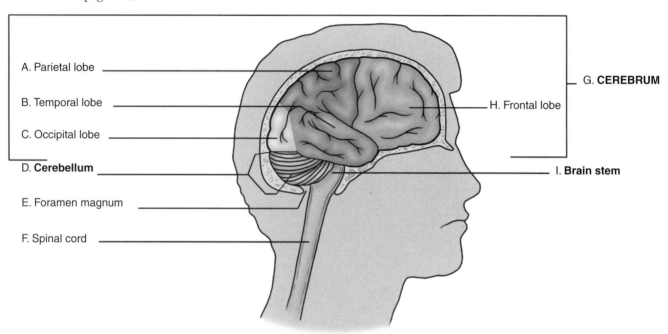

A. Parietal lobe

B. Temporal lobe

C. Occipital lobe

D. **Cerebellum**

E. Foramen magnum

F. Spinal cord

G. **CEREBRUM**

H. Frontal lobe

I. **Brain stem**

2. The Connecting Nerves in the Spinal Cord (page 875)

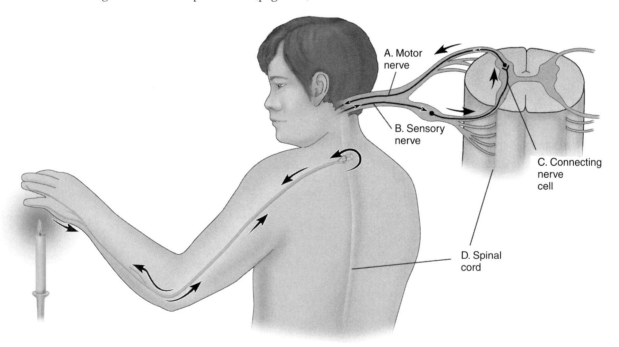

A. Motor nerve

B. Sensory nerve

C. Connecting nerve cell

D. Spinal cord

3. The Spinal Column (page 877)

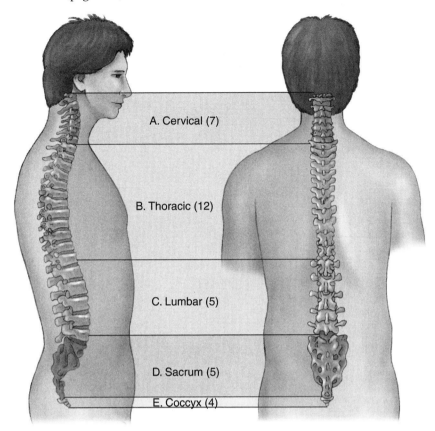

A. Cervical (7)

B. Thoracic (12)

C. Lumbar (5)

D. Sacrum (5)

E. Coccyx (4)

Fill-in

1. motor (page 874)
2. meninges (page 872)
3. central (page 872)
4. 31 (page 874)
5. cranial (page 874)
6. intervertebral discs (page 877)
7. cranium, face. (page 876)
8. arachnoid, pia mater (page 872)
9. sympathetic (page 875)
10. parasympathetic (page 876)

True/False

1. T (page 880)
2. F (page 875)
3. F (page 875)
4. T (page 875)
5. F (page 876)
6. F (page 890)
7. F (page 881)
8. T (page 891)
9. T (page 893)
10. T (page 898)

Short Answer

1. **1.** Does your neck or back hurt?

   2. What happened?

   3. Where does it hurt?

   4. Can you move your hands and feet?

   5. Can you feel me touching your fingers? Your toes? (page 881)

2. –Muscle spasms in the neck
   –Increased pain
   –Numbness, tingling, or weakness
   –Compromised airway or ventilations (page 890)

3. **1.** Concussion

   2. Contusion

   3. Intracranial bleeding (page 879)

4. –Lacerations, contusions, or hematomas to the scalp
   –Soft area or depression upon palpation
   –Visible fractures or deformities of the skull
   –Ecchymosis about the eyes or behind the ear over the mastoid process
   –Clear or pink cerebrospinal fluid leakage from a scalp wound, the nose, or the ear
   –Failure of the pupils to respond to light
   –Unequal pupil size
   –Loss of sensation and/or motor function
   –A period of unconsciousness
   –Amnesia
   –Seizures
   –Numbness or tingling in the extremities
   –Irregular respirations
   –Dizziness
   –Visual complaints
   –Combative or other abnormal behavior
   –Nausea or vomiting (page 880)

5. **1.** Establish an adequate airway.

   2. Control bleeding.

   3. Assess the patient's baseline level of consciousness. (page 890)

6. **1.** Is the patient's airway clear?

   2. Is the patient breathing adequately?

   3. Can you maintain the airway and assist ventilations if the helmet remains in place?

   4. How well does the helmet fit?

   5. Can the patient move within the helmet?

   6. Can the spine be immobilized in a neutral position with the helmet on? (page 900)

Word Fun

Ambulance Calls

1. There is significant mechanism of injury involved in this incident. The impact with the car, the lack of a helmet, the obvious head trauma, and the patient's decreased level of consciousness all indicate significant life-threatening injuries. You must quickly apply full spinal precautions and manage her airway. Provide prompt transport, high-flow oxygen and perform ongoing assessments while en route to the nearest appropriate medical facility.

2. This child landed face down on a concrete surface with significant force. She is now unconscious with a partially obstructed airway. You must move her to a supine position to manage her airway. Take note of any apparent injuries to her back as you reposition her. Ideally, you will have the appropriate equipment and adequate staffing to quickly and safely move her to a long backboard immediately. However, do not delay appropriately moving her to a supine position as you must do this to assess and manage her airway. This may be difficult as she likely has facial fractures and possibly broken teeth, blood, and secretions in her airway. Be prepared to suction her airway and apply positive pressure ventilations (this may be especially challenging in the presence of significant facial fractures). Use high-flow oxygen and promptly transport her to the nearest appropriate facility according to your local protocols.

3. –Leave the patient in his car seat.

   –Pad appropriately to immobilize the patient.

   –Use blow-by oxygen if the patient will tolerate it.

   –Monitor vital signs.

   –Continue assessment.

   –Rapid transport due to mechanism of injury and death in vehicle

Skill Drills

Skill Drill 30-1: Performing Manual In-Line Stabilization (page 889)

1. Kneel behind the patient and place your hands firmly around the **base** of the **skull** on either **side**.

2. Support the lower jaw with your **index** and **long** fingers, and the head with your **palms**. Gently **lift** the head into a **neutral, eyes-forward** position, aligned with the torso. Do not **move** the head or neck excessively, forcefully, or rapidly.

3. Continue to **support** the head manually while your partner places a rigid **cervical collar** around the neck. Maintain **manual support** until you have completely secured the patient to a backboard.

Skill Drill 30-2: Immobilizing a Patient to a Long Backboard (page 892)

1. Apply and maintain **cervical stabilization**.

   Assess **distal functions** in all extremities.

2. Apply a **cervical collar**.

3. Rescuers **kneel** on one side of the patient and place **hands** on the far side of the patient.

4. On command, rescuers **roll** the patient toward themselves, quickly examine the **back**, slide the backboard under the patient, and roll the patient onto the board.

5. **Center** the patient on the board.

6. Secure the **upper torso** first.

7. Secure the **chest, pelvis,** and **upper legs**.

8. Begin to secure the patient's head using a commercial immobilization device or **rolled towels**.

9. Place **tape** across the patient's forehead.

10. Check all **straps** and readjust as needed.

    Reassess **distal functions** in all extremities.

Skill Drill 30-3: Immobilizing a Patient Found in a Sitting Position (page 895)

1. Stabilize the head and neck in a neutral, in-line position.

   Assess pulse, motor, and sensory function in each extremity.

   Apply a cervical collar.

2. Insert a short spine immobilization device between the patient's upper back and the seat.

3. Open the side flaps, and position them around the patient's torso, snug around the armpits.

4. Secure the upper torso flaps, then the midtorso flaps.

5. Secure the groin (leg) straps. Check and adjust torso straps.

6. Pad between the head and the device as needed.

   Secure the forehead strap and fasten the lower head strap around the collar.

7. Wedge a long backboard next to the patient's buttocks.

8. Turn and lower the patient onto the long board.

   Lift the patient, and slip the long board under the spine device.

9. Secure the immobilization devices to each other.

   Reassess pulse, motor, and sensory functions in each extremity.

Skill Drill 30-4: Immobilizing a Patient found in a Standing Position (page 897)

1. While **manually** stabilizing the head and neck, apply a **cervical collar**. Position the board **behind** the patient.

2. Position EMT-Bs at **sides** and **behind** the patient.
   Side EMT-Bs reach under patient's **arms** and grasp **handholds** at or slightly above **shoulder** level.

3. Prepare to lower the patient. EMT-Bs on the sides should be **facing** the EMT-B at the head and **wait** for his or her **direction**.

4. On command, **lower** the backboard to the ground.

Skill Drill 30-5: Application of a Cervical Collar (page 899)

1. Apply **in-line** stabilization.

2. Measure the proper **collar size**.

3. Place the **chin support** first.

4. **Wrap** the collar around the neck and **secure** the collar.

5. Assure proper **fit** and maintain **neutral, in-line** stabilization.

Skill Drill 30-6: Removing a Helmet (page 903)

1. Kneel down at the patient's head with your **partner** at one side.
   Open the face shield to assess **airway** and **breathing**. Remove **eyeglasses** if present.

2. Prevent head movement by placing your **hands** on either side of the helmet and fingers on the **lower jaw**. Have your partner **loosen** the strap.

3. Have your partner place one hand at the **angle** of the **lower jaw** and the other at the **occiput**.

4. Gently slip the helmet about **halfway** off, then stop.

5. Have your partner slide the hand from the **occiput** to the **back** of the head to prevent it from snapping back.

6. Remove the helmet and **stabilize** the cervical spine.
   Apply a **cervical collar** and secure the patient to a **long backboard**.
   Pad as needed to prevent neck flexion or extension.

Assessment Review

1. D (page 906)
2. C (page 907)
3. B (page 907)
4. C (page 907)
5. A (page 907)

Emergency Care Summary
Immobilizing a Patient Found in a Sitting Position

neutral, in-line
cervical collar
short spine
lower head strap
long backboard
extremity
(page 907)

Chapter 31: Pediatric Emergencies

Matching

1. I (page 934)
2. F (page 935)
3. E (page 934)
4. J (page 918)
5. B (page 919)

6. A (page 920)
7. H (page 918)
8. G (page 918)
9. C (page 918)
10. D (page 916)

Multiple Choice

1. D (page 917)
2. D (page 917)
3. A (page 917)
4. C (page 917)
5. B (page 917)
6. D (page 917)
7. C (page 918)
8. D (page 918)
9. B (page 918)
10. C (page 919)
11. B (page 933)
12. C (page 933)
13. D (page 934)
14. A (page 935)

Fill-in

1. pediatrics (page 916)
2. tongue (page 917)
3. trachea (page 917)
4. 200 (page 917)
5. pale skin (page 917)
6. fontanels (page 917)
7. adolescents (page 920)

True/False

1. T (page 918)
2. T (page 918)
3. F (page 920)
4. F (page 918)

5. T (page 919)
6. T (page 933)
7. F (page 935)
8. T (page 935)

Short Answer

1. At first, infants respond mainly to physical stimuli. Crying is the main avenue of expression. Later, infants learn to coo, smile, roll over, and recognize caregivers. Usually, they demonstrate no stranger anxiety. Observe infants from a distance first and allow the caregiver to hold the baby as you perform the examination. (page 918)

2. –Children who were born prematurely and have associated lung disease problems
   –Small children or infants with congenital heart disease
   –Children with neurologic disease, such as cerebral palsy
   –Children with chronic disease or with functions that have been altered since birth (page 933)

3. Toddlers begin to walk and explore the environment. Because they are able to open doors, boxes, and so forth, injuries in this age group are more frequent. Stranger anxiety develops early in this period. They are also developing their own ideas about almost everything. Make as many observations as you can before touching the child. When appropriate, examine the child on the caregiver's lap and use a distraction. Allowing the child to examine your equipment may be enough distraction to allow you to complete an assessment. (page 918)

4. These children are beginning to act more like adults and can think in concrete terms, respond sensibly to direct questions, and help take care of themselves. Talk to the child, not just the parent, when obtaining a history. Give the child choices and explain any procedures. Give them simple explanations and carry on a conversation to distract them. Reward the child after a procedure. (page 920)

5. Adolescents are able to think abstractly, can participate in decision making, and are more susceptible to peer pressure. They are very concerned about body image and how they appear to peers and others. Respect their privacy at all times. They can understand complex concepts and treatment options and should be included on decision making concerning their care. Advise them in advance of painful procedures and talk with them about their interests to distract them. (page 920)

Word Fun

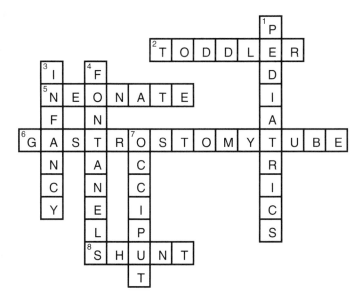

Ambulance Calls

1. Young children typically experience febrile seizures when their temperature rises rapidly, not simply the temperature of the fever itself. As with any call, you should assess airway, breathing, and circulation of this 2-year-old. Ensure that his airway is patent, assist him with breathing using a BVM device and airway adjunct as necessary, apply high-flow oxygen, and remove excessive clothing. The child's level of consciousness should improve, but if it doesn't or the child experiences additional seizure activity, this is a very serious sign that should be relayed to the receiving emergency department.

2. –Immediately open and suction the airway.

   –Assess breathing and apply high-flow oxygen by nonrebreathing mask or BVM device.

   –Assess patient further en route.

   –Rapid transport

   –Obtain history from grandmother en route.

   –Reassess patient's airway and vital signs en route.

3. When a patient cannot answer questions at all or can only respond in one- or two-word responses, this is a true emergency. She obviously has a history of respiratory disease or condition that requires her use of a metered-dose inhaler. You can ask her "yes" or "no" responses to aid in clarification of her medical history. Provide rapid transport and high-flow oxygen and assist her with her MDI in accordance with your local protocols.

Chapter 32: Pediatric Assessment and Management

Matching

1. D (page 958)
2. F (page 959)
3. H (page 958)
4. J (page 959)
5. L (page 959)
6. I (page 952)
7. K (page 976)
8. C (page 976)
9. E (page 981)
10. G (page 968)
11. B (page 978)
12. A (page 975)
13. M (page 944)

Multiple Choice

1. D (page 950)
2. B (page 952)
3. D (page 955)
4. D (page 957)
5. A (page 958)
6. D (page 959)
7. C (page 944)
8. D (page 975)
9. A (page 977)
10. D (page 977)
11. B (page 978)
12. D (page 978)
13. C (page 978)
14. D (page 943)
15. A (page 976)
16. C (page 963)
17. D (page 964)
18. D (page 970)
19. B (page 967)
20. D (page 968)
21. B (page 943)
22. C (page 971)
23. C (page 944)
24. B (page 977)
25. B (page 957)
26. C (page 945)

Fill-in

1. status epilepticus (page 977)
2. see, hear (page 942)
3. Dehydration (page 978)
4. Airway adjuncts (page 950)
5. gag reflex (page 951)
6. Skin condition (page 944)
7. tidal volume (page 955)

8. Infection (page 958)
9. seizure (page 977)
10. head (page 971)
11. shock (page 976)

True/False

1. T (page 955)
2. F (page 978)
3. T (page 978)
4. F (page 978)
5. T (page 978)
6. T (page 978)
7. F (page 978)
8. T (page 969)
9. T (page 976)

Short Answer

1. By the number of wet diapers: 6 to 10 per day is normal. (page 753)
2. –Nasal flaring
   –Grunting
   –Wheezing, stridor, or other abnormal airway sounds
   –Use of accessory muscles
   –Retractions
   –Tripod position (page 746)

Word Fun

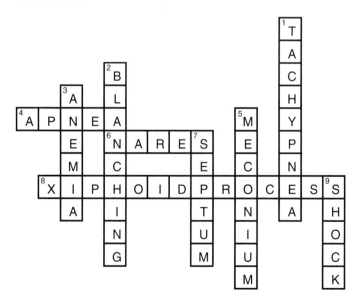

Ambulance Calls

1. This child has very likely aspirated vomitus into his lungs. You must immediately assess and manage his airway. This will require suctioning, use of airway adjuncts, and assisting his ventilations with high-flow oxygen and a BVM device. It appears that this boy has consumed alcohol and/or drugs, and the condition and situation of the scene require your notification of law enforcement. Ensure that no trauma has occurred or recently occurred to this patient; when in doubt, apply full c-spine precautions. You may find that obtaining information regarding medical conditions or history of recent trauma or illness is challenging in situations dealing with minors who fear punishment from their parents or guardians or consequences with law enforcement.

2. –Allow the child to remain in the mother's arms to decrease her anxiety.
–Offer oxygen via a nonrebreathing mask with the mother holding it.
–If she will not tolerate the nonrebreathing mask, use blow-by oxygen.
–Allow the mother to ride in the patient compartment to comfort the child.
–Provide rapid transport in position of comfort with as much oxygen as she will tolerate.
–Continually assess patient for signs of altered mental status and decreasing tidal volume; be prepared to assist ventilations.
–Obtain further history en route.

3. This infant is deceased, possibly as a result of sudden infant death syndrome (SIDS). After you have quickly assessed the infant and made this determination, you must communicate the condition of the baby to the mother and family. This may be difficult, and they will possibly request resuscitation attempts regardless of your findings. This becomes a judgment call, which can be clarified by utilizing online medical direction and/or standing orders. You should survey the scene and document any history of recent illness, congenital conditions, and so forth. Be supportive of family members and assist them as appropriate. Calls involving infants and children can be traumatic experiences for emergency medical providers as well. Request debriefing as necessary and follow local protocols.

Skill Drills

Skill Drill 32-1: Positioning the Airway in a Child (page 951)
1. Position the child on a **firm** surface.
2. Place a **folded** towel about 1 inch thick under the **shoulders** and **back**.
3. **Immobilize** the forehead to limit **movement** and use the head tilt-chin lift to open the airway.

Skill Drill 32-2: Inserting an Oropharyngeal Airway in a Child (page 953)
1. Determine the **appropriately sized** airway.
Confirm the correct size **visually**, by placing it next to the patient's **face**.
2. Position the patient's **airway** with the appropriate method.
3. Open the mouth.
Insert the airway until the **flange** rests against the **lips**.
Reassess the airway.

Skill Drill 32-3: Inserting a Nasopharyngeal Airway in a Child (page 954)
1. Determine the correct airway size by comparing its **diameter** to the opening of the **nostril** (naris).
Place the airway next to the patient's **face** to confirm correct **length**.
Position the airway.
2. **Lubricate** the airway.
Insert the **tip** into the right naris with the bevel pointing toward the **septum**.
3. Carefully move the tip forward until the **flange** rests against the **outside** of the nostril.
Reassess the **airway**.

Skill Drill 32-4: One-rescuer BVM Ventilation on a Child (page 957)
1. Open the airway and insert the appropriate airway adjunct.
2. Hold the mask on the patient's face with a one-handed head tilt-chin lift technique (E-C grip).
Ensure a good mask-face seal while maintaining the airway.
3. Squeeze the bag using the correct ventilation rate of 12–20 breaths/min. Allow adequate time for exhalation.
4. Assess effectiveness of ventilation by watching bilateral rise and fall of the chest.

Skill Drill 32-5: Removing a Foreign Body Airway Obstruction in an Unconscious Child (page 961)

1. Position the child on a **firm**, flat surface.
2. Inspect the **airway**. Remove any visible **foreign** object that you see.
3. Attempt **rescue** breathing. If unsuccessful, **reposition** the head and try again.
4. Locate the proper **hand** position on the chest of the child.
   If ventilation is still **unsuccessful**, begin CPR.
5. Administer **30** chest compressions and look inside the child's mouth. If you see the object, **remove** it.

Skill Drill 32-6: Performing Infant Chest Compressions (page 969)

1. Position the infant on a **firm** surface while **maintaining** the airway.
   Place two **fingers** in the **middle** of the sternum just below a line between the **nipples**.
2. Use two fingers to **compress** the chest about **one third to one half its depth** at a rate of 100 times/min.
   Allow the sternum to return **briefly** to its **normal** position between compressions.

Skill Drill 32-7: Performing CPR on a Child (page 970)

1. Place the child on a **firm** surface, open the airway, and deliver **two** rescue breaths.
2. Place the **heel** of one or both hands in the **center** of the chest, in between the nipples, avoiding the **xiphoid** process.
3. Compress the chest **one third** to **one half** the depth of the chest at a rate of **100** times/min. Coordinate compressions with ventilations in a 30:2 ratio (one rescuer) or 15:2 (two rescuers), pausing for **ventilations**.
4. Reassess for **breathing** and **pulse** after every **5** cycles (about 2 minutes) of CPR. If the child resumes **effective** breathing, place him or her in a position that allows **frequent** reassessment of the airway and vital signs during transport.

Skill Drill 32-8: Immobilizing a Child (page 972)

1. Use a towel under the back, from the **shoulders** to the **hips**, to maintain the head in a **neutral** position.
2. Apply an appropriately sized **cervical collar**.
3. **Log roll** the child onto the **immobilization** device.
4. Secure the **torso** first.
5. Secure the **head**.
6. Ensure that the child is **strapped** in properly.

Skill Drill 32-10: Immobilizing an Infant Out of a Car Seat (page 974)

1. Stabilize the head in neutral position.
2. Place an immobilization device between the patient and the surface he or she is resting on.
3. Slide the infant onto the board.
4. Place a towel under the back, from the shoulders to the hips, to ensure neutral head position.
5. Secure the torso first; pad any voids.
6. Secure the head.

Chapter 33: Geriatric Emergencies

Matching

1. E (page 989)
2. C (page 989)
3. G (page 994)
4. I (page 994)
5. J (page 992)

6. F (page 993)
7. D (page 993)
8. H (page 933)
9. A (page 988)
10. H (page 989)

Multiple Choice

1. B (page 988)
2. D (page 986)
3. D (page 986)
4. A (page 988)
5. C (page 988)
6. D (page 988)
7. B (page 989)
8. C (page 989)
9. B (page 989)
10. D (page 989)
11. A (page 989)
12. C (page 989)
13. A (page 989)
14. D (page 989)
15. B (page 990)

16. B (page 994)
17. A (page 994)
18. D (page 986)
19. C (page 991)
20. B (page 992)
21. A (page 991)
22. D (page 991)
23. D (page 991)
24. A (page 991)
25. D (page 992)
26. C (page 994)
27. D (page 995)
28. B (page 995)
29. B (page 996)
30. D (page 998)

Fill-in

1. 65 (page 986)
2. mask (page 986)
3. stereotypes (page 988)
4. sweat glands (page 988)
5. Cardiac output (page 989)
6. aneurysm (page 989)
7. kyphosis (page 990)

True/False

1. F (page 989)
2. T (page 988)
3. T (page 988)
4. F (page 988)

5. T (page 989)
6. T (page 989)
7. F (page 996)

Short Answer

1. –Physical
  –Psychological
  –Financial (page 996)

2. 1. Cardiac dysrhythmias/dysrhythmias/heart attack: The heart is beating too fast or too slowly, the cardiac output drops, and blood flow to the brain is interrupted. A heart attack can also cause syncope.

2. Vascular and volume: Medication interactions can cause venous pooling, and vasodilation, widening of the blood vessel, results in a drop in blood pressure and inadequate blood flow to the brain. Another cause of syncope can be a drop in blood volume because of hidden bleeding from a condition such as an aneurysm.

3. Neurologic: A transient ischemic attack (TIA) or "brain attack" can sometimes mimic syncope. (page 993)

3. –Repeated visits to the emergency department or clinic
–A history of being "accident prone"
–Soft-tissue injuries
–Unbelievable or vague explanations of injuries
–Psychosomatic complaints
–Chronic pain
–Self-destructive behavior
–Eating and sleep disorders
–Depression or lack of energy
–Substance and/or sexual abuse (page 997)

4. –Dyspnea
–Weak feeling
–Syncope/confusion/altered mental status (page 993)

Word Fun

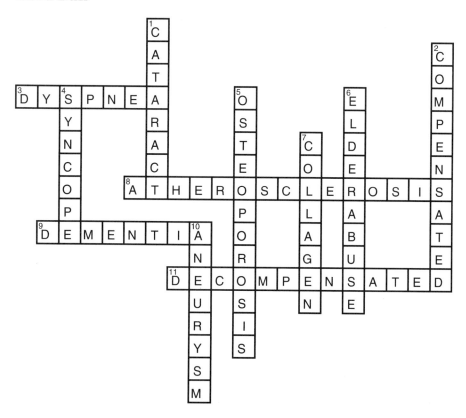

Ambulance Calls

1. –Survey the scene for any signs or clues to the patient's care.
   –Ask about patient's normal mental status.
   –Apply high-flow oxygen.
   –Obtain patient's medications to take to the hospital.
   –Place patient supine on stretcher with legs elevated.
   –Keep patient warm.
   –Rapid transport
   –Monitor vital signs en route.

2. This situation appears to be one of neglect and possible elder abuse. Perform a thorough assessment, especially if the patient is confused or is otherwise unable to express how or why she ended up on the floor. Ask for the patient's medical chart to obtain accurate information regarding her medical condition(s) and current medications. Immobilize her affected leg (and spine if needed) using the most comfortable methods possible. Geriatric patients need gentle care and many cannot tolerate conventional methods of splinting and immobilizing. Mechanisms of injury not viewed as significant in young, healthy patients can prove quite devastating or result in serious injury in older patients with frail skin and brittle bones.

3. Osteoporosis can be quite insidious in its onset. Individuals who were once relatively strong and healthy can suddenly find themselves with a fracture from something they may have done many times in the past. Postmenopausal, thin, Caucasian women are at higher risk of developing osteoporosis, and a spinal fracture in this scenario should be suspected. Perform an assessment to determine the presence of other injuries, provide full spinal immobilization, apply oxygen, and promptly transport this patient to the nearest appropriate facility. Always follow local protocols.

Chapter 34: Geriatric Assessment and Management

Matching

1. E (page 1012) **5.** C (page 1007)
2. G (page 1015) **6.** A (page 1015)
3. F (page 1010) **7.** D (page 1009)
4. B (page 1010) **8.** H (page 1015)

Multiple Choice

1. D (page 1004) **7.** A (page 1009)
2. D (page 1016) **8.** D (page 1010)
3. C (page 1005) **9.** B (page 1012)
4. A (page 1006) **10.** C (page 1013)
5. A (page 1007) **11.** C (page 1015)
6. D (page 1008) **12.** D (page 1016)

Fill-in

1. polypharmacy (page 1007)   **6.** compression fracture (page 1010)
2. stable (page 1009)   **7.** burst (page 1010)
3. unstable (page 1009)   **8.** seat belt fracture (page 1010)
4. acetabulum (page 1012)   **9.** septicemia (page 1015)
5. central cord syndrome (page 1010)   **10.** bacteremia (page 1015)

True/False

1. T (page 1015) **6.** T (page 1007)
2. T (page 1016) **7.** T (page 1007)
3. F (page 1004) **8.** F (page 1010)
4. T (page 1005) **9.** F (page 1012)
5. T (page 1006)

Short Answer

1. Apply and maintain cervical stabilization.
Apply a cervical collar.
Rescuers should kneel on one side of the patient.
Rescuers roll the patient toward themselves and quickly examine the patient's back.
Slide the backboard under the patient, and roll the patient onto the board.
Pad the void space produced by the kyphotic spine.
Secure the torso to the backboard with straps.
Secure the patient's head and padding. (page 1010)

2. Assess pulse, motor, and sensory function of the extremity.
Place the patient on a scoop stretcher or long backboard.
Place a blanket roll between the patient's legs.
Place blankets and pillows under the injured extremity.
Secure both legs and the padding to the backboard with at least three cravats or straps.
Reassess pulse, motor, and sensory function. (page 1013)

3. What is the patient's chief complaint today?
What initial problem caused the patient to be admitted to the facility? (page 1016)

Word Fun

```
              1                2
              S                B
              E                A
              P                C
              T                T
              I                E
3          4
B  U  R  S  T  F  R  A  C  T  U  R  E  S
         Y           E           E
         N           M           M
         C           I           I
      5
      P  O  L  Y  P  H  A  R  M  A  C  Y
         P
   6
   A  C  E  T  A  B  U  L  U  M
```

Ambulance Calls

1. Straining or "bearing down" during a bowel movement is done by trying to forcefully exhale against a closed glottis or vocal cords. This can cause a significant decrease in heart rate that is not always well-tolerated by geriatric patients. When the Valsalva maneuver just described results in a temporary loss of consciousness, it is referred to as vasovagal syncope. Attempt to determine whether or not the patient lost consciousness before and/or after the event. As well as attempting to determine the cause of the fall, you must now care for his resultant head injury and potential spinal fractures. You should control any bleeding, provide full spinal immobilization, apply high-flow oxygen, provide prompt transport, and perform a thorough assessment and history taking to determine the presence of any other injuries or medical conditions.

2. Hip fractures are a misnomer because true "hip fractures" are in fact fractures of the femur. You should determine if the mechanism of injury or suspicion exists for possible spinal fractures. Survey the scene to determine if there was a reason for the fall or if the break was spontaneous, giving the patient the impression of "tripping" on an object. If you believe that it was a spontaneous fracture, this patient suffers from extreme osteoporosis and likely suffered other fractures during the fall. If no spinal immobilization is deemed necessary, isolated hip fracture care requires an assessment of the patient's pulse, motor, and sensation of both lower extremities. You should also determine if the affected extremity is the same or shorter than the uninjured side. It is also helpful to note if the leg is rotated inward, outward, or not at all. Immobilize the injured leg using blankets, pillows, and cravats or straps after placing the patient on a long spine board or scoop stretcher. After immobilization has occurred, reassess the pulse, motor, and sensory functions of the injured extremity. Provide reassurance for the patient because many feel that hip fractures signal the loss of their independence. Always follow local protocols.

3. Patients with kyphosis are extremely uncomfortable when immobilized on long spine boards. You should pad all voids using pillows, blankets, towels, or other appropriate forms of padding. Failure to adequately pad a patient's spine can result in increased pain and the inability of the patient to remain still. This can result in exacerbated injuries as well as a very unhappy patient. Thoracic spine injuries and injuries to the ribs can make it difficult for patients to breathe without pain. Determine if her shortness of breath is related to pain during inspiration and/or if she has a history of respiratory disease such as emphysema, chronic bronchitis, or asthma. Local weather conditions have probably contributed to her fall. Depending upon the length of exposure to the elements, she could be suffering from hypothermia as well. Geriatric patients generally do not tolerate extremes in weather; therefore avoiding extended periods of time in cold temperatures will minimize the likelihood for hypothermia.

Chapter 35: Ambulance Operations

Matching

1. D (page 1048)
2. A (page 1044)
3. E (page 1043)
4. G (page 1040)
5. F (page 1024)
6. C (page 1040)
7. B (page 1040)

Multiple Choice

1. C (page 1024)
2. A (page 1024)
3. A (page 1027)
4. B (page 1025)
5. D (page 1027)
6. D (page 1028)
7. A (page 1028)
8. D (page 1028)
9. D (page 1028)
10. B (page 1029)
11. D (page 1032)
12. A (page 1032)
13. D (page 1034)
14. D (page 1034)
15. C (page 1035)
16. C (page 1036)
17. D (page 1040)
18. B (page 1036)
19. D (page 1041)
20. B (page 1038)
21. A (page 1042)
22. C (page 1044)
23. D (page 1046)
24. B (page 1046)
25. C (page 1046)
26. C (page 1046)
27. B (page 1047)
28. D (page 1036)
29. D (page 1037)
30. C (page 1038)
31. D (page 1040)
32. D (page 1049)

Fill-in

1. jump kit (page 1032)
2. Star of Life (page 1026)
3. hearse (page 1024)
4. First responder vehicles (page 1024)
5. nine (page 1026)
6. decontaminate (page 1040)
7. airway (page 1028)
8. CPR board (page 1030)

True/False

1. T (page 1027)   **6.** T (page 1042)

2. F (page 1030)   **7.** T (page 1049)

3. T (page 1039)   **8.** F (page 1047)

4. F (page 1036)   **9.** F (page 1049)

5. F (page 1044)

Short Answer

1. Type I: Conventional, truck cab chassis with modular ambulance body that can be transferred to a newer chassis as needed

Type II: Standard van, forward-control integral cab-body ambulance

Type III: Specialty van, forward-control integral cab-body ambulance (page 1025)

2.   **1.** Preparation for the call

   2. Dispatch

   3. En route to scene

   4. Arrival at scene

   5. Transfer of patient to the ambulance

   6. En route to the receiving facility (transport)

   7. At the receiving facility (delivery)

   8. En route to station

   9. Postrun (page 1026)

3.   **1.** Lack of experience of the dispatcher

   2. Inadequate equipment in the ambulance

   3. Inadequate training of the EMT

   4. Inadequate driving ability

   5. Siren syndrome (page 1044)

4.   **1.** To the best of your knowledge, the unit must be on a true emergency call.

   2. Both audible and visual warning devices must be used simultaneously.

   3. The unit must be operated with due regard for the safety of all others, on and off the roadway. (page 1046)

5.   **1.** At the time of dispatch, select the shortest and least congested route to the scene.

   2. Avoid routes with heavy traffic congestion.

   3. Avoid one-way streets.

   4. Watch carefully for bystanders as you approach the scene.

   5. Once you arrive at the scene, park the ambulance in a safe place.

   6. Drive within the speed limit while transporting patients, except in the rare extreme emergency.

   7. Go with the flow of the traffic.

   8. Use the siren as little as possible en route.

   9. Always drive defensively.

   10. Always maintain a safe following distance.

   11. Try to maintain an open space in the lane next to you as an escape route in case the vehicle in front of you stops suddenly.

   12. Use your siren if you turn on the emergency lights, except when you are on a freeway.

   13. Always assume that other drivers will not hear the siren or see your emergency lights. (page 1042)

6. Approach from the front of the aircraft using an approach area of between nine o'clock and three o'clock as the pilot faces forward. (page 1049)

7. –A clear site that is free of loose debris, electric or telephone poles and wires, or any other hazards that might interfere with the safe operation of the helicopter

–A minimum of 100 ft. by 100 ft. is recommended for the landing zone. (page 1049)

Word Fun

Ambulance Calls

1. Fire departments have a distinct chain of command. You should obtain permission to place yourself en route to this medical emergency from your shift captain or appropriate fire officer. Just because a call is dispatched does not mean you should automatically place yourself en route to that location. The incident commander of the confirmed structure fire may need your immediate assistance on the scene. Because you do not know the nature medical emergency at this point, it would be prudent to place that decision in appropriate hands. Always follow local protocols and chain of command.

2. Ideally, you should be very familiar with your response area, and you should take time every shift to study local roads. This can be done individually or as a group shift activity. Obviously, the larger your coverage area the more difficult it becomes to memorize addresses, especially if you are not native to the area. Regardless, you should be able to read maps quickly, and you should not rely on your memory to assist in locating a new address. To attempt to memorize all local streets is a good goal, but failing to consult with a map before leaving the station for reasons of pride is downright foolish. Ensure that each and every agency vehicle used in response to emergencies is equipped with local maps and other resources to ensure you arrive promptly at emergency scenes. Utilize your local dispatch center for directions if necessary.

3. –Park your ambulance in a safe area.
   –Ensure personal safety.
   –Either you or your partner should handle traffic control until the police arrive.
   –The person not handling traffic control should assess the patient.
   –Provide patient care while ensuring personal safety and patient safety.

Chapter 36: Gaining Access

Matching

1. E (page 1059)
2. A (page 1062)
3. D (page 1062)
4. F (page 1060)
5. B (page 1065)
6. G (page 1068)
7. C (page 1068)

Multiple Choice

1. D (page 1060)
2. B (page 1060)
3. C (page 1060)
4. A (page 1060)
5. D (page 1059)
6. A (page 1062)
7. B (page 1061)
8. D (page 1063)
9. D (page 1065)
10. C (page 1065)
11. D (page 1066)
12. B (page 1067)

Fill-in

1. Removal (page 1059)
2. safety (page 1058)
3. communication (page 1058)
4. stable (page 1061)
5. safest (page 1063)
6. harm (page 1064)
7. self-contained breathing apparatus (page 1068)

True/False

1. F (page 1060)
2. T (page 1059)
3. T (page 1058)
4. F (page 1062)
5. T (page 1061)

Short Answer

1. –Firefighters: responsible for extinguishing any fire, preventing additional ignition, ensuring that the scene is safe, and washing down spilled fuel.
   –Law enforcement: responsible for traffic control and direction, maintaining order at the scene, investigating the crash or crime scene, and establishing and maintaining lines so that bystanders are kept at a safe distance and out of the way of rescuers.
   –Rescue group: responsible for properly securing and stabilizing the vehicle, providing safe entrance and access to patients, extricating any patients, ensuring that patients are properly protected during extrication or other rescue activities, and providing adequate room so that patients can be removed properly.
   –EMS personnel: responsible for assessing and providing immediate medical care, triage and assigning priority to patients, packaging the patient, providing additional assessment and care as needed once the patient has been removed, and providing transport to the emergency department. (page 1060)

2. Is the patient in a vehicle or in some other structure?
   Is the vehicle or structure severely damaged?
   What hazards exist that pose risk to the patient and rescuers?
   In what position is the vehicle? On what type of surface? Is the vehicle stable or is it apt to roll or tip? (page 1062)

3. **1.** Provide manual stabilization to protect c-spine, as needed.

   2. Open the airway.

   3. Provide high-flow oxygen

   4. Assist or provide for adequate ventilation.

   5. Control any significant external bleeding.

   6. Treat all critical injuries. (page 1063)

4. The extent of injury and whether there is a possibility of hidden bleeding
   Evaluate sensation in the trapped area to discover if an object is pressing on or impaled in the patient. (page 1063)

5. Ensure that each EMT-B can be positioned so that he/she can lift and carry at all times.
   Move the patient in a series of smooth, slow, controlled steps, with stops designed between to allow for repositioning and adjustments.
   Plan the exact steps and pathway that you will follow.
   Choose a path that requires the least manipulation of the patient or equipment.
   Make sure that sufficient personnel are available.
   Make sure that you move the patient as a unit.
   While moving the patient, continue to protect him/her from any hazards. (page 1064)

Word Fun

Ambulance Calls

1. You cannot assume that the man has left the premises, and you must act in the best interest of the patient. If the man is unconscious or otherwise incapacitated, time is of the essence. You should look inside the residence to see if the patient (in whole or part) is visible. Law enforcement should be immediately summoned to the scene for legal purposes if you feel that forced entry is justified. Always "try before you pry" meaning check open windows and doors to gain access, and if this is not possible, choose a small window to break rather than a door. It is less expensive to repair. Sometimes windows adjacent to doors can be broken, and access to the interior lock can be gained. You must be careful when entering, as there is a family dog that will most likely protect the property. Animal control should be requested if you feel the dog presents a safety hazard. It is prudent to err on the side that the patient needs immediate assistance rather than waiting for the third party caller to arrive with keys. Always follow local protocols.

2. Entering the water could be a death sentence for you both. If you have a cellular phone and/or department radio with you, your best course of action would be to inform incoming units of the boy's location and situation. It would be advisable to direct some responding units down stream so that if the boy should lose his grasp on the log, other responders will be available to retrieve him. If your department has areas that contain the possibility of swift water rescue (even if only seasonal), training should be conducted and appropriate helmets, throw bags and life jackets should be made available for safe response. You must assess the scene before trying to effect immediate rescue operations. Many responders have been killed by "jumping into" all types of rescues without performing a scene size-up or using the proper gear. Don't become a victim. Doing so will only make yourself part of the problem.

3. Try to learn as much about the chemical as possible by having the dispatcher contact CHEMTREC or another agency to find out about possible effects to the patient.

   Prepare necessary equipment to manage the airway and ventilation.

   Be prepared to do CPR if necessary.

   Have all equipment within reach.

   Once patient is brought to you, rapidly begin to manage the ABCs and prepare for rapid transport.

Chapter 37: Special Operations

Matching

1. J (page 1079)
2. K (page 1074)
3. L (page 1089)
4. D (page 1076)
5. H (page 1091)
6. E (page 1092)
7. B (page 1084)
8. I (page 1084)
9. A (page 1085)
10. F (page 1079)
11. G (page 1077)
12. C (page 1077)

Multiple Choice

1. D (page 1074)
2. D (page 1075)
3. C (page 1075)
4. A (page 1076)
5. D (page 1079)
6. C (page 1077)
7. D (page 1083)
8. B (page 1076)
9. D (page 1079)
10. D (page 1091)
11. B (page 1091)
12. D (page 1090)
13. A (page 1090)
14. B (page 1087)
15. D (page 1087)
16. D (page 1087)
17. D (page 1086)
18. D (page 1087)
19. B (page 1086)
20. D (page 1086)
21. A (page 1083)
22. B (page 1083)
23. D (page 1083)
24. C (page 1083)
25. A (page 1083)
26. D (page 1083)
27. A (page 1083)
28. B (page 1083)
29. D (page 1083)
30. C (page 1083)

Fill-in

1. incident command system (page 1074)
2. command post (page 1074)
3. safety officer (page 1075)
4. hazardous materials incident (page 1085)
5. extrication (page 1076)
6. disaster (page 1084)
7. Triage (page 1076)
8. contaminated (page 1091)
9. airway, breathing (page 1090)
10. Decontamination (page 1089)
11. relocate (page 1087)
12. toxic (page 1086)
13. placard (page 1086)

True/False

1. T (page 1085)
2. F (page 1088)
3. F (page 1089)
4. F (page 1092)
5. T (page 1083)
6. F (page 1083)
7. T (page 1086)
8. F (page 1089)
9. F (page 1090)
10. T (page 1076)

Short Answer

1. Level 0: Materials that would cause little, if any, health hazard if you came into contact with them
   Level 1: Materials that would cause irritation on contact, but only mild residual injury, even without treatment
   Level 2: Materials that could cause temporary damage or residual injury unless prompt medical treatment is provided
   Level 3: Materials that are extremely hazardous to health
   Level 4: Materials that are so hazardous that minimal contact will cause death (page 1089)

2. Level A: Fully encapsulated, chemical-resistant protective clothing; SCBA; special, sealed equipment
   Level B: Nonencapsulated protective clothing, or clothing designed to protect against a particular hazard; SCBA; eye protection
   Level C: Nonpermeable clothing; eye protection; facemasks
   Level D: Work uniform (page 1092)

3. 1. Command
   2. Staging
   3. Extrication
   4. Triage
   5. Treatment
   6. Supply
   7. Transportation
   8. Rehabilitation (page 1075)

4. First priority (red): Patients who need immediate care and transport
   Second priority (yellow): Patients whose treatment and transportation can be temporarily delayed
   Third priority (green): Patients whose treatment and transportation can be delayed until last
   Fourth priority (black): Patients who are already dead or have little chance for survival (page 1081)

5. (page 1088)

Class	Type
Class 1	Explosives
Class 2	Gases
Class 3	Flammable liquids
Class 4	Flammable solids
Class 5	Oxidizers
Class 6	Poisons
Class 7	Radioactive
Class 8	Corrosives
Class 9	Miscellaneous

6. The process of removing or neutralizing and properly disposing of hazardous materials from equipment, patients, and rescue personnel. The decontamination area is the designated area where contaminants are removed before an individual can go to another area. (page 1089)

Word Fun

Ambulance Calls

1. –The 4-year-old should be triaged as first priority (red).
–The 27-year-old should be triaged as third priority (green).
–The 42-year-old should be triaged as fourth priority (black).

2. It's fortunate that this is occurring on shift change, as members from two shifts are available to respond to this mass casualty incident. Since this is a remote area, implying response times are extended, you should use this time to gather information from the crop duster pilot/responsible parties regarding the substance and notify all local hospitals. Area HazMat team members should be requested, and once the chemicals have been identified, appropriate first aid and other instructions should be relayed to the affected patients on-scene through the dispatch center. Law enforcement should be requested to cordon off the area to prevent more individuals such as co-workers and family members from entering the contaminated area. Consult HazMat team members and/or CHEMTREC for appropriate information including medical treatment, PPE and minimum distances that are required to avoid exposure to the identified substance. Obviously, do not enter the area unless you have been trained and have the proper equipment to do so.

3. The good news is that you are uphill from the possible hazardous materials. You should be uphill and upwind from contaminated areas (especially when dealing with an unidentified substance). Each chemical reacts differently to outside air, temperature and other ambient conditions. You should immediately instruct all civilians along the road to move out of the immediate area and/or instruct them to wait in a specific location if you feel that they have already been contaminated by the substance. Do not allow the driver to contaminate the passersby. Law enforcement should be immediately notified as well as the local HazMat team. Attempt to gain information regarding the shipment, keeping a safe distance (using binoculars, PA system to relay instructions, etc.) Prevent exposure to yourself and your crew and the members of the public. Always follow local protocols.

Chapter 38: Response to Terrorism and Weapons of Mass Destruction

Matching

1. I (page 1107)
2. G (page 1107)
3. E (page 1112)
4. A (page 1108)
5. J (page 1117)
6. C (page 1115)
7. B (page 1106)
8. F (page 1112)
9. D (page 1106)
10. H (page 1116)

Multiple Choice

1. D (page 1100)
2. A (page 1101)
3. B (page 1102)
4. B (page 1103)
5. A (page 1104)
6. A (page 1105)
7. D (page 1106)
8. C (page 1107)
9. A (page 1108)
10. A (page 1109)
11. A (page 1110)
12. B (page 1111)
13. A (page 1112)
14. B (page 1113)
15. D (page 1114)
16. D (page 1115)
17. C (page 1116)
18. A (page 1117)
19. A (page 1118)
20. B (page 1119)
21. B (page 1120)
22. D (page 1122)

Fill-in

1. domestic terrorism (page 1100)
2. weapon of mass destruction (page 1101)
3. State-sponsored terrorism (page 1102)
4. orange (page 1103)
5. Cross-contamination (page 1104)
6. Route of exposure (page 1106)
7. contact hazard (page 1107)
8. Nerve agents (page 1108)
9. Off-gassing (page 1109)
10. Dissemination (page 1112)
11. Virus (page 1113)
12. viral hemorrhagic fevers (page 1114)
13. Anthrax (page 1115)
14. lymph nodes (page 1116)
15. botulinum (page 1117)
16. Points of distribution (page 1119)
17. radiological dispersal device (page 1120)

True/False

1. F (page 1100)
2. T (page 1101)
3. F (page 1102)
4. T (page 1103)
5. T (page 1104)
6. F (page 1105)
7. F (page 1106)
8. F (page 1107)
9. T (page 1108)
10. F (page 1109)
11. F (page 1110)
12. T (page 1111)
13. T (page 1112)
14. F (page 1113)
15. F (page 1114)
16. T (page 1115)
17. T (page 1116)
18. F (page 1117)
19. F (page 1118)
20. F (page 1119)
21. T (page 1120)
22. T (page 1122)

Short Answer

1. What are your initial actions?
   Who should you notify, and what should you tell them?
   What type of additional resources might you require?
   How should you proceed to address the needs of the victims?
   How do you ensure your own and your partner's safety, as well as the safety of the victims?
   What is the clinical presentation of a victim exposed to a WMD?
   How are WMD patients to be assessed and treated?
   How do you avoid becoming contaminated or cross-contaminated with a WMD agent? (page 1100)

2. Vesicants
   Respiratory agents
   Nerve agents
   Metabolic agents (page 1102)

3. Type of location
   Type of call
   Number of patients
   Victim's statements
   Pre-incident indicators (page 1103)

4. Skin irritation, burning, and reddening
   Immediate intense skin pain
   Formation of large blisters
   Gray discoloration of skin
   Swollen and closed or irritated eyes
   Permanent eye injury (page 1107)

5. Shortness of breath
   Chest tightness
   Hoarseness and stridor due to upper airway constriction
   Gasping and coughing (page 1108)

6. Salivation, sweating
   Lacrimation
   Urination
   Defecation, drooling, diarrhea
   Gastric upset and cramps
   Emesis
   Muscle twitching (page 1110)

7. Shortness of breath and gasping respirations
   Tachypnea
   Flushed skin color
   Tachycardia
   Altered mental status
   Seizures
   Coma
   Apnea
   Cardiac arrest (page 1112)

8. Fever
   Chills
   Headache
   Muscle aches
   Nausea
   Vomiting
   Diarrhea
   Severe abdominal cramping
   Dehydration
   Gastrointestinal bleeding
   Necrosis of liver, spleen, kidneys, and gastrointestinal tract (page 1118)

9. Hospitals
   Colleges and universities
   Chemical and industrial sites (page 1120)

10. Time
    Distance
    Shielding (page 1122)

Word Fun

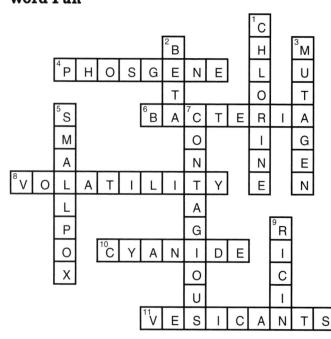

Ambulance Calls

1. There is no confirmation regarding the nature of the explosion. This could have been caused by a number of hazards, but any time there is the chance for a terrorist attack, precautions should be taken. No one should rush into the scene, because terrorists have been known to deliberately target first responders by placing secondary explosive devices. Request all available resources including local, state, and federal specialized HazMat and law enforcement agencies to assist in this call. Notify all local and regional hospitals of the situation; assess the scene from a distance. Establish command and designate a staging area for incoming units. Enter the scene only when it has been determined to be safe. Regularly scheduled mock drills to include expected agencies to respond in this type of situation can greatly improve communications and the overall effectiveness and efficiency of response.

2. The presence of numerous people with the same signs and symptoms, who were in the same location at the same time, should immediately send up red flags. These patients were exposed to contaminated food, air, or water in which they ingested some sort of toxin. Ingestion of ricin can produce these signs and symptoms and will usually be seen 4 to 8 hours after exposure. It can be difficult to initially determine what substances caused these signs and symptoms, but through careful history taking and investigation of the patients' commonalities, the exposure can be found. Unfortunately for those exposed to ricin, there is no vaccination or other specific treatment available. Only supportive measures for airway, breathing, and circulation can be applied.

3. Cyanide is colorless and has an odor similar to bitter almonds. It interferes with the body's ability to utilize oxygen and can result in headache, shortness of breath, tachypnea, altered levels of consciousness, apnea, seizures, and even death. Antidotes can be given, but are rarely carried on ambulances. Patients' clothing must be removed to avoid exposure to the cyanide because it is released as a gas from the clothing fibers. For most patients, simply removing them from the source of cyanide and providing supportive therapy for their ABCs will be all that is needed. However, those with significant exposure may require aggressive airway intervention.

Chapter 39: Advanced Airway Management

Matching

1. K (page 1132)
2. G (page 1132)
3. L (page 1132)
4. E (page 1132)
5. J (page 1132)
6. C (page 1133)
7. B (page 1132)
8. M (page 1146)
9. H (page 1135)
10. A (page 1142)
11. I (page 1137)
12. D (page 1139)
13. F (page 1139)

Multiple Choice

1. D (page 1132)
2. B (page 1132)
3. C (page 1132)
4. C (page 1132)
5. A (page 1132)
6. B (page 1132)
7. B (page 1132)
8. A (page 1132)
9. D (page 1132)
10. C (page 1133)
11. D (page 1134)
12. B (page 1134)
13. A (page 1135)
14. B (page 1135)
15. D (page 1137)
16. B (page 1137)
17. D (page 1138)
18. A (page 1139)
19. B (page 1139)
20. C (page 1141)
21. A (page 1141)
22. C (page 1141)
23. B (page 1142)
24. B (page 1142)
25. A (page 1141)
26. D (page 1138)
27. C (page 1145)
28. A (page 1145)
29. C (page 1146)
30. D (page 1150)
31. D (page 1150)
32. B (page 1137)
33. D (page 1151)
34. B (page 1152)
35. D (page 1153)
36. D (page 1154)
37. C (page 1155)
38. D (page 1156)

Labeling

The Upper and Lower Airways (page 1133)

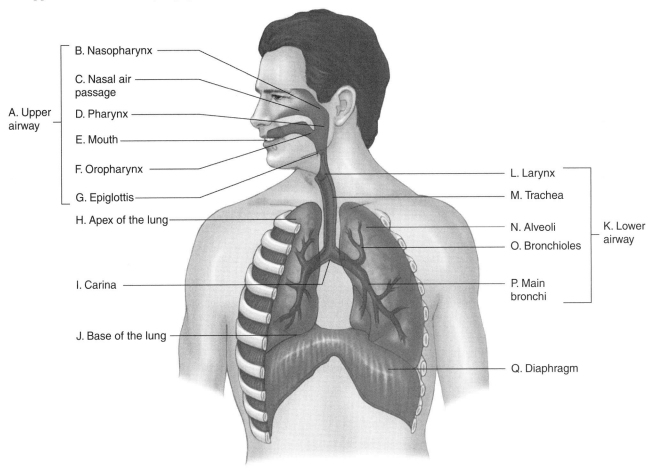

Fill-in

1. inhalation (page 1132)
2. bronchi (page 1132)
3. passive (page 1132)
4. oxygen (page 1132)
5. Carbon dioxide (page 1132)
6. 4 to 6 (page 1133)
7. bronchioles (page 1132)
8. capillaries (page 1132)

True/False

1. T (page 1140)
2. T (page 1137)
3. F (page 1140)
4. F (page 1141)
5. T (page 1142)
6. F (page 1142)

Short Answer

1. To perform the Sellick maneuver, place a thumb and index finger on either side of the midline of the cricoid cartilage. Apply firm—but not excessive—pressure, as too much pressure could collapse the larynx. Maintain this pressure until the patient is intubated. (page 1136)

2. 1. Completely controls and protects the airway
   2. Delivers better minute volume without the difficulty of maintaining an adequate mask seal
   3. If prolonged ventilation is required, it may be left in place for a long time.
   4. Prevents gastric distention
   5. Minimizes risk of aspiration of stomach contents into the respiratory system

6. Allows for direct access to the trachea for suctioning

7. Allows for delivery of high volumes of oxygen at higher than normal pressures

8. Provides a route for administration of certain medications (page 1137)

3. 1. Intubating the right mainstem bronchus

   2. Intubating the esophagus

   3. Aggravating a spinal injury

   4. Taking too long to intubate

   5. Patient vomiting

   6. Soft-tissue trauma

   7. Mechanical failure

   8. Patient intolerance of ETT

   9. Decrease in heart rate (page 1150)

4. 1. Conscious or semi-conscious patients with a gag reflex

   2. Children younger than age 16 years

   3. Adults shorter than 5 feet tall

   4. Patients who have ingested a caustic substance

   5. Patients who have a known esophageal disease (page 1153)

5. Benefits: Ease of proper placement; no mask seal necessary; requires minimal skill and practice to maintain; easily used in spinal injury patients; may be inserted blindly; protects the airway from upper airway secretions; stays in place well (ETC); sturdy balloon (ETC).

   Complications: Loses effectiveness (cuff malfunction); requires deeply comatose patient; requires constant balloon observation; cannot be used on patients shorter than 5' tall; requires great care in listening to breath sounds; large balloon is easily broken and tends to push the PtL out of the mouth when inflated. (page 1152)

Word Fun

Ambulance Calls

1. If successful intubation has occurred, you will hear equal, bilateral breath sounds and no sounds over the epigastrium. This endotracheal tube should be removed, and the patient should be ventilated with high-flow oxygen via BVM device and the placement of an oral airway. Use of secondary placement devices should be utilized when assessing placement of the ET tube. Direct visualization of the ET tube passing through the cords along with the use of esophageal detector devices and/or end-tidal carbon dioxide detectors can ensure successful placement has been accomplished. If endotracheal intubation is not possible, consider the use of multilumen airways such as the Esophageal Tracheal Combitube (ETC), Pharyngeotracheal Lumen Airway (PtL) or Laryngeal Mask Airway (LMA). Always follow local protocols.

2. In this case, either nasogastric or orogastric tubes can be placed. You should measure the tube, lubricate it if necessary, place the patient in the proper position, pass the tube through the mouth or nose, confirm proper placement, aspirate gastric contents using a large syringe or suction unit and secure the tube in place. It is important to remember that this technique should be performed when other airway and cardiac needs have been addressed. Do not delay the securing of the patient's airway and/or application of the AED in the cardiac arrest patient in order to place an OG or NG tube. Also, remember to utilize the Sellick Maneuver until gastric contents have been aspirated and the abdomen is no longer distended. Always follow local protocols.

3. –Maintain c-spine control.
   –Use portable suction if needed.
   –Ventilate with BVM and 100% oxygen while c-spine and jaw thrust is maintained.
   –Insert a combitube to protect the airway (great because it is a blind insertion).
   –Continue to ventilate through combitube until patient is extricated.

Skill Drills

Skill Drill 39-1: Performing the Sellick Maneuver (page 1136)

1. Visualize the **cricoid** cartilage.
2. **Palpate** to confirm its location.
   Apply **firm** pressure on the cricoid **ring** with your thumb and index finger on either side of the **midline**. Maintain pressure until the patient is **intubated**.

Skill Drill 39-2: Performing Orotracheal Intubation (page 1148)

1. Open and clear the airway.
   Place an oropharyngeal airway, and preoxygenate with a BVM device.
2. Assemble and test intubation equipment as your partner continues to ventilate.
3. Confirm adequate preoxygenation, and remove the oral airway.
   If available, have another rescuer perform the Sellick manuever to improve visualization of the cords.
   Use the head tilt-chin lift maneuver to position a nontrauma patient for insertion of the laryngoscope.
4. In a trauma patient, maintain the cervical spine in-line and neutral as your partner lies down or straddles the patient's head to visualize the vocal cords.
5. Insert the laryngoscope from the right side of mouth, and move the tongue to the left. Lift the laryngoscope away from the posterior pharynx to visualize the vocal cords. **Do not pry or use the teeth as a fulcrum.**
   Insert the ET tube from the right side until the ET tube cuff passes between the vocal cords. Remove the laryngoscope and stylet. Hold the tube carefully until it is secured.
6. Inflate the balloon cuff, and remove the syringe as your partner prepares to ventilate.
7. Begin ventilating, and confirm placement of the ET tube by listening over the stomach and both lungs. Also confirm placement with an end-tidal carbon dioxide detector or EDD, if available.
8. Secure the tube, and continue to ventilate.
   Note and record depth of insertion (centimeter marking at the teeth), and reconfirm position after each time you move the patient.

Chapter 40: Assisting With Intravenous Therapy

Matching

1. C (page 1170)
2. G (page 1171)
3. A (page 1172)
4. H (page 1167)
5. D (page 1172)
6. B (page 1174)
7. F (page 1170)
8. E (page 1172)

Multiple Choice

1. D (page 1166)
2. B (page 1167)
3. D (page 1168)
4. A (page 1170)
5. B (page 1171)
6. D (page 1172)
7. B (page 1173)
8. C (page 1174)
9. B (page 1175)
10. C (page 1176)

Fill-in

1. administration set (page 1167)
2. Macrodrip sets (page 1168)
3. Saline locks (page 1170)
4. Jamshedi needle (page 1171)
5. phlebitis (page 1172)
6. Catheter shear (page 1174)

True/False

1. T (page 1166)
2. F (page 1167)
3. F (page 1168)
4. T (page 1170)
5. F (page 1171)
6. T (page 1172)
7. T (page 1173)
8. F (page 1174)
9. T (page 1175)
10. F (page 1176)

Short Answer

1. Follow BSI precautions.
   Obtain solution and check the bag for clarity, expiration date, and correct solution.
   Choose appropriate administration set.
   Obtain the requested catheter.
   Spike the bag.
   Remove air from the tubing.
   Prepare tape for securing tubing.
   Open alcohol prep.
   Have 4 x 4 ready for bleeding.
   Dispose of sharps.
   Hook up tubing and adjust flow. (page 1166)

2. Infiltration
   Phlebitis
   Occlusion
   Vein irritation
   Hematoma (page 1172)

3. Allergic reactions
   Air embolus
   Catheter shear
   Circulatory overload
   Vasovagal reactions (page 1173)

4. Place patient in shock position.
   Apply high-flow oxygen.
   Monitor vital signs.
   ALS provider should insert an IV catheter in case fluid resuscitation is needed. (page 1174)

5. IV fluid
   Administration set
   Height of bag
   Type of catheter
   Constricting band (page 1175)

Word Fun

Ambulance Calls

1. You should immediately slow the IV to a TKO rate, raise the patient's head, and apply high-flow oxygen (or increase the current LPM). Notify the receiving facility of the error. Healthy adults can handle the sudden influx of intravenous fluids without detrimental effects, but individuals who have diseased or otherwise weakened hearts, lungs, or kidneys do not possess the ability to cope with the fluid overload. To avoid this occurrence, check and double check the drip chamber/flow rate to prevent accidental fluid boluses.

2. There are many things that can hinder the flow of IV fluids. One of the most obvious is failure to remove the tourniquet. Other contributing factors can include the condition of the IV fluids, the height of the bag, and perforation of the vein. Look for signs of these situations in order to determine the source of the problem. Do not automatically suggest the discontinuance of the IV, because the problem may be easily corrected.

3. Swelling around the insertion site indicates the infiltration of IV fluids into the surrounding tissues. This will also be identified with slowing or stopping of the IV fluid flow as well as pain and tightness around the IV site. When this occurs, the ALS provider must remove the IV and perform the venipuncture at another location. Direct pressure should be applied to minimize further swelling and bleeding.

Chapter 41: Assisting With Cardiac Monitoring

Matching

1. C (page 1187)
2. F (page 1188)
3. A (page 1182)
4. E (page 1187)
5. B (page 1187)
6. D (page 1188)
7. G (page 1185)

Multiple Choice

1. D (page 1182)
2. D (page 1183)
3. B (page 1184)
4. A (page 1185)
5. B (page 1187)
6. A (page 1188)
7. B (page 1189)
8. C (page 1191)
9. C (page 1192)

Fill-in

1. electrical conduction system (page 1182)
2. Sinus rhythm (page 1185)
3. Arrhythmia (page 1187)
4. Ventricular fibrillation (page 1188)
5. Asystole (page 1189)
6. limb leads (page 1191)
7. Chest leads (page 1192)

True/False

1. F (page 1182)
2. F (page 1183)
3. T (page 1185)
4. F (page 1187)
5. T (page 1188)
6. T (page 1189)
7. T (page 1191)
8. F (page 1192)

Short Answer

1. The sinoatrial node
   Internodal pathways
   Atrioventricular node
   The bundle of His
   The left bundle
   The right bundle
   The left anterior superior fascicle
   The right posterior fascicle
   The Purkinje system (page 1182)
2. Baseline period
   SA node fires.
   P wave forms.
   QRS begins to form.
   T wave forms.
   The final phase (page 1185)

Word Fun

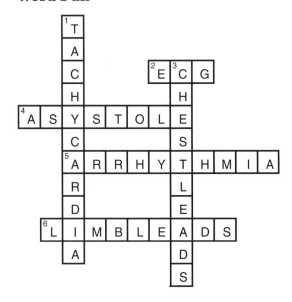

Ambulance Calls

1. Do not be afraid to inform the paramedic that the leads have not been appropriately attached. You can save valuable time, by simply relaying what you have seen. Sometimes EMTs can feel intimidated by the presence of paramedics and become afraid to speak out when they see something that doesn't appear right. Do not underestimate your ability to help, because sometimes you may notice things that the paramedic has not.

2. This arrhythmia is sinus bradycardia. When a patient is experiencing symptomatic bradycardia, the paramedic will take steps to correct this arrhythmia. Notifying the paramedic of the slow heart rate and being familiar with basic arrhythmias such as sinus bradycardia will make you a more helpful team member.

3. In order for the 12-lead ECG machine to work properly, the patient's chest must be clean, dry, and with minimal hair in order to ensure the electrodes can adhere to the skin. This requires that the patient be wiped down to remove excessive sweat and that benzoin be used (if necessary) and chest hair be removed. Some protocols do not allow for the shaving of chest hair (due to the potential of skin abrasions and cuts and their potential impact on the use of thrombolytics). It is also important to note that brassieres of female patients must be removed. It is helpful to carry gowns on the ambulance and ask any unnecessary personnel to exit the room in order to preserve a patient's modesty. Always follow local protocols.

Appendix A: BLS Review

Matching

1. E (page A-4)
2. D (page A-12)
3. B (page A-3)
4. C (page A-5)
5. A (page A-3)

Multiple Choice

1. D (page A-3)
2. C (page A-5)
3. B (page A-5)
4. D (page A-5)
5. A (page A-6)
6. D (page A-6)
7. D (page A-6)
8. D (page A-7)
9. C (page A-8)
10. D (page A-9)
11. A (page A-10)
12. B (page A-11)
13. C (page A-17)
14. B (page A-18)
15. A (page A-20)
16. D (page A-20)
17. A (page A-21)
18. D (page A-21)
19. B (page A-28)
20. D (page A-28)
21. D (page A-29)
22. B (page A-30)
23. D (page A-11)
24. D (page A-12)
25. B (page A-13)
26. C (page A-14)

Fill-in

1. 4, 6 (page A-3)
2. barrier (page A-5)
3. ABCs (page A-6)
4. opening (page A-4)
5. Advance directives (page A-7)
6. firm (page A-8)
7. airway (page A-9)

True/False

1. T (page A-6)
2. F (page A-6)
3. T (page A-6)
4. F (page A-20)
5. F (page A-10)
6. T (page A-7)
7. T (page A-23)
8. F (page A-23)
9. F (page A-20)
10. T (page A-19)
11. F (page A-19)

Short Answer

1. 1. Rigor mortis, or stiffening of the body after death
   2. Dependent lividity (livor mortis), a discoloration of the skin due to pooling of blood
   3. Putrefaction or decomposition of the body
   4. Evidence of nonsurvivable injury, such as decapitation (page A-7)

2. (page A-4)

Procedure	Infants (younger than age 1 y)	Children (1 y to onset of puberty)
Airway	Head tilt–chin lift; jaw thrust if spinal injury is suspected	Head tilt–chin lift; jaw thrust if spinal injury is suspected
Breathing		
Initial breaths	2 breaths with duration of 1 second each with enough volume to produce chest rise	2 breaths with duration of 1 second each with enough volume to produce chest rise
Subsequent breaths	1 breath every 3 to 5 seconds (12 to 20 breaths/min)	1 breath every 3 to 5 seconds (12 to 20 breaths/min)
Circulation		
Pulse check	Brachial artery	Carotid or femoral artery
Compression area	Just below nipple line	Center of chest, between nipples
Compression width	2 fingers or 2-thumb hands-encircling technique	Heel of one or both hands
Compression depth	$\frac{1}{3}$ to $\frac{1}{2}$ depth of chest	$\frac{1}{3}$ to $\frac{1}{2}$ depth of chest
Compression rate	100/min	100/min
Ratio of compressions to ventilations	30:2 (1 rescuer); 15:2 (2 rescuers)	30:2 (1 rescuer); 15:2 (2 rescuers)
Foreign body obstruction	Conscious: back slaps and chest thrusts Unconscious: CPR	Conscious: abdominal thrusts Unconscious: CPR

3. S–The patient starts breathing and has a pulse.
   T–The patient is transferred to another person who is trained in BLS or ALS, or another emergency medical responder.
   O–You are out of strength or too tired to continue.
   P–A physician who is present assumes responsibility for the patient. (page A-8)

4. To perform the head tilt-chin lift maneuver, make sure the patient is supine. Place one hand on the patient's forehead, and apply firm backward pressure with your palm to tilt the head back. Next, place the tips of your fingers of your other hand under the lower jaw near the bony part of the chin. Lift the chin forward, bringing the entire lower jaw with it, helping to tilt the head back. (page A-10)

5. To perform the jaw-thrust maneuver, kneel above the patient's head. Place your index or middle finger behind the angle of the lower jaw on both sides. Forcefully move the jaw forward, and tilt the head back. Use your thumbs to pull the patient's lower jaw down to allow breathing through the mouth and nose. (page A-11)

6. Ensure that the patient is on a firm, flat surface. Place your hands in the proper position. Lock your elbows with your arms straight and your shoulders directly over your hands. Give 30 compressions at a rate of about 100 beats/minute for an adult. Using a rocking motion, apply pressure vertically from your shoulders down through both arms to depress the sternum 1½" to 2" in the adult, then rise up gently. Count the compressions aloud. (page A-23)

7. Rescuer one begins another cycle of CPR while rescuer two moves to the opposite side of the chest and moves into position to begin compressions. Rescuer two begins compressing until 30 compressions have completed. Rescuer one delivers two ventilations. (page A-28)

8. –Standing: Stand behind the patient and wrap your arms around his or her waist. Press your fist into the patient's abdomen in quick inward and upward thrusts until the object is expelled.

   –Supine: Straddle the hips or legs. Place the heel of one hand against the patient's abdomen and the other hand on top of the first. Press your hands into the patient's abdomen in a series of five quick inward and upward thrusts. (page A-12)

9. –Standing: Stand behind the patient and wrap your arms under the armpits and around the patient's chest. Press your fist into the patient's chest and perform backwards thrusts until the object is expelled or the patient becomes unconscious.

   –Supine: Kneel next to the patient. Place your hands as you would to deliver chest compressions. Deliver slow chest thrusts until the object is expelled. (page A-13)

10. **1.** "Sandwich" the infant between your hands and arms.

    2. Deliver five quick back slaps between the shoulder blades, using the heel of your hand.

    3. Turn the infant face up.

    4. Give five quick chest thrusts on the sternum at a slightly slower rate than you would give for CPR. (pages A-16)

Word Fun

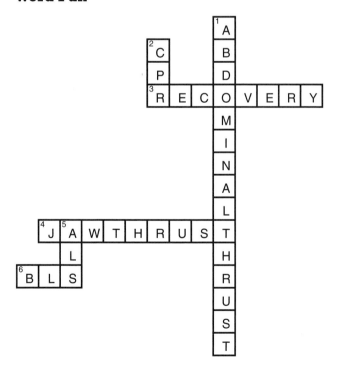

Ambulance Calls

1. –Question the family about the last time they spoke with her.
   –Explain that she has been down too long for CPR to be effective.
   –Comfort family members.
   –Notify your dispatcher to alert the supervisor and either law enforcement, the coroner, or a funeral home according to local protocols.

2. With the FDA's approval of Automated External Defibrillators for home use, it will become more common to see them being used by the lay provider. Simply purchasing an AED does not ensure appropriate usage during real life events. Training by knowledgeable, skilled instructors is needed to minimize confusion and inappropriate AED use. Remove the AED and explain to them that AED's are meant to be used only when a person is not breathing and has no pulse. Emphasize that it is appropriate to have the AED nearby in the event that a person goes into cardiorespiratory arrest, but it is not to be applied until then. After the patient has been transported to the hospital, tell the friends and family that you appreciate their willingness to be prepared for emergencies, and you would like to see them be successful in their usage of their AED. If you or your department offers CPR/AED courses, offer to train them and/or point them in the right direction to receive the appropriate training.

3. Given the scene, you must assume that there is the likelihood for trauma. This means that you must assess and maintain his airway without manipulating his spine. Use the jaw thrust maneuver and apply full c-spine precautions. Also, consider possible causes of his unconscious state, including any scene hazards and potential medical conditions.

Skill Drills

Skill Drill A-1: Positioning the Patient (page A-9)

1. Kneel beside the patient, leaving room to roll the patient toward you.
2. Grasp the patient, stabilizing the cervical spine if needed.
3. Move the head and neck as a unit with the torso as your partner pulls on the distant shoulder and hip.
4. Move the patient to a supine position with legs straight and arms at the sides.

Skill Drill A-2: Performing Chest Compressions (page A-21)

1. Place the **heel** of one hand on the sternum, **between the nipples**.
2. Place the heel of your other hand over the **first hand**.
3. With your arms straight, **lock** your elbows, and position your shoulders directly **over** your hands. Depress the sternum 1½ inches to 2 inches using a direct downward movement.

Skill Drill A-3: Performing One-rescuer Adult CPR (page A-24)

1. Establish unresponsiveness, and call for help.
2. Open the airway.
3. Look, listen, and feel for breathing. If breathing is adequate, place the patient in the recovery position and monitor.
4. If not breathing, give two breaths of 1 second each.
5. Check for a carotid pulse.
6. If no pulse is found, apply your AED. If there is no AED, place your hands in the proper position for chest compressions. Give 30 compressions at a rate of about 100/min.
   Open the airway, and give two ventilations of 1 second each.
   Perform five cycles of chest compressions and ventilations.
   Stop CPR, and check for return of the carotid pulse.
   Depending on patient condition, continue CPR, continue rescue breathing only, or place the patient in the recovery position and monitor breathing and pulse.

Skill Drill A-4: Performing Two-rescuer Adult CPR (page A-26)

1. Establish **unresponsiveness**, and take positions.
2. Open the airway.
3. Look, listen, and feel for breathing. If breathing is adequate, place the patient in the **recovery** position and **monitor**.
4. If not breathing, give **two** breaths of 1 second each.
5. Check for a **carotid** pulse.
6. If there is no pulse but an AED is available, apply it now. If no AED is available, begin **chest compressions** at about 100/min (30 compressions to **two** ventilations).
   After every **five** cycles, switch rescuer positions in order to minimize fatigue. Keep switch time to **5 to 10** seconds. Depending on patient **condition**, continue CPR, continue ventilations only, or place in the recovery position and monitor.